AF837093

BREAKING IMAGES

Breaking Images

Iconoclastic Analyses of Mathematics and its Education

Edited by
Brian Greer, David Kollosche, and Ole Skovsmose

OpenBook Publishers

https://www.openbookpublishers.com

©2024 Brian Greer, David Kollosche, and Ole Skovsmose (eds). Copyright of individual chapters remains with the chapter's author(s).

This work is licensed under a Creative Commons Attribution-NonCommercial-NoDerivs 4.0 International license (CC BY-NC 4.0). This license allows re-users to copy and distribute the material in any medium or format in unadapted form only, for non-commercial purposes only, and only so long as attribution is given to the creator. Attribution should include the following information:

Brian Greer, David Kollosche, and Ole Skovsmose (eds), *Breaking Images: Iconoclastic Analyses of Mathematics and its Education*. Cambridge, UK: Open Book Publishers, 2024, https://doi.org/10.11647/OBP.0407

Copyright and permissions for the reuse of many of the images included in this publication differ from the above. This information is provided in the captions and in the list of illustrations. Where no licensing information is provided in the caption, the figure is reproduced under the fair dealing principle. Every effort has been made to identify and contact copyright holders and any omission or error will be corrected if notification is made to the publisher.

Further details about the CC BY-NC license are available at http://creativecommons.org/licenses/by-nc/4.0/

All external links were active at the time of publication unless otherwise stated and have been archived via the Internet Archive Wayback Machine at https://archive.org/web

Any digital material and resources associated with this volume will be available at https://doi.org/10.11647/OBP.0407#resources

Volume 2 | Studies on Mathematics Education and Society Book Series
ISSN Print: 2755-2616
ISSN Digital: 2755-2624

ISBN Paperback: 978-1-80511-321-8
ISBN Hardback: 978-1-80511-322-5
ISBN Digital (PDF): 978-1-80511-323-2
ISBN Digital eBook (EPUB): 978-1-80511-324-9
ISBN HTML: 978-1-80511-325-6

DOI: 10.11647/OBP.0407

Cover image: *Fall* by Tara Shabnavard
Cover design: Jeevanjot Kaur Nagpal

Published with the support of the Open Access Publishing Fund of the University of Klagenfurt.

Contents

1. Beginning

Brian Greer, Ole Skovsmose, and David Kollosche

The contributions in this book focus on critically analysing the relationship between mathematics as a discipline and mathematics as a school subject. The discontents of school mathematics are universally acknowledged and include questions such as: Why do so many people, however intelligent and successful, have feelings of inadequacy and alienation towards the subject? Why does mathematics education in school not seem to improve despite all the effort put into it? Our collective attempt to address such questions through radical rethinking begins by arguing that it is more productive to speak in terms of doing mathematics, *in a variety of senses, rather than using words that imply that mathematics* exists as some kind of entity. *In particular, we reject the notion of mathematics being independent of human agency. Such a reformulation is in line with recent developments in mathematics and the philosophy of mathematics that problematise the quest for a definitive and timeless definition of mathematics. Related developments in history of mathematics, anthropology, and related fields make it imperative to acknowledge historical, cultural, social, ethical, and political – in short, human – dimensions of mathematics and mathematics education. Multiple important themes that are generated by this perspective are summarised.*

The purpose of this book is to examine, critically and in their full complexity, relationships between conceptions of mathematics (mainly presented in Part 1 of this book) and the teaching/learning of mathematics in schools (mainly presented in Part 2 of this book).

The reader of this introductory chapter, and of the book as a whole, can hardly fail to become aware of the tension produced by the attempt to keep within reasonable length a discussion that involves negotiating a minefield of exploding concepts while trying to avoid omission of

©2024 Brian Greer et al., CC BY-NC 4.0 https://doi.org/10.11647/OBP.0407.01

essential aspects. We have made certain decisions necessary to keep the scope within manageable bounds, such as essentially limiting the contexts to those of 'the West'. Thus, we do not address, for example, Indian philosophies of mathematics, or mathematics education in China. Discussion on mathematics education relates predominantly to that which happens in schools, as opposed to university mathematics education and learning in out-of-school contexts. The following sections outline some of the main themes of the book.

Conceptions of mathematics

In his book *What is Mathematics, Really?*, Reuben Hersh makes the following observation:

> The working mathematician is a Platonist on weekdays, a formalist on weekends. On weekdays, when doing mathematics, he's a Platonist, convinced he's dealing with an objective reality whose properties he's trying to determine. On weekends, if challenged to give a philosophical account of the reality, it's easiest to pretend he doesn't believe it. He plays formalist, and pretends mathematics is a meaningless game. (Hersh, 1997, p. 39)

We refer to such a formulation as a working philosophy of mathematics. It need not be well articulated, and, as indicated by Hersh, it need not even be consistent. Formal philosophies of mathematics have been elaborated in all directions (for overviews, see Benacerraf & Putnam, 1964; Hacking, 2014; Shapiro, 2000). As a term avoiding a sharp distinction between the two, we tend to refer to 'conceptions of mathematics'.

'What is mathematics?'

A short sampling of answers:

> Mathematics may be defined as the subject in which we never know what we are talking about, nor whether what we are saying is true. (Russell, 1901, p. 1)

> Mathematics is the art of giving the same name to different things. (Attributed to Poincaré)

> Mathematics is the language in which God has written the universe. (Attributed to Galileo)

> Mathematics is the study of all possible patterns. (Sawyer, 1955, p. 12)

Answers to the question 'What is Mathematics?' may be enigmatic, aphoristic, religious, hubristic, aesthetic. However, one also finds answers that have been elaborated through deep philosophical investigations. Let us briefly recapitulate the positions of logicism, formalism, and intuitionism.

Since Antiquity, Platonism has been carefully articulated and further developed. Gottlob Frege (e.g., 1967) reworked Platonism in relation to mathematics into a completely new format, claiming that the idealised and permanent mathematical objects are sets. In this way, he launched the logicist programme, which tries to show that mathematical entities in fact are logical entities, and that mathematical statements are logical statements. In *Principia Mathematica*, Alfred Whitehead and Bertrand Russell (1910–1913) elaborated this programme to the extreme. As already quoted, Russell characterised mathematics as a subject in which 'we never know what we are talking about'. In a more serious mood, he declared George Boole's *Laws of Thought* to be about formal logic, adding 'and this is the same thing as mathematics' (Russell, 1901, p. 1).

David Hilbert, wanting to systematically address mathematical theories, advocated formalising mathematics so that they could be investigated with respect to, for instance, consistency and completeness. This programme led directly to the idea that mathematics can be identified with formal systems. In *Outlines of a Formalist Philosophy of Mathematics*, Haskell Curry (1970) provides a comprehensive presentation of formalism and what it means to identify mathematics with formalism.

A third answer to 'What is mathematics?' comes from L. E. J. Brouwer (e.g., 1913), who formulated intuitionism as a philosophy of mathematics. According to Brouwer, formalism represents a complete misunderstanding of mathematics and formal structures. While formalists see formal structures as being precise expressions of mathematics, intuitionists view formal structures as imprecise and, in many cases, inappropriate approximations to mathematics. Wagner (2017) summarises intuitionism as questioning 'any mathematics that cold not be finitely constructed starting with counting a sequence of moments (in a Kant-like framework of temporality)' (p. 17). In this way, Brouwer characterised mathematics as a human, mental, activity.

Philosophical answers to the question 'What is mathematics?' reveal multiple conceptions of mathematics. Some see mathematics as an essential constituent of our world while others consider it as man-made. Such diversity and contrasts suggest that a search for a definitive characterisation of the *essence* of mathematics is a chimera, albeit one that, like the quest for the Philosopher's Stone, stimulates productive inquiry.

Posing a better question

We suggest that 'What *is* mathematics?' is not a good question. 'Mathematics' means a lot of different things for school students, for engineers, for philosophers in contemporary times, in the late nineteenth century, in Antiquity. The very grammar of the question tempts us to search for a universal essence of mathematics. However, how such an essence could be found and verified constitutes an unsolved, arguably unsolvable, philosophical problem. Every attempt to capture the essence of mathematics entails the danger of generalising a particular perspective at the expense of others. Mathematics has, and will continue to evolve, a history, and the families of activity systems that involve mathematics are diverse.

Semantically speaking, in terms of the discourse theory of Ernesto Laclau and Chantal Mouffe (1985/2001), 'mathematics' constitutes a floating signifier, a concept whose strength in combining with other concepts, activities, and expectations depends on its conceptual flexibility, on its openness to assume different facets of meaning in different discourses. It is also worth questioning to what extent 'mathematics' is best regarded as a noun. Could it be interpreted more like a verb? In fact, we are going to suggest a shift to thinking about *what can be done through mathematics*. We contend that the characterisation of mathematics is more usefully framed not in terms of an entity, but in terms of what humans, individually and collectively, *do* when they engage with mathematics.

Hans Freudenthal, inspired by Brouwer and by intuitionism, highlighted again and again that 'mathematics is a human activity'

(Chapters 7 and 8, this volume). Let us clarify what we mean by 'human activity'. For emphasis, we may instead use the phrase 'social activity', reflecting an orientation that envisages collective mental activity, not just what Brouwer took to be an activity of a single mind. Such an individualistic formulation may seem natural when we consider a child having a breakthrough insight, or a solitary mathematician struggling with a proof. Even in such circumstances, however, the social nexus is still there. The mathematician is part of a community with well-established norms (Chapter 9, this volume) that has worked on the problem; the child is in an educational setting. Indeed, to the extent that thought may be considered internal communication (a question we will not attempt to address), it is inherently socially grounded, in particular linguistically.

School and its associated practices (and not just learning and teaching) constitute a very particular form of historically evolved social activity. In other cultural settings, there are very different forms of learning and teaching, including those in which 'doing' and 'learning' are embedded in the same activity. Contrast that with the familiar answer from mathematics teachers to the question 'Why are we doing this?', namely some variant of 'Because it will be useful to you later'.

For all of these reasons, we contend that a better question than 'What is mathematics?' is to ask something like 'What do people do when they use mathematics within an activity system?'. Such a shift away from essentialist to performative paradigms is not unique to the philosophy of mathematics. For example, there is a parallel with Ludwig Wittgenstein's (1997/1953) later work of interpreting language through its use in what John Searle (1969) called 'speech acts'. The conception of language shifted from a descriptive perspective to a performative perspective. In a similar way, then, we want to pay particular attention to performative features of mathematics, which are highlighted by Ole Ravn and Ole Skovsmose (2019) through their formulation of a four-dimensional philosophy of mathematics. We may point to ethnographic studies of people doing mathematics in workplace contexts, for example. George Pólya's (e.g., 1962) emphasis on how mathematicians behave stems from a similar motivation.

Seeing mathematics as a social activity has profound implications. It shifts the balance away from 'mathematics' as something that *exists* (in whatever sense) to something that is *done by people*. It makes it natural to adopt both the historical and diversity lenses and prompts many other considerations that are relevant to both mathematics as an academic discipline and mathematics education. It becomes natural to consider how conceptions of mathematics have changed over historical time and to acknowledge that differently situated people might mean different activities when they refer to mathematics, even one and the same person might refer to different activities. For example, when we refer to mathematics in academic situations, activities such as defining concepts, testing hypotheses, and formulating proofs are central activities, but often they are not typical activities in school mathematics.

To signal and emphasise that one aspect among many is being highlighted, we use 'as' rather than 'is' in phrases such as 'Mathematics as a process of discovery' (see Ravn & Skovsmose, 2019). So, in addition to mathematics as academic discipline, we will also talk about, for example, mathematics as cultural constructions, mathematics as practices in work, mathematics as engineering techniques, mathematics as school subject, and so on. We allow ourselves to be unsystematic in our use of 'mathematics as...' and we fully recognise that we have to cope with a fuzzy way of using the words. Clearly, this phrasing in terms of families of practices in which mathematics is embedded is closely aligned with the concept of Ethnomathematics (Chapters 10 and 17, this volume).

Evolution of academic mathematics

As a human activity, that set of practices that we term 'academic mathematics' has a long history (Chapter 2, this volume). In the course of that history, radical conceptual restructuring has taken place, and continues to take place. To use the most familiar example, what is meant by 'number' stretches from the 'natural numbers' 1, 2, 3, ... to the equation $e^{i\pi} = -1$ and beyond.

A first central question regarding such developments is: What are the processes through which conceptual restructuring occurs? Answers to this question minimally include the following:

- In response to human needs. For example, because of its late development, we have a relatively clear historical picture of how probability theory was initially motivated by the needs of gamblers, and developed in close proximity to situations such as jury trials, risk assessment, social theories of the nature of man, and so on (Hacking, 1990).

- By asking 'What if...?' questions, such as 'What if we don't assume Euclid's fifth axiom?', a question that led to revolutionary developments in geometry.
- Through the symbiotic development of tools, including representational tools, for example, coordinate geometry based on the Cartesian representation.
- Through making connections between apparently disjoint fields, notably the translatability between geometry and algebra achieved by René Descartes (discussed at length by Hacking, 2014).
- Through internal crises, disequilibria, a famous example being the realisation that the diagonal of a square is incommensurable with its side.
- Through the detachment of mathematical structures from their origins in systematised situations. A clear example is the concept of 'group' which eventually came to be defined as a set, together with an operation on ordered pairs thereof, having certain properties. Given this definition, mathematicians could pursue their researches independently of any particular examples or applications of group structures.
- Through the reconceptualisation of conceptual entities within mathematics. The case study by Imre Lakatos (1976) on a theorem about polygons is a prime example; changed ideas of the nature of mathematical proof given the advent of computers is another.

A second key question is 'To what extent is the development of mathematics necessary, and to what extent contingent?' Rafael Núñez (2000) argues that it is not a binary choice, stating that that 'mathematics is not transcendentally objective, but it is not arbitrary either (not the result of pure social conventions)' (p. 3). There are mathematical developments that feel like they could not have happened otherwise

– for example, the extension from natural numbers to rational numbers and directed numbers. It is not so obvious, however, when it comes to the question posed by Núñez: 'Have you ever thought why (I mean, really *why*) the multiplication of two negative numbers yields a positive one?' (p. 3)

That the development of academic mathematics proceeds in a way that is absolutely predetermined is arguably disproved by the diversity within it. For example, Raju (2007, p. 413) declared that within European mathematics there are two streams:

1. from Greece and Egypt a mathematics that was spiritual, anti-empirical, proof-oriented, and explicitly religious, and
2. from India via Islamic countries a mathematics that was pro-empirical, and calculation-oriented, with practical objectives.

Raju's (2007) work is also an important contribution to one aspect of Ethnomathematics, namely the construction of a counter-narrative to the myth that academic mathematics is a purely European achievement.

Is doing mathematics inherently beneficial to humankind?

In the European context, since Antiquity, mathematics has been admired and celebrated, while, in academia, a critical conception of mathematics has only been articulated within the last century. Plato admired mathematics, which showed what it could mean to enter the world of ideas. Via the human senses such access was not possible, but through rationality, it was assumed, we can explore properties of idealised objects. The Platonist admiration of mathematics turned into a celebration of Euclid's *Elements*, which brought together an axiomatisation of geometry that right up to the late nineteenth century was considered to be perfect, serving as the epitome of the systematisation of mathematics within formal structures, and taken as the role model for how to build theories in science.

The admiration of mathematics acquired more fuel through the so-called scientific revolution. The people contributing to this were deep believers in God, as, for instance, Isaac Newton. They saw the world as created by God, meaning that insight and understanding of nature meant insight and understanding of God's creation. God had inserted laws of nature that could be captured by mathematics, truly an overwhelming insight. Through mathematics we human beings become able to grasp the rationality of God! When the natural sciences, following a protracted ideological struggle, separated from religious beliefs, the celebration of mathematics continued, and mathematics became nominated as the language of science. The celebration of mathematics has also become an integral part of much philosophy of science (e.g., Shapiro, 2000).

In contemporary circumstances, practitioners and proponents of mathematics (more generally the fashionable complex of Science, Technology, Engineering, and Mathematics, STEM) enjoy a great deal of political and cultural capital. In political and economic media discourse, statements to the effect that high achievement in STEM education is essential for economic competitiveness in the global marketplace are pervasive. A preponderance of what is written or spoken about mathematics in public, political, and academic discourses reflects an unexamined belief in what Paola Valero (2004) called 'the unquestioned intrinsic goodness of both mathematics and mathematics education [that represents] the core of its "political" value' (p. 13).

In this book, we leave behind the blind admiration of mathematics and consider the emergence of a critical stance towards mathematics, in particular its dehumanising effects (Chapter 5, this volume). The most concerted critique has emanated from within the group of critical mathematics educators (Chapter 11, this volume; and see Greer & Skovsmose, 2012, for a history of that movement). Relatively few mathematicians have expressed a critical attitude towards what people have done using mathematics. Writers commenting on the human condition who have done so include, notably, Charles Dickens, who was repelled by the class oppression that was exacerbated by the Industrial Revolution (Chapter 12, this volume).

Perhaps we should make clear that we by no means discount the very many ways in which mathematics has been, and can be, used to benefit our lives both practically and intellectually. However, given that there is no lack of writing in praise of mathematics, we feel the need to emphasise rather its problematic uses, including in the service of imperialism, for advancing the techniques of war, and its inextricable links with capitalism.

Ubiratan D'Ambrosio concluded his paper introducing Ethnomathematics as follows:

> Ideology [...] takes a more subtle and damaging turn, with even longer and more disrupting effects, when built into the formation of the cadres and intellectual classes of former colonies, which constitute the majority of so-called Third World countries. We should not forget that colonialism grew together in a symbiotic relationship with modern science, in particular with mathematics, and technology. (D'Ambrosio, 1985, p. 47)

Beyond the material military contributions to colonial conquest through technology, we have to consider the symbolic violence of suppressing other forms of knowledge and replacing them with European epistemologies and practices.

Mathematics has long been used in the service of war, and many mathematicians have devoted their talents to the design of more effective ways of killing people. Others have used mathematics for the more efficient management of warfare. A very strong statement was made by Zygmunt Bauman (1989) that the Holocaust was not an anomaly within modernity but, in its monstrous effectivity, depended on the most modern practices of organisation, including mathematics (Chapter 5, this volume).

Again, the use of mathematics in the service of capitalism constitutes a vast subject and here we merely draw attention to some specific aspects. Most fundamental, perhaps, are the connections between the great abstractions of number and capital, intermediated through money as represented materially and, increasingly, in virtual forms. Economic and political theorists can present various dynamic system analyses of the possibly irreversible development of the particular pathological form of capitalism currently in the United States and beyond. We may consider to what extent contemporary mathematics education within particular political regimes plays a role in preparing children to be active proponents or passive citizens within capitalist systems.

In view of the discussion above, we take the position that it is no longer possible for mathematicians (or scientists, or any scholars) to claim ethical/political neutrality, such a claim in itself being a kind of ideology (Chapter 4, this volume). Specifically, in Chapter 3, Skovsmose discusses the views of G. H. Hardy as presented in *A Mathematician's Apology*, in which Hardy (1967) suggests that a mathematician can operate as a pure intellectual, with no responsibility for what is done with her/his work. Another mathematician, Chandler Davis (2015) issued a different kind of apology – not in the sense of 'apologia' – when he regretted that he and other mathematicians had not done more to oppose war, including the Mutually Assured Destruction (MAD) principle that guided policy during the Cold War, and was substantially based on the work of John von Neumann and others on game theory.

In direct opposition to Hardy's stance, Ubiratan D'Ambrosio, in the manifesto for 'Non-killing Mathematics',[1] asserts that it is not enough for mathematicians to do good work, they must pay attention to what will be done using that work, and that it is not enough for mathematics educators to teach students well, they must pay attention to what those students will do with what they have been taught.

Development of mathematical understanding under instruction

In considering relationships between the development of mathematics by humankind and the development of mathematical knowledge and understanding in a contemporary student, the most obvious point is that the former occurred over millenia as opposed to a small number of years. A child today is expected to deal, at least procedurally, with mathematical content that historically took multiple good brains collectively a very long time to figure out.

As stated by Freudenthal (1991), 'we know nearly nothing about how thinking develops in individuals, but we can learn a great deal from the development of mankind' (p. 48). In response to his own question as to whether the learner should repeat the learning process of mankind, his response is 'of course not'. Instead, his recommendation is that 'the learner should reinvent mathematizing rather than mathematics; abstracting rather than abstractions; schematizing rather than schemes; formalising rather than formulas; alogrithmising rather than algorithms; verbalising rather than language', which chimes with

1 Ethics/Nonkilling/Mathematics (2024, April 5). *Wikiversity*, https://en.wikiversity.org/wiki/Ethics/Nonkilling/Mathematics

our emphasis on actions. In the same spirit, Pólya argued for children having the opportunity to experience problem-solving for themselves: 'How can you know if you like raspberry pie if you have never tasted it?' (Pólya, 1945, p. v).

Another important point has been clearly stated thus:

> Teaching is one of the immense social influences that can affect a child, but its effects can be out of proportion to any other kind of social influence once the first beginnings of a child's life are past. In it once again knowledge builds on knowledge, but the form of experience that makes it possible is really quite unlike those forms of experience that come the individual's way when teaching is not involved. (Hamlyn, 1978, p. 144)

Multidiversity

In terms of mathematics as a human activity, 'multidiversity' relates to differences among and within families of mathematical activities emergent from their cultural and historical underpinnings, including forms of life, worldviews, cognition, language, value systems, and so on. In terms of school mathematics, it relates to the myriad of differences, interacting in complex ways, among students (and also among teachers, a story in itself). These include, notably, ethnic diversity (Chapter 18, this volume) and gender (Chapter 19, this volume). Within mathematics education, much of the foundational work addressing diversity has been concerned with 'equity' and 'access'. The sloganising of these terms demands more careful analysis (e.g., Martin, 2019; Pais, 2012) and we pinpoint the following preliminary questions and comments:

- Access to what? Many if not most of the exhortations to improve access takes mathematics-as-school-subject as an unexamined given.
- Equity on whose terms? Is it merely assimilation, involving the denial of cultural identity?
- Beyond equity and access lie identity and agency.

All of these, of course, are intensely political in nature.

In current circumstances, we can observe a hegemonical struggle between acknowledgment and valorisation of diversity in all its aspects, and multifaceted forces tending towards homogenisation, linked with globalisation (Westernisation), corporatisation, metrification, and so on. Such homogenisation is certainly prominent within mathematics education. Perhaps the most obvious manifestation is in curricular documents, for which the Common Core State Standards within the United States may serve as an example. We draw attention to its stated principle of benchmarking with similar projects from other countries, contributing to a process of convergence towards global uniformity, exacerbated by the effects of the international comparison industry (Chapters 15 and 16, this volume).

Parenthetically, as a parallel, think of the onward march of English as a global language, among the consequences of which is a significant distortion of our field. This book, in English, has been written by speakers of many languages and edited by two people for whom English is a foreign language and one who grew up speaking English because of early colonisation and linguicide.[2]

Epistemological pluralism is another central issue, including from the perspective of mathematics-as-discipline. Rik Pinxten, Ingrid van Dooren, and Frank Harvey (1983), who studied the fundamentally different epistemology of the Navajo people, in particular in relation to space, commented that:

> Through a systematic superimposition of the world view and thought system of the West on traditional non-Western systems of thought and action all over the world, a tremendous uniformization is taking hold [...] The risks we take on a worldwide scale, and the impoverishment we witness is – evolutionarily speaking – quite frightening. (pp. 174–175)

As a closing comment, we observe that in terms of families of mathematical practices, there is obvious diversity within mathematics as cultural constructions, mathematics in work practices, mathematics of everyday life, and, indeed, within academic mathematics (e.g., Hersh, 2006). Yet this diversity is not generally manifest in school mathematics; we regard that as a problem.

2 To respect authors' linguistic preferences and cultural identities, authors of each chapter have opted to follow British or American English in spelling and punctuation.

Mathematics education as a research field

A survey volume edited by Anna Sierpinska and Jeremy Kilpatrick (1998) is tellingly titled *Mathematics Education as a Research Domain: The Search for Identity*. The emergence and development of mathematics education as a field has seen a diversification of influential disciplines and methodologies – broadly speaking, the balancing of technical disciplines by human disciplines such as sociology and anthropology, and formal statistical methods by interpretative methods of research and analysis.

The desire to have clearcut methodologies avoiding complex human judgments has passed through many manifestations from the early alignment with logical positivism and related positions. In his address to the first International Conference on Mathematics Education in 1969, Edward Begle explicitly recommended the empirical-scientific approach through a program of identifying the important variables and systematically studying the relations between them. In Begle (1979), he confessed to feeling depressed that a decade of experimental work had produced little progress. In fact, a range of theoretical frameworks may be characterised as attempts to apply scientific precision to the complexity of understanding and improving mathematics education – behaviourism, information-processing theory, Artificial Intelligence, neurocognition – aligned with a reliance on narrowly defined standards of empirical research and statistical modelling. Kilpatrick (1981), in a paper entitled 'The Reasonable Ineffectiveness of Research in Mathematics Education', cited Irving Kristol (1973), who raised the question why we can send a man to the moon, but cannot improve mathematics education, and answered it by pointing out that the former is a technical problem, the latter is a human problem.

Academic mathematicians' claims over mathematics education

The most obvious difference between mathematics-as-discipline and mathematics-as-school-subject lies in the nature of the populations involved. Picture a pyramid representing all those who are taught mathematics in school. A very small peak corresponds to those who will become academic mathematicians. A rather larger zone beneath that corresponds to those, such as engineers, that will use significant technical mathematics. The largest part of the pyramid represents people who secure material support for those at the peak and who do, indeed, use mathematics, but most often learned in context as needed, using situated procedures unrelated to what they learned in school, and mediated by tools (Lave, 1988).

Accordingly, we ask 'To what extent, and in what ways, should academic mathematicians be accorded control over school mathematics education?' Mathematicians have vested interests in the reproduction of their kind, and so may be suspected of bias, as well as developmental ignorance, by which we mean that, in their expertise, they forget what it is like to struggle with mathematics. We put forward two propositions for consideration. The first is that mathematicians should not dominate school mathematics – simply put, mathematics education is far too important to be left to mathematicians. The second is that mathematics education is about much, much more than the transmission of a subset of accumulated and systematised mathematical knowledge and techniques. We take issue with the position that the predominant role of those who work in mathematics education should be simply to study and implement better ways to effect this transmission. For a clear statement of that position, broadly speaking, see the book edited by Michael Fried and Tommy Dreyfus (2014).

The most obvious manifestation of mathematicians shaping mathematics education is through the formulation of curricula. The Common Core State Standards for Mathematics in the United States, mentioned above, may be taken as representative of the search for the perfect model. It was primarily designed by three mathematicians, albeit with an advisory group that included mathematics educators. But there are many, many other actors that have direct and indirect roles in shaping mathematics education in the United States, as analysed in great detail by Mark Wolfmeyer (2014).

We suggest that the uses of the term 'mathematics' in political discourse support an unreasonable sway over the policies and administration of mathematics education. Both reflecting and influencing what politicians do, the images of mathematics and mathematics education among the public in general (Chapter 20, this volume) matter greatly.

School mathematics as an instrument of the state

The advancement and perfection of mathematics are immediately connected with the prosperity of the state. (Attributed to Napoleon, 1800)

The functioning of the modern state presupposes a variety of mathematical technologies – accounting, statistics, and much more. Mathematics, on its part needs the institutions of the state (schools, universities, research institutions, etc.) to secure financing, recruitment and the rearing of competence. (Høyrup, 2019, p. 635)

Ian Hacking (1990) has documented, in painstaking detail, the ways in which the formal mathematics of probability and statistics developed within socio-political contexts, in close relationship to changing views of the nature of humans, and in the service of states. In a rare overtly political statement, he trenchantly observed that:

> We obtain data about a governed class whose deportment is offensive, and then attempt to alter what we guess are relevant conditions of that class in order to change the laws of statistics that the class obeys. (Hacking, 1990, p. 119)

The two most obvious mechanisms through which states control school mathematics are curriculum (in concert with mathematicians, see above) and standardised testing (in concert with psychometricians and others). We assert that curriculum, historically, has been characterised by inertia and stasis in terms of content and pedagogy, and as argued within this book, accords little weight to the needs of people in general.

Arguably, however, the sharpest tool for state control of school mathematics lies within the proliferation of standardised testing, locally, nationally and globally, within which mathematics has a particular importance. On the one hand, mathematics is implicated because it underpins the models used to construct such testing and interpretations of the results and, at a deeper level, the culture of affording unjustified authority to numbers (e.g., Porter, 1975) and mathematical models (O'Neil, 2016; Skovsmose, 2005). And the imposition of such testing constrains and distorts mathematics teaching and learning (for a detailed historical survey by a battle-scarred participant, see Chapter 14, this volume).

Formative assessment, in the sense of assessment by a teacher in the course of interactions with students, forms an integral part of learning and teaching within a long-term relationship. Such a process has at least the potential of affording an effective form of communication. By contrast, summative assessment, in its typical forms, is a form of communication whose flaws are compounded across many stages (Miller-Jones & Greer, 2009). In the United States, the standard use of the term 'achievement gap', implying a deficit model, instead of 'differences in test scores' is another pernicious use of language. And accreditation in mathematics creates a barrier to educational and financial opportunities through imposing requirements unrelated to the actual needs of chosen career paths, as has been particularly well documented by Hacker (2016).

Turning to the escalating power of international comparative assessment exercises, Christine Keitel and Kilpatrick (1998) concluded a critique with the following damning assessment:

> The studies rest on the shakiest of foundations – they assume that the mantel of science can cover all weaknesses in design, incongruous data and errors of interpretation. They not only compare the incomparable, they rationalize the irrational. (p. 254)

In their edited volume, *Education by the Numbers and the Making of Society*, Sverker Lindblad, Daniel Pettersson, and Thomas Popkewitz (2018) analyse the dominance of international educational assessments (in which mathematics has a pre-eminent place in terms of its role in constructing models and in terms of its prominence as subject-matter of tests) in shaping educational policymaking on a global scale, to the extreme of shaping the right kind of people and the right kind of countries. Most fundamentally, they present arguments about the harmful effects of uncritical obeisance to the authority of numbers, and about the use of statistical and modelling techniques in furthering the rise of neoliberal hegemony in education.

While curriculum and testing are the most blatant instruments, there are more subtle ways in which mathematics education may both reflect and frame forms of life and worldviews. Here we exemplify core elements of the standard school mathematics curriculum and their possible effects:

- In many systems of mathematics education, considerable emphasis is given to procedural fluency with algorithms. Might it be that this helps form a disposition for following rules, and abdicating responsibility for making personal judgments? (Skovsmose, 1994).
- It has been amply documented (e.g., Verschaffel, Greer, & De Corte, 2000) that children manifest suspension of sense-making when solving word (or story) problems in mathematics. Is it going too far to suggest that this kind of experience over years of schooling contributes to inculcating a frame of mind whereby a person uncritically accepts an unproblematic mapping of situations in the world onto equations? (see, e.g., Porter, 1975).
- More generally, it could be argued that the nature of mathematical modelling in general is poorly conveyed in mathematics education, failing to address a critical attitude to modelling that takes into account the motivations of the modellers, the limitations of representational and physical modelling tools available, the reliance of models on assumptions made, the difficulty of gauging the effects of simplification, the complexities of interpretation, and the nuances of communicating conclusions. Accordingly, mathematics education typically fails to prepare students to become citizens with a critical disposition and a desire to achieve and wield agency.
- A specific aspect of viewing the world that teachers and users of mathematics may unwittingly promote is the implicit rule that anything can be measured on a single dimension (Horkheimer & Adorno, 1944/1997). Once that is done, there are numerous implications, such as that averages can be worked out for different populations and compared (the history of measurements of intelligence provides an obvious example).

Final comments

Commentaries on mathematics education in schools – from students, parents, teachers, mathematics educators, researchers, politicians, and people in general – tend to be dominated by discontents and a sense of puzzlement about why such education seems to be unsuccessful in many ways despite the efforts put into improving it. In this book we argue that one starting point in addressing these discontents and their causes is a back-to-basics analysis of what is meant by 'doing mathematics', in particular by people designated as 'mathematicians', and how that vast diversity of activities contributes to shaping what happens in school classrooms. Throughout, we emphasise that the doing and teaching and learning of mathematics are situated in historical, cultural, social, and political – in short, human – contexts.

References

Bauman, Z. (1989). *Modernity and the Holocaust*. Polity Press.

Begle, E. G. (1979). *Critical variables in mathematics education*. Mathematical Association of America.

Benacerraf, P., & Putnam, H. (Eds.). (1964). *Philosophy of mathematics: Selected essays*. Cambridge University Press.

Brouwer, L. E. J. (1913). Intuitionism and formalism. *Bulletin of the American Mathematical Society*, *20*(2), 81–96. (Reprinted in Benacerraf & Putnam, 1964, pp. 77–89)

Brouwer, L. E. J. (1964). *Consciousness, philosophy and mathematics*. In P. Benacerraf & H. Putnam (Eds.), *Philosophy of mathematics* (pp. 90–96). Prentice-Hall.

Curry, H. B. (1970). *Outlines of a formalist philosophy of mathematics*. North-Holland.

D'Ambrosio, U. (1985). Ethnomathematics and its place in the history and pedagogy of mathematics. *For the Learning of Mathematics*, *5*(1), 41–48.

Davis, C. (2015). Choosing our future. *Pakula lecture*. http://www.dlsph.utoronto.ca/2015/04/waging-peace-the-annual-pakula-lecture-by-professor-chandler-davis

Frege, G. (1967). *Begriffsschrift: A formula language, modelled upon that of arithmetic, for pure thought*. In J. van Hiejenoort (Ed.), *From Frege to Gödel: A source book in mathematical logic, 1879–1931* (pp. 1–82). Harvard University Press.

Freudenthal, H. (1991). *Revisiting mathematics education*. Kluwer.

Fried, M. N., & Dreyfus, T. (2014). *Mathematics and mathematics education: Searching for common ground*. Springer. https://doi.org/10.1007/978-94-007-7473-5

Greer, B., & Skovsmose, S. (2012). Seeing the cage? The emergence of critical mathematics education. In O. Skovsmose & B. Greer (Eds.). *Opening the cage: Critique and politics of mathematics education* (pp. 1–20). Sense. https://doi.org/10.1007/978-94-6091-808-7_1

Hacker, A. (2016). *The math myth and other STEM delusions*. New Press.

Hacking, I. (1990). *The taming of chance*. Cambridge University Press. https://doi.org/10.1017/CBO9780511819766

Hacking, I. (2014). *Why is there philosophy of mathematics at all?* Cambridge University Press. https://doi.org/10.1017/CBO9781107279346

Hamlyn, D. W. (1978). *Experience and the growth of understanding*. Routledge and Kegan Paul.

Hardy, G. H. (1967). *A mathematician's apology*. Cambridge University Press.

Hersh, R. (1997). *What is mathematics, really?* Oxford University Press.

Hersh, R. (Ed.). (2006). *18 unconventional essays on the nature of mathematics*. Springer. https://doi.org/10.1007/0-387-29831-2

Horkheimer, M., & Adorno, T. W. (1997) *Dialectic of enlightenment*. Verso. (Original work published 1944)

Høyrup, J. (2019). *Selected essays on pre- and early modern mathematical practice*. Springer. https://doi.org/10.1007/978-3-030-19258-7

Keitel, C., & Kilpatrick, J. (1998). The rationality and irrationality of international comparative studies. In I. Huntley, G. Kaiser, & E. Luna (Eds.), *International comparisons in mathematics education* (pp. 241–255). Falmer. https://doi.org/10.4324/9780203012086-19

Kilpatrick, J. (1981). The reasonable ineffectiveness of research in mathematics education. *For the Learning of Mathematics*, 2(2), 22–29.

Kristol, I. (1973, January 8). Some second thoughts. *New York Times*. 55, 62.

Laclau, E., & Mouffe, C. (2001). *Hegemony and the socialist strategy*. Verso. (Original work published 1985)

Lakatos, I. (1976). *Proofs and refutations: The logic of mathematical discovery*. Cambridge University Press. https://doi.org/10.1017/CBO9781139171472

Lave, J. (1988). *Cognition in practice: Mind, mathematics and culture in everyday life*. Cambridge University Press. https://doi.org/10.1017/CBO9780511609268

Lindblad, S., Pettersson, D., & Popkewitz, T. S. (Eds.). (2018). *Education by the numbers and the making of society*. Routledge.

Martin, D. B. (2019). Equity, inclusion, and antiblackness in mathematics education. *Race Ethnicity and Education, 22*, 459–478. https://doi.org/10.1080/13613324.2019.1592833

Miller-Jones, D., & Greer, B. (2009). Conceptions of assessment of mathematical proficiency and their implications for cultural diversity. In B. Greer, S. Mukhopadhyay, S. Nelson-Barber, & A. B. Powell (Eds.). *Culturally responsive mathematics education* (pp. 165–186). Routledge. https://doi.org/10.4324/9780203879948-14

Núñez, R. E. (2000). Mathematical idea analysis: What embodied cognitive science can say about the human nature of mathematics. In *Proceedings of the 24th International Conference for the Psychology of Mathematics Education* (Vol. 1, pp. 3–22). IGPME.

O'Neil, C. (2016). *Weapons of math destruction: How big data increases inequality and threatens democracy*. Crown Books.

Pais, A. (2012). A critical approach to equity. In O. Skovsmose & B. Greer (Eds.). *Opening the cage: Critique and politics of mathematics education* (pp. 49–92). Sense. https://doi.org/10.1007/978-94-6091-808-7_3

Pinxten, R., van Dooren, I., & Harvey, F. (1983). *The anthropology of space: Explorations into the natural philosophy and semantics of the Navajo*. University of Pennsylvania Press.

Pólya, G. (1945). *How to solve it*. Princeton University Press.

Pólya, G. (1962). *Mathematical discovery: On understanding, learning, and teaching probem solving*. Wiley.

Porter, T. M. (1975). *Trust in numbers: The pursuit of objectivity in science and public life*. Princeton University Press.

Raju, C. K. (2007). *Cultural foundations of mathematics: The nature of mathematical proof and the transmission of the calculus from India to Europe in the 16th c. CE*. PearsonLongman.

Ravn, O., & Skovsmose, O. (2019). *Connecting humans to equations: A reinterpretation of the philosophy of mathematics*. Springer. https://doi.org/10.1007/978-3-030-01337-0

Russell, B. (1901). Recent work on the principles of mathematics. *International Monthly, 4*, 83–101.

Russell, B. (1918). *Mysticism and logic and other essays*. Longmans, Green, & Company.

Sawyer, W. W. (1955). *Prelude to mathematics*. Pelican.

Searle, J. (1969). *Speech acts*. Cambridge University Press. https://doi.org/10.1017/CBO9781139173438

Shapiro, S. (2000). *Thinking about mathematics: The philosophy of mathematics*. Oxford University Press.

Sierpinska, A., & Kilpatrick, J. (Eds.). (1998). *Mathematics education as a research domain: A search for identity*. Kluwer. https://doi.org/10.1007/978-94-011-5470-3

Skovsmose, O. (1994). *Towards a philosophy of critical mathematics education*. Kluwer. https://doi.org/10.1007/978-94-017-3556-8

Skovsmose, O. (2005). *Travelling through education: Uncertainty, mathematics, responsibility*. Sense.

Valero, P. (2004). Socio-political perspectives on mathematics education. In P. Valero & R. Zevenbergen (Eds.). *Researching the socio-political dimensions of mathematics education: Issues of power in theory and methodology* (pp. 5–23). Kluwer. https://doi.org/10.1007/1-4020-7914-1_2

Verschaffel, L., Greer, B., & De Corte, E. (2000). *Making sense of word problems*. Swets & Zeitlinger.

Wagner, R. (2017). *Making and breaking mathematical sense: Histories and philosophies of mathematical practice*. Princeton University Press.

Whitehead, A., & Russell, B. (1910–1913). *Principia mathematica I–III*. Cambridge University Press.

Wittgenstein, L. (1997). *Philosophical investigations*. Blackwell. (Original work published 1953)

Wolfmeyer, M. (2014). *Math education for America? Policy networks, big business, and pedagogy wars*. Routledge.

PART 1

2. Why and how people develop mathematics

Brian Greer

The development of mathematics by humans has a long and unfinished history. In this, necessarily highly selective, overview, the discussion is framed in terms of the environments – physical, cultural, socio-political, specialised – within which people, including those designated as 'mathematicians' do what is called 'mathematics' in all its many forms. These forms include the traditional divide between 'pure' and 'applied'. A distinction is drawn between internal and external processes driving the development, and within internal drivers between those of creation and those of systematisation. The links between this chapter and Chapter 13 are stressed throughout.

Introduction

> Philosophers, like most other people who think about it at all, tend to take 'mathematics' for granted (Hacking, 2014, p. 41).

Arguably, Hacking's observation also holds true for most mathematicians, mathematics teachers, researchers on mathematics education – and everyone else. A major thrust of this book is to combat this tendency.

One of the most important and powerful antidotes to taking mathematics for granted is to examine the history of people – in particular the special kinds of people who are designated as 'mathematicians' – creating, chronicling, developing, systematising, applying what people call 'mathematics' or 'doing mathematics'.

A historian of mathematics faces the problem faced, *mutatis mutandis*, by the anthropologist, the child psychologist, the therapist, and many

©2024 Brian Greer, CC BY-NC 4.0 https://doi.org/10.11647/OBP.0407.02

others, namely how to understand, from within one's one cultural and epistemological frameworks, those of the Other. As pointed out by scholars who have done the hard work, notably Jens Høyrup (Greer, 2021), some historians of mathematics address this challenge better than others.

In the context of China, but with general application, Christopher Cullen (2009) made a fundamental point in ruling out

> the idea that there is a priori a universal ahistorical, cross-cultural 'natural kind' called 'mathematics' that can simply be located and studied once one can penetrate the linguistic barrier to see what it is called in Chinese, and on which one can simply impose all the structures and expectations that a modern person finds in the subject called 'mathematics' in twenty-first-century English. (p. 592)

And, as with all history, the history of mathematics is complexified by gaps, errors of translation and interpretation, ideologically motivated falsifications, and other imperfections in the record. As a particularly striking instance, if you, the reader, would agree with the statement 'Pythagoras was a mathematician', you are recommended to read the entry on Pythagoras in the *Stanford Encyclopaedia of Philosophy*, available online (Huffman, 2018). And the entry on Socrates, in which it is stated that: 'Each age, each intellectual turn, produces a Socrates of its own' (Nails & Monoson, 2022).

It will be obvious that, in the service of writing this chapter, draconian selection was inevitable. The range of educational systems considered is limited. Topics are chosen with an eye to the arguments advanced in Chapter 13 in this volume. Thus, the preponderance of mathematical content addressed does not go beyond that of school mathematics. There is heavy reliance on what I judge to be load-bearing examples.

As a simple but convenient scheme, I frame the discussion by asking what are the 'drivers' of mathematical development, choosing that word to connote both impulsion and steering, the 'why' and the 'how' of the chapter title. I distinguish between *external* and *internal* drivers. The former are framed in terms of adaptations to environments – physical, cultural, political. A theme throughout is the relationship between the two faces of mathematics – on the one hand, the decontextualised codifications of accumulated mathematical knowledge and, on the other, the contextualised applications of mathematics to aspects of physical and human reality.

I further divide the discussion of internal drivers into those relating to acts of creating mathematics, and those relating to acts of systematising – again an obvious simplification that bears on discussion of a number of important issues, such as the fluid relationship between diversification and unification within mathematics, ways in which the development may be considered as following an inevitable trajectory or being contingent, and the relative contributions of individuals and collectives.

Internal drivers shape the discipline that emerged as a self-aware field of human activity in diverse milieux, with their own subcultures and norms, as 'constructed environments'. There are also special-purpose constructed environments relating to particular activity systems that mathematics can serve, such as military engineering.

The history of mathematics makes it abundantly clear that its development is a long and difficult process, and that constitutive of that development are epistemological crises and their resolutions. Periods of relatively steady elaboration and consolidation are punctuated by discontinuities.

In discussing mathematical creativity, I do not focus on the stories of individual triumphs that are often prominent in superficially 'popular' histories; instead, the emphasis is on collective aspects and on some of the salient factors conducive to the gaining and dissemination of new insights. In this respect it is difficult to overstate the importance of material representations, including the revolutionarily new resources made available through computer technology.

Turning to systematisation of mathematical knowledge, it is argued that while many aspects of the development of mathematics are contingent and subject to cultural diversity, that development is not arbitrary, since mathematics is an activity of humans existing in bodies, within social groupings, on a planet that affords underpinnings for mathematics, notably countable entities. Thus, any systematisation will reflect the balance between contingency and constraints. In particular, those constraints are manifest in general mechanisms of development, variously described in terms of hierarchical levels with each succeeding level building on its predecessor, well articulated by Hans Freudenthal

(1991), or Piagetian notions of successions of local equilibria – permeated throughout by the dialectical imperative. Along the same lines, the mathematician William Thurstone (1994) invoked recursion:

> As mathematics advances, we incorporate it into our thinking. As our thinking becomes more sophisticated, we generate new mathematical concepts and new mathematical structures: the subject matter of mathematics changes to reflect how we think. (p. 162)

Next, with a narrowing of focus to 'pure' or 'theoretical' mathematics, attention is given to the emphasis within 'modern mathematics' on elusive, temporary and local, aspirations for certainty such as absolutely precise definitions, irrefutable proofs, impeccable structures. A particular example examined in some detail is the Bourbaki enterprise that enjoyed considerable influence within academic mathematics for much of the twentieth century. That analysis illuminates tensions between mathematics as an academic discipline and mathematics as a school subject, and debate over the extent and nature of the influence of the former over the latter.

The chapter concludes with a brief summary and look ahead to Chapter 13.

External drivers

Mathematical practices may originate in the interactions between the human species and their physical environments, but humans, from a very early stage, have felt needs beyond the necessities of staying alive, including needs that may be described as spiritual, aesthetic, ludic, and the need for explanations and understanding. Thus, astronomy, which has been prominent for so long in so many cultures, has practical aspects relating to navigation, and has also been one of the salient areas for the metanotion that the physical world is governed by laws that can be mathematically framed, and it also has deeply religious connotations.

Then I briefly address the roles of mathematical activities within socio-political environments, with particular attention to how the discipline exists in a symbiotic relationship with the state, reflected in what might be called 'the unreasonable political effectiveness of "mathematics"' (where the quotation marks signal that what is being referenced is the propagandistic use of the word.)

Looking ahead to the next two sections, I consider some of the ways in which, as mathematics emerged as a recognised discipline with its acknowledged experts, creators, systematisers, practitioners, and teachers, external drivers have interacted with the drivers internal to the discipline. And, as an overarching theme, it is proposed that mathematics has 'two faces', one abstract and formal, the other relating to 'the real world' (a concept that I will not attempt to define, but assume to be meaningful in some way to the reader).

Physical environments, practical needs

Humans originally developed practices involving mathematics as part of adapting to their physical environments and the practicalities of survival. There are many experiences underpinning aspects of mathematics that are universal – birth and death, the force of gravity, cycles of day and night, seasons, and tides, observations of the night sky, objects and other entities that afford counting (fingers, prenatally listening to the maternal heartbeat), the approximate symmetry of the human body, and on and on. At this point in history, we should add finiteness in its multiple manifestations as an inherent aspect of the planet we inhabit.

As a counterpoint to universality (there is always a counterpoint) there is diversity in physical environments. It might be expected, for example, that the spatial epistemology, in interaction with visual perception, of people living in a dense forest would differ from that of people living on a treeless plain. One school of thought attributes diversity within the human race to climatic and environmental variation.

As already alluded to, a natural starting point is the human body, with obvious relevance to counting, measuring, perception, movement... The use of the vocal tract, mouth etc. for communication, evolving into language, was foundational for social development, and there followed the emergence of writing which enables, to a significantly greater extent that oral transmission, the extension of communication across space and time (Kaput & Schaffer, 2002). Writing also exemplifies the essentially human (though not exclusively so) characteristic of the use of tools extending the functionality of the body, underlying the emergence of cultural evolution beyond biological evolution.

Practices involving mathematics have been implicated in all forms of interactions of our species with the physical environment: adapting, observing and predicting, recording and organising data, understanding and explaining, controlling and changing to the point of destruction. Alan Bishop (1988) listed six families of practices significantly imbued with mathematical connotations that are found in essentially all cultures, namely counting, locating, measuring, playing, designing, and explaining. The first three, broadly speaking, represent ways of interacting with the physical environment in service of practical requirements, while the last three entail aspects that transcend, to a greater or lesser extent, the immediate needs of survival, as taken up in the next section.

Cultural environments, supra-utilitarian desires

As humans came to live within increasingly complex social/cultural environments, practices involving mathematical elements transcended issues of survival and day-to-day life. The study of mathematics may have been significantly motivated by contemplation of an immortal soul in the face of the ephemerality of bodily death. For many of the recognised greats of European mathematics, even into relatively recent times, the links to (broadly speaking) religious beliefs have been extremely strong (and often overlooked in histories that emphasise the rationality of the 'great men [sic] of mathematics'). Perhaps the hope of finding non-tautologous absolute certainty through mathematics in recent centuries is related to the loss, with the growth of scientific worldviews, of the feeling of absolute certainty attainable through blind religious faith.

Aesthetic impulses run deep. Franz Boas (1927/1955) concluded that:

> No people [...] however hard their lives may be, spend all their time, all their energies in the acquisition of food and shelter [...] Even the poorest tribes have produced work that gives them aesthetic pleasure [...] [They] devote much of their energy to the creation of works of beauty. (p. 9)

Jens Høyrup (2019) discusses the relationship between the geometrical structures (symmetries, in particular) that can be found in pottery, weaving, and other artefacts, and the development of formal geometry. With particular reference to the studies by Paulus Gerdes and his colleagues into the decorative art of Subsaharan Africa, he asserted that 'the decorations of many cultures [...] can be regarded in full right as expressions of formal investigation and experiment' (p. 202). Nevertheless, he cautioned that 'no necessity leads from an aesthetics of forms to formal investigation of forms' (p. 203). In any case, common to aesthetically motivated creations and formal mathematics is the idea of pattern (Mukhopadhyay, 2009).

The ludic impulse ('playing', in Bishop's list) likewise may be invoked as a wellspring of mathematical activity. In the earlier known recordings of mathematical activity, in such forms as cuneiform and papyri, are inscribed mathematical puzzles as well as data and practical problems. And besides puzzles, games of strategy and chance are also found across cultures. The attraction of intellectual play may be seen both in the popularity among general populations of puzzles such as crosswords (I can claim expertise in that field) and in the pursuit of 'pure' mathematics for its own sake. However, as Volker Runde (2003) reminds us, 'mathematicians live in the real world and their mathematics interacts with the real world in one way or another'. Which takes us to the next section...

State environments, socio-political constraints

> The functioning of the modern state presupposes a variety of mathematical technologies – accounting, statistics, and much more. Mathematics, on its part, needs the institutions of the state (schools, universities, research institutions, etc.) to secure financing, recruitment and the rearing of competence. (Høyrup, 2019, p. 635)

In a footnote, Høyrup further comments that, in the last four decades or so, 'without information technology, the immense increase of administrative control of citizens (to mention but that) would never have been possible'.

As societies became more complex, mathematics became a major resource for governance and statecraft. For example, Gary Urton (2009) discussed the complex mathematical resources that served administration of the Inkan Empire. In such examples, we see early examples of what Houman Harouni (2015) terms 'Commercial-Administrative Mathematics' (p. 59), dealing with finance, trade, censuses, labour, and citizenship.

Within Europe, as the Industrial Revolution gathered steam and thereafter, mathematics education was progressively tailored to produce a minimally trained workforce and to prepare people to live as practitioners or consumers of capitalism. Beyond Europe, it was 'the secret weapon of imperialism' (Bishop, 1990), and implicated in White supremacy, so cogently expressed in Høyrup's (2020) phrase 'the ideological shroud assigning the right to conquer and kill in the name of moral superiority' (p. 8).

At the beginning of the nineteenth century, Napoleon wrote that 'the advancement and perfection of mathematics are immediately connected with the prosperity of the state' (cited in Moritz, 1958). We find an echo in the *Executive Summary of the Final Report of the National Mathematics Advisory Panel* (2008), where it is stated that:

> During most of the 20th century, the United States possessed peerless mathematical prowess [...] But without substantial and sustained changes to its educational system, the United States will relinquish its leadership in the 21st century. Much of the commentary on mathematics and science in the United States focuses on national economic competitiveness and the economic well-being of citizens and enterprises. There is reason enough for concern about these matters, but it is yet more fundamental to recognize that the safety of the nation and the quality of life – not just the prosperity of the nation – are at issue. (p. xi)

This quotation exemplifies what President Eisenhower, in a draft of his retirement speech, referred to as the 'military-industrial-academic complex'. Mathematicians benefit from the perceived importance of their discipline, typically with scant acceptance or even awareness of moral responsibilities; enabling 'the unreasonable political effectiveness of "mathematics"'.

Another category proposed by Harouni, that of 'social-analytical mathematics' (p. 67) is exemplified in economics and social statistics.

A genealogical account of the development of mathematics of this kind, based on collecting vast amounts of data and creating conceptual and mathematical means for their analysis, inextricably intertwined with views on the nature of collective human behaviour within societies, was provided by Hacking (1990). Advances in information technology have immensely increased the ability to accumulate and process data and to build models that format many aspects of our lives (as pointed out by Ole Skovsmose for decades), models that are generally beyond the control of those affected and typically not even accessible to their inspection.

As for the relationship between mathematics education and governance, space does not permit even a minimal discussion, so I restrict myself to the following (adapted) aphorism: 'All education tends to control, and mathematics education tends to control absolutely.'

Internal/external drivers, and the two faces of mathematics

> Three main models have been traditionally used to explain scientific development and change. According to one, scientists respond to the results of earlier science and to questions raised by these results ('internalism'); according to another, general (mostly technological) social needs are the moving force, and their absence a brake (one brand of 'externalism'). The third approach [...] looks into the general history of ideas more specifically into the history of philosophy, for the causes that make scientists organise their search and shape their theories as they do. (Høyrup, 1994, p. 124)

Høyrup characterises the above as a simplistic, nevertheless convenient, scheme, and it is so applied in this chapter, simplified further by omitting explicit discussion of 'the third approach' though that does appear *passim* in relation to 'general history of ideas', in particular:

- the emergence of empirical science;
- Eurocentrism – more specifically, Grecocentrism;
- logical positivism and its extended family (discussed in many chapters of this book);
- structuralism (see below).

The two faces of mathematics mentioned in the introduction are reflected in the conventional opposition of 'pure' and 'applied'. In general, the external drivers bear more on applications, and the internal relate more to pure mathematics. Again, there are many interactions, such as the familiar observation that the 'purest' of mathematics turns out to have applications, often decades after its development – even for Hardy, for example (see Chapter 3, this volume). Hacking (2014, pp. 146–148) refers to the older term 'mixed mathematics' reflecting an area such as theoretical physics that is dependent on a combination of empirical investigations and mathematical modelling. Runde (2003) also offered an improvement on 'pure':

> Pure mathematics isn't pure: neither in the sense that it is removed from the real world, nor in the sense that its practitioners can ultimately avoid the moral questions faced by more applied scientists. It would much better be called 'theoretical mathematics'. (p. 3)

This also covers the point made by Hacking (2014, p. 9) that mathematics can be applied to (theoretical) mathematics also.

Modelling acts constitute the interface between the two faces of mathematics. The modelling cycle is often simplistically represented in terms of mathematisation of a situation, derivation of results within theoretical mathematics, interpretation back into the context, and a reality check possibly followed by revision of the model. To those elements should be added (at least) the motivations of the modellers, the adequacy of the assumptions on which the modes is based, the range of applicable mathematics to hand or derivable for the task at hand, communication of interpretations to interested groups.

Historically, modelling was first applied to physical phenomena, notably in cosmology and physics; more recently, particularly through harnessing the power of computer simulations, the modelling of social and political phenomena has become prevalent. For such phenomena, the assumptions on which the model is based become critical, are often extremely tenuous, and ideologically porous. Modelling physical and social phenomena may be broadly contrasted as manifesting 'unreasonable effectiveness' (Wigner, 1960) and 'reasonable ineffectiveness'.

Internal drivers: Creating

For internal drivers, another convenient distinction may be drawn between acts of creating, addressed in this section, and acts of organising, addressed in the next. As throughout the chapter, there is a concentration on strategically chosen aspects laying groundwork for arguments advanced in Chapter 13.

Extending the general notion of environments – physical, cultural, and political – appealed to in the section on external drivers, the first part here deals with the constructed environments within which people designated as 'mathematicians' carry on the activities that are recognised as 'doing mathematics'. Now primarily universities (historically also religious institutions, royal courts, intellectual salons, and other milieux), these also include settings outside the academy, notably military establishments, industry, and the corporate world.

Any study of the history of mathematics makes clear the importance of people running up against the puzzles created when their current ways of thinking cannot cope with what they are noticing. Galileo, for example, was intrigued that there are as many squared natural numbers as there are natural numbers, but it took nearly three centuries before Georg Cantor proposed a reconceptualisation that resolved the issue – and famously commented that 'I see it, but I don't believe it'.

The next focus is on mathematical creativity, highlighting certain intellectual aspects and mental processes, such as those described by George Pólya based on his observations of the behaviour of mathematicians, including himself. A particularly powerful weapon in the mathematician's armamentarium is a sensitive antenna for the perception of structure, in particular the same underlying structure in apparently different contexts. Formally such insights are termed isomorphisms, aphoristically by Henri Poincaré's characterisation of mathematics as the art of giving the same name to different things (Verhulst, 2012, p. 157).

Running through the whole story of human interplay between biological and cultural evolution is the role of material representations (Kaput & Schaffer, 2002). In particular, the impact of computers represents a fifth stage; for a seminal analysis, see Kaput, 1992.

Constructed environments, disciplinary norms

Over the last two millennia or more, with cultural variations, formal mathematics has emerged as a discipline. Analysis in any detail of how this happened in different cultures would require another book. Here I merely stipulate some 'boundary conditions' for such a work, beginning with a caution from Cullen (2009):

> Can we identify an activity in ancient China with a family resemblance to what would nowadays be called 'mathematics'? Or was there a self-conscious and publically recognized group of people in ancient China with a family resemblance to what would be called nowadays 'mathematicians'? (p. 593)

In a similar vein, Høyrup (2013) discussed the criteria that might be used to judge the appropriateness of the term 'Babylonian mathematicians' (concluding that there were some, even if a small minority). Fast-forwarding to the modern era, Karen Parshall (2009) traced the internationalisation of mathematics between 1800 and 1960.

Within the academy, the niches established/occupied by individual mathematicians are naturally diverse (the image comes to mind of Hardy at high table enjoying port and walnuts). In general, a mathematician with a university position has enough financial security to devote her/his time to research and teaching, and most of them do. Further, such an individual enjoys the support of a local and extended community – a very full discussion of such collective aspects will be found in Hersh and John-Steiner (2011). The specific case of the Bourbaki collective, an extreme example of a norm-dense subculture, is discussed below.

There is also the issue of how mathematics relates to other disciplines – most obviously physics, statistics, and computer science but also social sciences – through statistical and other forms of modelling (see Chapter 8, this volume). Further, what I term special-purpose constructed environments exist outside universities. Highly specific constructed environments that come to mind are the Manhattan Project to develop nuclear weapons, and the code-breaking team led by Alan Turing at Bletchley Park; current military applications include a great deal of Artificial Intelligence, for example to program drones so that they can, without human intervention, 'decide' to kill people.

The corporate business world also provides environments for mathematical work. Two examples that spring to mind are William Sealy Gossett, developer of the t-test while employed as the Head Experimental Brewer by Guinness, and Claude Shannon, who developed Information Theory while working for Bell Labs. In both cases, work initially driven by situated problems proved to be of much wider significance (as any psychology student knows).

Epistemological crises, conceptual change

Expanding mathematical knowledge is much more than mere accumulation; it is driven by conceptual restructuring. The history of mathematics is replete with examples of puzzlement. Epistemological crises may break in moments (relatively speaking) when the unthinkable becomes thinkable and the ineffable effable, but the ramifications can extend across centuries (for example, from Galileo to Cantor, referred to above), indeed millennia.

I begin by sketching the fascinatingly complex history of what people have meant by 'number'. Every (or at least, essentially every) culture makes use of counting, and does so in natural ways reflecting the affordances of the environment; beyond that complications ensue. Here I minimally comment on four epistemologically revolutionary extensions of what is meant by 'number', intimately tied to the basic arithmetical operations.

Natural numbers to positive rationals

It appears that for a long time, the conceptualisation of positive rationals remained tied to that of natural numbers. For the Greeks, for example, fractions intimately related to ratios and proportions, often in geometrical contexts. Cultural diversity is evident – why, for example, did the Egyptians and others restrict themselves almost entirely to unit fractions? The Mesopotamians developed procedures for division by fractions equivalent to the rule not infrequently taught to students today to 'invert and multiply', using table of reciprocals. And so on...

Positive numbers to directed numbers

I cannot resist beginning with the quotation:

> 3 – 8 is an impossibility, it requires you to take from 3 more than there is in 3, which is absurd.

The source of the above statement was neither someone writing centuries ago, nor a mathematical ignoramus. It was Augustus De Morgan (1806–1871), an eminent English mathematician, in his extremely interesting book called *Study and Difficulties of Mathematics* (De Morgan, 1831/1910).

While it is relatively easy to expand the domain of application of numbers to directed numbers for addition and subtraction, it took a very long time to agree on an explanation for something the poet W. H. Auden, in his *A Certain World* (1970), remembered from school:

> Minus times minus makes a plus
> The reason for this we need not discuss.

Rational numbers to real numbers

The realisation that, for example, the exact length of the diagonal of a unit square cannot be expressed as the ratio of two natural numbers required a reconceptualisation of number; the details are unclear in a historical record complicated by mythical stories. It is generally considered that a rigorous theory of irrational numbers was accomplished in the nineteenth century by Richard Dedekind, Cantor, and Karl Weierstrass.

Real numbers to complex numbers

The story of how complex numbers came to be accepted is even more fascinating. A key part was the invention of diagrams providing a representation for the numbers and arithmetical operations on them. And then there are quaternions, surreal numbers, on and on ... *And they are all called 'numbers'*!

The above sketch primarily relates to the expansion of numbers within theoretical mathematics. Another perspective is that numbers are embedded within cultural matrices – in Urton's (1997) phrase, they have a 'social life'. In contrast to the formal structural analysis of numbers and the operations upon them (see Bourbaki discussion below), in human contexts multiplication and division are polysemous (Greer, 1992). Emphasis on what people *do with numbers*, whether for practical purposes, societal functioning, or for intellectual pleasure, contrasts with what I assert, without further elaboration, is the unproductive, arguably even meaningless, question 'Do numbers (of a specified type, especially negative, irrational, complex) *exist*?'.

Beyond arithmetic, parallel examples can easily be found illustrative of the points attempted in this chapter from the histories of other components of school mathematics: algebra, geometry, calculus, probability. Space allows only the briefest hints of how those discussions might go:

- For 2500 years, formal algebra ('rich in structure but weak in meaning' as René Thom put it) had little or no practical purpose (Høyrup, 2013). The familiar school algebra of today (satirised as 'the intensive study of the last three letters of the alphabet') is the product of a representationally driven development over millennia.
- As the familiar story goes, Euclid's *Elements* provided a model of the axiomatic method in mathematics until flaws were discovered and rectified by David Hilbert – at the cost of losing the simplicity of the original five axioms. And the problem of the fifth axiom, that bothered mathematicians (such as Omar Khayyam) for a very long time, finally was resolved (at least temporarily) by the emergence of non-Euclidean geometries. Further liberating reconceptualisations ensued, with the escape from a mere three dimensions to many, and on to Mandelbrot's exposition of fractal geometry. For some mathematicians (notably the Bourbakists), geometry became detached from its roots in locating and spatial cognition, and was absorbed into formalism.
- The story of calculus is long, and profoundly illustrates the importance of representations (Kaput, 1994). While its roots lie deep in intuitions of time and movement as continuous, it became a major topic for the nineteenth-century drive for

rigour. And it raises issues of intellectual priority between 'Europe' and India.

- Probability is a very special case, since its explicit mathematisation is relatively recent and hence more open to historical documentation and analysis. Hacking's (1990) remarkable work shows how its development was related to the most general social and political issues of statehood, mass collections of data, conceptions of the nature of humanity in the mass.

Across all of the branches of mathematics, there has occurred a revolutionary shift from the conception of mathematical formulations as providing, in some sense, a direct picture of the world, to the reconceptualisation that they model the world, in some sense.

There are many theoretical frameworks that may be invoked to explicate the above. In terms of theoretical mathematics, a pervasive need is for closure, in the technical sense. It is a prime driver in the expansion of 'number' to more and more complex structures. When addition and its inverse, subtraction, and multiplication and its inverse, division, arise through contemplation and applications of the natural numbers, the fact that subtraction and division are not always possible drives consideration of the possibility of negative and rational numbers and so on, for each expansion. At each stage, a local equilibrium is achieved (the real numbers, with a coherent representation in the number line, the complex numbers underpinning the fundamental theory of algebra) which itself harbours the germ of a disequilibrium. The parallel with a central aspect of Jean Piaget's account of cognitive development should be obvious.

Mathematical creativity

Throughout the history of mathematics, there have been individuals who have realised remarkable insights in posing and solving mathematical problems. The completion of a proof of Fermat's Last Theorem by Andrew Wiles and others arouses intense admiration among the general public, although, or perhaps because, the technical details are beyond all but a very small number of mathematicians; yet the pleasure of solving a problem insightfully is open to everyone, including children. On the assumption that the reader knows some or many of the canonical examples, the focus of this section is on how the individual's intellectual feats are embedded within the collective activities of communities of mathematicians.

To begin with a statement of the obvious, mathematicians approaching a creative challenge come forearmed with a great deal of resources – including methods, representations, proofs, structural analysis, problem-solving strategies, and so on (and they differ fundamentally from schoolchildren in these respects). These resources have been recorded, accumulated, critiqued, and systematised over centuries and multiple cultures. Mathematicians operate within the circles of their forebears and contemporaries, which, due to communicational advances, are now globally and speedily accessed.

These resources are activated by deploying a range of routine methods, heuristics, strategies, combined with mental flexibility. Above all is the disposition to look for and exploit structure. One of my teachers used to say 'Good mathematicians are lazy' by which he meant that they would look for an insightful rather than routine but laborious technique. (The apocryphal story of the young Gauss finding a 'smart' way to sum the integers from 1 to 100 is the classic example.)

Through observation and analysis of the behaviour of mathematicians, including himself, and a great variety of examples, Pólya inductively taxonomised some of these strategies. The following are among the most salient aspects of mathematicians' armamentaria:

- In its most explicit form, the exploitation of structure involves an isomorphism (Greer & Harel, 1998). A famous example is this account of a sudden insight:

> The idea came to me, apparently with nothing whatever in my previous thoughts having prepared me for it, that the transformations which I had used to define Fuchsian functions were identical with those of non-Euclidean geometry. (Poincaré, quoted in Newman, 1956, Vol. 4, p. 2020)

- By a kind of pattern recognition, before thinking about the details of a solution it is often possible to recognise problem/solution types, e.g., 'this kind of problem may well hinge on finding an invariant', 'clearly this can be handled by induction', and so on.

- Intuition is very frequently invoked by mathematicians. An agreed-upon definition, let alone a convincing theoretical explanation remains elusive (Fischbein, 1987). For present purposes, I take it to mean 'any immediate inference in which there is no conscious reasoning' (Hacking, 2014, p. 17).

Beyond the deployment of these resources, in ways that may be more or less routinised, there are the most elevated forms of creativity when an individual or group achieves a conceptual restructuring, finds a hitherto unknown proof that goes to the structural heart of a big idea – or designs a transformative representation.

Material representations

Material representations have been of crucial importance in the creation, accumulation, organisation, and communication of mathematical results. As throughout this chapter, an attempt is made to use space efficiently through powerful examples. In this section, the focus is on inscriptions on paper and other materials (in particular notations), and diagrams. A separate section, which follows, outlines the revolutionarily new resources afforded by advances in computer-based representations (Kaput, 1992). No attempt is made to address the vast topic of natural language and mathematics or that of mental representations.

A human starting point is the body; it is no accident that the most common bases for numerical systems are 5, 10, 20; some cultures go beyond manual and pedal digits. Body parts are also ubiquitous in measurement (hand, foot, cubit...). And in recent years much attention has been given to embodied cognition.

If it is helpful – which I doubt – to speak of mathematics as a language, then it is one that draws on natural languages, with enhancements, and with particular notations. A glance at the encyclopaedic work of Florian Cajori (e.g., 1928–1929/1993) is enough to make clear how rich and complex, messy and arbitrary, has been the evolution of such. A familiar example of how instrumental a good notational representation can be is the contrast between the user-friendliness of decimal numbers for purposes of calculation and the system used by the Romans. For a more advanced example, De Morgan (1910, p. 185), citing Pierre-Simon Laplace and referring to notation for powers, wrote:

> Newton extended to fractional and negative powers the analytical expression which he had found for whole and positive ones. You see in their extension one of the great advantages of algebraic language which expresses truths much more general than those which were at first contemplated [...]

(which stands in marked contrast to his blinkered view on 3 – 8 cited above). Thus, the power of the notation x^n is that it opens up the possibility of conceiving of other values of n, eventually leading to the remarkable equation $e^{i\pi} = -1$ (Lakoff & Núñez, 2000, p. 433).

Graphical representations, naturally enough, are central to geometry, combined with the conventions for using letters to label elements. Here may be mentioned the distinctive position of Reviel Netz, emphasising what Bruno Latour (2008, p. 3) called 'scripto-visual inventions':

> I will argue that the two main tools for the shaping of deduction were the diagram, on the one hand, and the mathematical language on the other hand. Diagrams – in the specific way they are used in Greek mathematics – are the Greek mathematical way of tapping human visual cognitive resources. Greek mathematical language is a way of tapping human linguistic resources [...] But note that there is nothing universal about the precise shape of such cognitive methods. They are not neural; they are a historical construct. [...] One needs studies in cognitive history, and I offer here one such study. (Netz, 2003, pp. 6–7)

The fusion of geometry and algebra was, of course, a revolutionary passage in the history of mathematics, heavily dependent on the invention of Cartesian graphs. In similar vein, an extended analysis of the long history of representations in the development of calculus was provided by James Kaput (1994). And, as Kaput (1992) has pointed out, a fundamental level-shift in material representations lies between those which record and those which are manipulable, for example for executing calculations (e.g., the abacus or the Quechuan *yupana*). Computers have taken representational resources to new levels, as outlined next.

Computers: Opening new representational windows

Computers and associated technologies have significantly changed the doing of, and the conception of, mathematics in multiple ways, including the following:

- Most obviously, increase in brute computational power as exploited, for example, in testing conjectures and, generally, leading to the acceptance that an empirical element may enter mathematics.
- The theoretical notion of computability, captured in the conceptual device of the Turing machine, leading inexorably to analyses of the limitations of computability.
- Changes in the conception of proof prompted by computer proofs and discussions – philosophical and practical – about their status. Hacking (2014) discusses the debate over whether a totally computerised proof machine will ever be possible.

Perhaps most importantly, computers have provided more powerful representational resources. Benoît Mandelbrot, the originator of fractal geometry (a creative feat hard to imagine possible before the computer era), commented that 'computers have put the eye back into computing'. A small sample:

- Being able to represent continuous change in a perceptually direct way, thereby moving 'past the algebra bottleneck' (Kaput, 1998, p. 278) *en route* to calculus.
- The representation of geometrical procedures, not just diagrams. For example, consider the theorem that joining the midpoints of the sides of any quadrilateral produces a parallelogram. Using Geometer's Sketchpad, the user can store a procedure (not a static image) corresponding to that result. Then, any vertex of the quadrilateral can be 'grabbed' by the cursor and moved, and the whole configuration moves accordingly; it should be clear that this gives a whole new insight into the invariance at the centre of the theorem.
- Generativity, as shown *par exemple* in the simplicity of the Turing machine, relative to the huge mathematical edifice that can be built on that foundation. Another example is the Logo programming language built on the two primitives of moving forward a certain distance, and rotating through a certain angle. The language affords construction of a hierarchy of procedures building on procedures.

- Computers allow the display of much more complex data, qualitatively as well as quantitatively for example data varying over time, and for realistic modelling, for example the software STELLA, which enables school students to build, run, and evaluate system dynamical models (e.g., Fisher, 2021).

Internal drivers: Systematising

> A great part of mathematical activity today is organizing [...] When compared with creating, organizing scientific cognition seems to be an inferior activity. Yet [...] in no science are these two activities so densely interwoven as they are in mathematics. (Freudenthal, 1973, p. 414)

The sheer volume of established mathematical knowledge now is such that consolidating it as a coherent body of knowledge and techniques is a daunting task, even if restricted to 'pure' mathematics, as is the focus of this section. Material resources deployed in the attempt include inscriptions, notational systems, taxonomies, books that survey the field, classical textbooks. Internally, there are definitions, axioms, theorems, visual representations, structures...

Does it make sense to speak, as Nicolas Bourbaki (1950) did, of 'the architecture of mathematics'? Arguments against that are advanced below. Admittedly, there is considerable agreement on what is accredited within theoretical mathematics (using that term instead of 'pure'). Rejecting teleology as I do, thereby refusing to accept a forced choice between the development of mathematics being inevitable or contingent, the position of Rafael Núñez (2000) seems appropriate:

> Mathematics is not transcendentally objective, but it is not arbitrary either (not the result of pure social conventions). (p. 3)

In support of this position, examples are cited where some aspects of codified mathematics seem inevitable, being tied to the human condition, and reflecting a hard-to-deny internal coherence. Other aspects are contingent, reflecting environmental and cultural diversity, the impact of external events, technological developments, specific individual and collective creative acts. Relevant also is the evolutionary perspective; mathematics generated is subject to selection processes, only the fittest surviving.

Three aspects central to the systematisation of theoretical mathematics are abstraction from its human roots, rigour, and structure. The drive for rigour is perhaps the most defining characteristic of European mathematics in the nineteenth and twentieth centuries. And the complex notion of structure (with loose and disputed ties to the amorphous movement called 'structuralism'), does provide a systematic summarising of a great deal of theoretical mathematics, which in turn constitutes a very powerful resource for mathematicians advancing the field.

All of those aspects are clearly exemplified in the Bourbaki movement of the twentieth century which made a heroic, but arguably doomed, attempt to define the architecture of which Jean Dieudonné spoke.

Historico-genetic development of mathematics: Inevitable and contingent

> A repeating process, an interplay of form and content, which characterizes mathematical thought (Freudenthal, 1991, p. 10)

Explaining why he adopted the term 'anthropology of mathematics' to characterise his scholarly field, Høyrup (1994) stated:

> What I looked for was a term which suggested neither crushing of the socially and historically particular nor the oblivion of the search for possible more general structures: a term which neither implied that the history of mathematics was nothing but the gradual but unilinear discovery of ever-existing Platonic truths nor [...] a random walk [among] an infinity of possible systems of belief. A term, finally, which involved the importance of cross-cultural comparisons. (p. xi)

(And see the quotation from Núñez above.)

The question 'Is the development of mathematics inevitable or contingent?' presents, in my view, a false choice. There are, indeed, aspects of the development of mathematics that it is hard to imagine happening otherwise. Arguably the clearest example is 'number' as the usages of the word developed over many centuries, from the naturally termed 'natural numbers'. As Freudenthal (1991) put it:

> The first non-trivial structure as such, i.e. whole number as the product of the process of counting, begot rich process and product content which,

> organised by ever new structures, in turn begot new contents – a never ending cyclic process. (p. 10)

Mathematics has to be generative for the same reasons that language is generative. As more complex societies evolved, the practical need for dealing with large numbers meant that structural intervention became necessary in order to avoid the fate of Jorge Luis Borges' character 'Funes the Memorious', who had a separate image and name for every natural number. The specifics may be contingent, but the emergence of *some* such construction seems inevitable.

Further, there are many aspects of mathematics that make it difficult to disagree with Freudenthal (1991) when he stated that 'mathematics grows, as it were, by a self-organizing momentum' (p. 15). Again, think of numbers, and the simple example of going beyond whole numbers to fractions, motivated by so many practical situations. Of course, this took many centuries, with great cultural variation. The extension to directed (negative as well as positive) took even longer to bring to the point of formal respectability, although people managed much earlier to deal with practicalities such as debt.

In a very thorough and nuanced discussion of the issue, Hacking (2014) declares his support for what he calls 'the Latin model', the name being suggested by an analogy of the evolution of Latin into Romance languages – contingent in detail, but subject to significant constraints. He also argued that, while 'our notion of the infinite was not inevitable [...], our notion of complex numbers was inevitable' (pp. 117–121). It is hard to see the development of numbers beyond the counting numbers as other than inevitable, but that is clear only insofar as *some* such development was driven by practical, and also supra-utilitarian, needs. But inevitable to what degree? Fractions and negative numbers, surely, but complex numbers? Quaternions and octonions? Conway's surreal numbers? And when studying the natural numbers as a system became an interest, was it inevitable that prime numbers should take such a central role? Perhaps, but how about other named numbers with special properties given poetically suggestive names – 'perfect', 'amicable', and so on?

The formalisation and systematisation of mathematics accomplishes a great deal in terms of generativeness. From the five axioms of Euclid, a huge edifice can be constructed (albeit Hilbert pointed out cracks a long time later). The definition of a group is simple (see next section) yet on that foundation, again, so much can be constructed, including the recently completed classification of finite groups.

Have you noticed that when people want to argue that mathematics is universal and certain, they use simple examples, such as 'the angles of a triangle add up to 180°' or 2 + 2 = 4? With appropriate clarification, it's hard to argue with either statement. But similar statements about, for example, non-standard analysis are not at all clear. And there is no such simplicity or obviousness about probability, in particular *subjective* probability (Devlin, 2014). It is a norm within academic mathematics to take proof as central, yet Hacking (2014) was prepared to argue that 'deep mathematics could have developed without proof at all' (p. 115).

'Self-organising' may be interpreted in terms of each local equilibrium containing within itself the germ of disequilibrium. Studying the real numbers, mathematicians, from at least the Babylonians, became interested in quadratic and cubic equations. Throw in the apparently very strong psychological need for closure, in the mathematical sense, and eventually the need for positing the square root of –1 became tempting though frightening, then it appeared to work, eventually it became formally ratified.

Thus, time-dependency must be acknowledged. What appears inevitable in hindsight was certainly not so during the struggles for epistemological coherence. And the temptation to believe in teleology, implying the possibility of a definitive characterisation of mathematics, does not hold up. In his critique of the Bourbaki-Piaget axis (see Chapter 13), Freudenthal (1973) stated as follows:

> Piaget is not a mathematician, so he could not know how unreliable mathematical system builders are [...] Mathematics is never finished – anyone who worships a certain system of mathematics should take heed of this advice. (p. 46)

Or, as Høyrup (1995) put it:

> No critique is ever definitive. What seemed at one moment to be an absolute underpinning [...] turns out with historical insight to make other 'naïve' presuppositions which in their turn can be 'criticized'. (p. 5)

The discipline of the discipline: Abstraction, rigour, proof, structures

Dominant themes of (European and extended-European) theoretical mathematics in the nineteenth and twentieth centuries were: increased abstraction; the drive for rigour, yoked to the desire to establish unassailable proofs; specification of a structural architecture – all manifestations of a craving for absolute certainty, as was the dream of establishing mathematics on logic, discussed in various chapters of this volume. Elements of this quest include the chimerical search for absolute definitional precision, the power of axioms, impeccably formal proofs, and a network of abstract structures, replacing the problematic metaphor 'mathematical objects' with webs of relations among undefined entities. Discussion of abstraction and rigour will be found in various chapters of this book (and see the next section, on Bourbaki); some key points about proof and on general and specific notions of structure, follow.

A great deal of Hacking (2014) is concerned with proof. In particular, he makes a clear distinction between 'two visions of proof' (p. 11) which he labels with the names of two mathematical greats:

> There are proofs that, after some reflection and study, one totally understands, and can get in one's mind 'all at once'. That's Descartes.
>
> There are proofs in which every step is meticulously laid out, and can be checked, line by line, in a mechanical way. That's Leibniz.

Leibnizian proof is the dominant image of how people do proofs, reinforced by the norm of publishing mathematical papers whereby all traces of how the proof was found are expunged.

One characterisation of mathematics is as 'the study of all possible patterns'; patterns may be thought of as partial manifestations of the rigorously defined structures of modern mathematics. The pattern of addition and subtraction of even and odd integers (even + even = even, etc.) is accessible to quite young children; formally, this pattern is a feature of one instantiation of a group with two elements. The concept of a group is simple to define, yet with immense ramifications both in terms of modelling situations and in terms of the architecture of formal mathematics. A group is defined as the coupling of a set, S (which may have a finite or infinite number of elements), and an operation, S, applicable to any two elements of S and having certain properties (the list is redundant relative to minimal definitional requirements), including the following:

- Closure: For any two elements of S, x and y, $x \circ y$ exists and is an element of S.
- Identity element: S contains a unique element, e, such that for any x, $x \circ e = x$ and $e \circ x = x$.
- Inverse: for any x, there is another element x^{-1} in S such that $x \circ x^{-1} = x^{-1} \circ x = e$.

All of these properties relate to extremely pervasive aspects of mathematics.

The history of how this axiomatisation crystallised out of multiple, apparently unrelated, situations that could be modelled by groups is a fascinating episode in the history of mathematics. In arithmetic, the (positive and negative) integers, rationals, real numbers, and complex numbers, with the operation of addition, form groups, for example. In geometry, systems of transformations form groups. Galois theory in algebra is based on group theory. Groups are central to the theory of crystallography. They have been invoked by Piaget and Claude Lévi-Strauss, and are pervasive in the work of M. C. Escher. Rubik's cube was designed to help teach group theory. Groups and other structures such as rings and fields are central to the proposed architecture of mathematics as envisaged by the Bourbaki collective, to which we next turn.

The case of Bourbaki

> The most spectacular example of organizing mathematics is, of course, Bourbaki. (Freudenthal, 1973, p. 46)

There are two central reasons for including this section. First, as expressed in the quotation above, the Bourbaki project stands as the supreme attempt to deliver an organisation for *selected* parts of mathematics (excluding applied mathematics, probability theory, and much else). Second, as is taken up in Chapter 13, the influence of Bourbaki (not always emanating from Bourbaki itself) spread into mathematics education, with continuing and arguably harmful ramifications.

Bourbaki also serves as probably the most extreme example of a constructed environment within which self-described 'working mathematicians' could do mathematics, one whose origins can be traced back to the aftermath of the First World War, when the ranks of French academic mathematicians were depleted due to many of them falling in the war, and survivors wished to restore the standing of *French* mathematics. The very distinctive organisation of Bourbaki as a kind of secret society (Mashaal, 2006), or club, is well summarised in the Wikipedia entry (and see Hersh & John-Steiner, 2011, pp. 181–191). What is clear is that their effort to systematise mathematics in an uncompromisingly formalist style based on defining mathematical structures was extremely influential on the field through much of the twentieth century, with residual influence to this day. The following, from one of the most distinguished mathematicians of the recent past, strikes me as a balanced view:

> All mathematicians of my generation, and even those of subsequent decades, were aware of Nicolas Bourbaki, the Napoleonic general whose reincarnation as a radical group of young French mathematicians was to make such a mark on the mathematical world. His memory may now have faded, the books are old and yellowed, but his influence lives on. Many of us were enthusiastic disciples of Bourbaki, believing that he had reinvigorated the mathematics of the twentieth century and given it direction. But others believed that Bourbaki's influence had been pernicious and narrow, confining mathematics behind walls of rigour, and cutting off its external sources of inspiration. (Atiyah, 2007, p. 1150)

Atiyah neatly underlines the last point by pointing out that 'had Euler worried too much about rigour, mathematics would have suffered' (p. 1151).

While most emphasis is on the collective aspect of the Bourbaki mathematicians, Gerhard Heinzmann and Jean Petitot (2020) clarify that Bourbaki 'was at the same time the collective author of a monumental and long-lasting treatise [...] and a pleiad of individual geniuses [...] who were at the cutting edge of innovation and creativity' (pp. 187–188). Heinzmann and Petitot also emphasise the view within the collective persona of Bourbaki that they were providing a powerful toolbox to facilitate the creativity of 'working mathematicians'. They also point to a central Bourbakian tenet of the unity of mathematics, as implied by the singular form in the title of their treatise *Elements de Mathématique*, and as manifest in so many examples of structural connexions across diverse branches of mathematics. Hacking (2014, p. 13) stated that 'the history of mathematics is one of diversification and unity', so that when, for example, Descartes brought together geometry and algebra, they turned out to be 'the same stuff' (p. 11).

Other mathematicians diverged from Bourbaki, including Thom (1971) and Mandelbrot (2002). Alexander Grothendieck proposed a new organisation around category theory that was not taken on board. Mandelbrot (2002) convincingly argues that Bourbaki's history was shaped by a series of historical accidents that they never acknowledged, believing themselves to be 'the necessary and inevitable response to the call of history' (p. 31).

The attenuated but continuing impact of Bourbaki on school mathematics is discussed in Chapter 13. Here, for the sake of brevity, I point to some facets of what I see as the supreme irony of Bourbaki – the contrast between its adherence to *mathematical* rigour and the irrationality and contradictions of its philosophical and socio-political stances:

- *Universal versus chauvinistic mathematics*: The image of mathematics venerated within Bourbaki is universal, yet the organisation of Bourbaki as a constructed environment within which to systematise mathematics was decidedly French in terms of original motivation, membership (predominantly), and style.
- *Cavalier attitude to philosophy*: As expressed by Reuben Hersh most 'working mathematicians' do not fret over philosophical issues. Dieudonné (1970) made a similar comment:

> On foundations we believe in the reality of mathematics, but of course when philosophers attack us with their paradoxes we rush to hide behind formalism [...]. Finally, we are left in peace to go back to our mathematics and do it as we have always done, with the feeling each mathematician has that he is working with something real.

- Mandelbrot (2002) stated that Bourbaki had 'only contempt for the logical foundations of mathematics', such as the work of Kurt Gödel and Turing (p. 31).

- *Bizarre claim to include all of mathematics*: in their series of textbooks (more accurately described as an encyclopaedia, as Leo Corry (2009) has pointed out) they claimed to be surveying the whole of modern mathematics, despite totally excluding applied mathematics, any connection with physics, and also probability.

On a more specific point, their taboo against diagrams (while allowing themselves poetic licence in choosing technical vocabulary) is hard to understand, and I have found no clear explanation for that. It is hardly surprising, then, that mathematicians such as Thom and Mandelbrot were alienated.

Their claimed liaisons with other manifestations of the general cultural movement, structuralism (or, to be careful, other uses of the term) seem opportunistic – in particular, their rather one-sided romance with Piaget. Hacking (2014) proposed a clear distinction between 'mathematician's structuralism' and the structuralism of recent analytic philosophy (p. 237). Kantor (2011) unequivocally characterised the supposed relationship of Bourbaki's structures to structuralism as 'pure intellectual fraud' and he elaborated that 'referring to Bourbaki in structuralist essays was a way of giving some scientific credit and weight to works of variable quality' (and see Aubin, 1997).

The Bourbakists were, of course, entitled to define 'mathematics' as they wished, essentially ignoring one of its faces, as long as the definition was clear, which it was. However, it is arguable that they influenced the image of mathematics among mathematicians and non-mathematicians in an unbalanced way, which had harmful effects on mathematics education during the New Math period and continuing to this day.

Looking back and forward

Central to this book is disruption of the tendency to take for granted mathematics-as-discipline, mathematics-as-school-subject, and the relations between them; to that end, this chapter is intended to support arguments advanced in the intimately related Chapter 13. A necessarily broad-brush sketch of the history of people developing mathematics has been attempted, with a simplifying framework of distinguishing between external and internal drivers, and between acts of creating and acts of systematising.

The history of mathematics continues to happen. Looking into the future, a clearer picture of the powers and limitations of computers and Artificial Intelligence will emerge. There is no lack of unfinished business from the past. When Hilbert, in 1900, set out twenty-three mathematical problems to be solved in the twentieth century, the continuum hypothesis was the first. This unproved hypothesis relates to the cardinality of the real numbers, which, since Cantor, is known to be greater than that of the natural numbers (or the rationals) but it remains unknown whether there are any intermediate cardinalities. A major theorem which may help to settle the issue was recently published (Asperó & Schindler, 2021). This example serves admirably to show that many questions within mathematics remain open, as does the Wikipedia summary of the current consensus among mathematicians in relation to the Hilbert's problems as to whether they have been solved, remain unsolved, or were not stated with sufficient precision.

Among the overarching themes in this chapter selected for the framing of Chapter 13 are: the conception of environments from physical through socially constructed, the emergence of mathematics and mathematicians as identifiable collective activities and actors, the two faces of mathematics, the centrality of epistemological shocks and their resolutions, the defining characteristic of mathematicians to look for and exploit patterns, and the role of material representations.

Key issues to be addressed in Chapter 13, with references back to this chapter, include:

- The relevance of history of mathematics to school mathematics, rejecting any simplistic interpretation of 'ontogeny recapitulates phylogeny'. In particular, what can be learned from epistemological crises and their resolutions to figure out how to help children through the radical reconceptualisations they need to negotiate.
- The embeddedness of mathematics in culture, despite historical disembeddings, which has massive implications for school mathematics.

- Absolutely central point: the child learns *under instruction* – in a constructed environment, insofar as school learning is concerned. This is, perhaps, the fault-line between Freudenthal and Piaget. I will argue that the latter's idea of some sort of 'natural development', and variations on that theme by radical constructivists, do not bear examination. Piaget was impressed by someone figuring out that the cardinality of a set of objects is independent of the order of counting but offered no account, as far as I can see, of how to get from there to, say, the solution of a quadratic equation.
- In particular, the supposed correspondence between the mother-structures of Bourbaki and the structures of Piaget's developmental theory, and the educational damage that resulted, will be addressed. The waning direct influence of Bourbaki does not mean that it is dead. A contemporary curricular framework (for which I suggest the Common Core State Standards in the US affords a representative example) could be characterised as 'Bourbaki light' – a sequence structured in a superficially 'logical' form.
- School mathematics is predominantly presented as pre-systematised, with little opportunity for students to experience systematising, let alone creating.
- In relation to mathematical modelling, it will be argued that school mathematics, in general, fails to deal with the core issues. A long-term research program on word problems feeds directly into this discussion, and a thread can be followed from there to all the work on formatting and so on.
- Discussion of Bourbaki naturally raises the question of selection. Out of all the mathematics now assembled and organised, what should be selected for children to learn in school? Some possibly iconoclastic ideas for curriculum and pedagogy will be presented.
- It will be argued that academic mathematicians enjoy too much power to influence how school mathematics is framed and done.

- Elementary mathematics education is foundational, not just for later mathematics education, but in the framing of an individual's worldview; it will be argued that school mathematics, as typically practiced, tends to produce a destructive image of mathematics.

References

Asperó, D., & Schindler, R. (2021). Martin's Maximum^{++} implies Woodin's P_{max} axiom (∗). *Annals of Mathematics, 193*(3), 793–835. https://doi.org/10.4007/annals.2021.193.3.3

Atiyah, M. (2007). *Bourbaki, A Secret Society of Mathematicians* and *The Artist and the Mathematician* [Book review]. *Notices of the American Mathematical Society, 54*(9), 1150–1152.

Aubin, D. (1997). The withering immortality of Nicolas Bourbaki: A cultural connector at the confluence of mathematics, structuralism, and the Oulipo in France. *Science in Context, 10*(2), 297–342. https://doi.org/10.1017/S0269889700002660

Bishop, A. J. (1988). *Mathematical enculturation: A cultural perspective on mathematics education*. Kluwer. https://doi.org/10.1007/978-94-009-2657-8

Bishop, A. J. (1990). Western mathematics: The secret weapon of cultural imperialism. *Race & Class, 32*(2), 51–65. https://doi.org/10.1177/030639689003200204

Boas, F. (1955). *Primitive art*. Dover. (Original work published 1927)

Bourbaki, N. (1950). The architecture of mathematics. *American Mathematical Monthly, 57*(4), 221–232.

Cajori, F. (1993). *A history of mathematical notations*. Dover. (Original work published 1928–1929)

Corry, L. (2009). Writing the ultimate mathematics textbook: Nicholas Bourbaki's Éléments *de Mathématique*. In E. Robson & J. Steadall (Eds.), *Oxford handbook of history of mathematics* (pp. 565–588). Oxford University Press.

Cullen, C. (2009). People and numbers in early imperial China. In E. Robson & J. Steadall (Eds.), *Oxford handbook of history of mathematics* (pp. 591–618). Oxford University Press.

De Morgan, A. (1910). *Study and difficulties of mathematics*. University of Chicago Press. (Original work published 1831)

Devlin, K. (2014). The most common misconception about probability? In E. J. Chernoff & B. Sriraman (Eds.), *Probabilistic thinking: Presenting plural perspectives* (pp. ix–xiii). Springer.

Dieudonné, J. A. (1970). The work of Nicholas Bourbaki. *American Mathematical Monthly, 77*, 134–145.

Fischbein, E. (1987). *Intuition in science and mathematics: An educational approach.* Reidel.

Fisher, D. (2021). Global understanding of complex systems problems can start in pre-college education. In F. K. S. Leung, G. A. Stillman, G. Kaiser, & K. L. Wong (Eds.), *Mathematical modeling education in East and West*. Springer. https://doi.org/10.1007/978-3-030-66996-6_3

Freudenthal, H. (1973). *Mathematics as an educational task*. Reidel.

Freudenthal, H. (1991). *Revisiting mathematics education*. Kluwer.

Greer, B. (1992). Multiplication and division as models of situations. In D. Grouws (Ed.), *Handbook of research on mathematics education* (pp. 276–295). Macmillan.

Greer, B. (2021). Learning from history: Jens Høyrup on mathematics, education, and society. In D. Kollosche (Ed.), *Exploring new ways to connect: Proceedings of the Eleventh International Mathematics Education and Society Conference* (Vol. 2, pp. 487–496). Tredition. https://doi.org/10.5281/zenodo.5414119

Greer, B., & Harel, G. (1998). The role of isomorphisms in mathematical cognition. *Journal of Mathematical Behavior*, *17*(1), 5–24. https://doi.org/10.1016/S0732-3123(99)80058-3

Hacking, I. (1990). *The taming of chance.* Cambridge University Press. https://doi.org/10.1017/CBO9780511819766

Hacking, I. (2014). *Why is there philosophy of mathematics at all?* Cambridge University Press. https://doi.org/10.1017/CBO9781107279346

Harouni, H. (2015). Toward a political economy of mathematics education. *Harvard Educational Review, 85*(1), 50-74. https://doi.org/10.17763/haer.85.1.2q580625188983p6

Heinzmann, G., & Petitot, J. (2020). The functional role of structure in Bourbaki. In E. H. Reck & G. Schiemer (Eds.), *The prehistory of mathematical structuralism* (pp. 187–214). Oxford University Press. https://doi.org/10.1093/oso/9780190641221.003.0008

Hersh, R., & John-Steiner, V. (2011). *Loving and hating mathematics*. Princeton University Press.

Høyrup, J. (1994). *In measure, number, and weight*. State University of New York Press.

Høyrup, J. (1995). *The art of knowing: An essay on epistemology in practice* [Lecture notes]. https://ojs.ruc.dk/index.php/fil1/article/view/1947

Høyrup, J. (2013). *Algebra in cuneiform* [Preprint]. Max Planck Institute for the History of Science. https://www.mpiwg-berlin.mpg.de/Preprints/P452.PDF

Høyrup, J. (2019). *Selected essays on pre- and early modern mathematical practice.* Springer. https://doi.org/10.1007/978-3-030-19258-7

Høyrup, J. (2020). *From Hesiod to Saussure, from Hippocrates to Jevons: An introduction to the history of scientific thought between Iran and the Atlantic* [Preprint]. Max Planck Institute for the History of Science.

Huffman, C. (2018). Pythagoras. In E. N. Zalta (Ed.), *The Stanford Encyclopedia of Philosophy* (Winter 2018 Edition). https://plato.stanford.edu/archives/win2018/entries/pythagoras

Kantor, J.-M. (2011). Bourbaki's structures and structuralism. *The Mathematical Intelligencer, 33*(1). https://doi.org/10.1007/s00283-010-9173-4

Kaput, J. J. (1992). Technology and mathematics education. In D. A. Grouws (Ed.), *Handbook of research on mathematics teaching and learning* (pp. 515–556). Macmillan.

Kaput, J. (1994). Democratizing access to calculus: New routes to old roots. In: A. H. Schoenfeld (Ed.), *Mathematical thinking and problem solving* (pp. 77–156). Lawrence Erlbaum Associates.

Kaput, J. J. (1998) Representations, inscriptions, descriptions and learning: A kaleidoscope of windows. *Journal of Mathematical Behavior, 17*(2), 283–301. https://doi.org/10.1016/S0364-0213(99)80062-7

Kaput, J. J., & Schaffer, D. W. (2002). On the development of human representational competence from an evolutionary point of view. In K. Gravemeijer, R. Lehrer, B. van Oers, & L. Verschaffel (Eds.), *Symbolizing, modelling and tool use in mathematics* (pp. 277-293). Kluwer. https://doi.org/10.1007/978-94-017-3194-2_17

Lakoff, G., & Núñez, R. E. (2000). *Where mathematics comes from: How the embodied mind brings mathematics into being*. Basic Books.

Latour, B. (2008). The Netz-works of Greek deductions. *Social Studies of Science, 38*, 441–449.

Mandelbrot, B. B. (2002). Mathematics and society in the 20th century. In M. L. Frame & B. B. Mandelbrot (Eds.), *Fractals, graphics, and mathematics education* (pp. 29–32). Mathematical Association of America.

Mashaal, M. (2006). *Bourbaki: A secret society of mathematicians.* American Mathematical Society.

Moritz, R. E. (1958). *On mathematics: A collection of witty, profound, amusing passages about mathematics and mathematicians.* Dover.

Mukhopadhyay, S. (2009). The decorative impulse: Ethnomathematics and Tlingit basketry. *ZDM Mathematics Education*, *41*, 117–130. https://doi.org/10.1007/s11858-008-0151-7

National Mathematics Advisory Panel (2008). *Foundations for success: The final report of the National Mathematics Advisory Panel*. US Department of Education. https://files.eric.ed.gov/fulltext/ED500486.pdf

Nails, D., & Monoson, S. (2022). Socrates. In E. N. Zalta (Ed.), *The Stanford Encyclopedia of Philosophy* (Summer 2022 Edition). https://plato.stanford.edu/archives/sum2022/entries/socrates/

Netz, R. (2003). *The shaping of deduction in Greek mathematics: A study in cognitive history*. Cambridge University Press. https://doi.org/10.1017/CBO9780511543296

Newman, J. R. (Ed.). (1956). *The world of mathematics*. Simon and Schuster.

Núñez, R. (2000). Mathematical idea analysis: What embodied cognitive science can say about the human nature of mathematics. In T. Nakahara & M. Koyama (Eds.), *Proceedings of the 24th Conference of the International Group for the Psychology of Mathematics Education* (Vol. 1, pp. 3–22). IGPME.

Parshall, K. H. (2009). The internationalization of mathematics in a world of nations, 1800–1960. In E. Robson & J. Steadall (Eds.), *Oxford handbook of history of mathematics* (pp. 85–104). Oxford University Press.

Runde, V. (2003). Why I don't like 'pure' mathematics. *Pi in the Sky*, *7*, 30–31. https://arxiv.org/abs/math/0310152

Thom, R. (1971). 'Modern' mathematics: An educational and philosophical error? *American Scientist*, *59*(6), 695–699.

Thurstone, W. P. (1994). On proof and progress in mathematics. *Bulletin of the American Mathematical Society*, *30*, 161–177.

Urton, G. (1997). *The social life of numbers: A Quechua ontology of numbers and philosophy of arithmetic*. University of Texas Press.

Urton, G. (2009). Mathematics and authority: A case study in Old and New World accounting. In E. Robson & J. Steadall (Eds.), *Oxford handbook of history of mathematics* (pp. 27–55). Oxford University Press.

Verhulst, F. (2012). Mathematics is the art of giving the same name to different things: An interview with Henri Poincaré. *Niew Archief voor Wiskunde*, *13*(5), 154–158.

Wigner, E. (1960). The unreasonable effectiveness of mathematics in the natural sciences. *Communications in Pure and Applied Mathematics*, *13*, 1–14.

3. Hardy's deep sigh

Ole Skovsmose

In his book A Mathematician's Apology, *Godfrey H. Hardy presents a conception of mathematics according to which real mathematics can be considered harmless and innocent. By 'real' mathematics, Hardy has in mind, for instance, advanced number theory. He contrasts real mathematics with different examples of applied mathematics and cases of elementary mathematics. Hardy argues for the thesis of innocence by asserting that the utilitarian value of real mathematics is nil. Real mathematics does not have any useful applications. By assuming a utilitarian perspective on ethics, Hardy can claim that real mathematics operates at a comfortable distance from any ethnical and political controversies. However, number theory, that Hardy considered the epitome of real mathematics, has tremendous applications itself within war technology. Hardy's explicit justification of the thesis of innocence is simply fallacious. Most ironically, the doctrine of neutrality continues to operate. According to this doctrine, mathematics can be researched and developed while ignoring any kind of ethical and socio-political considerations. The doctrine of neutrality becomes acted out through mathematical research paradigms, dominating the vast majority of university departments in mathematics the world over.*

By the turn of the nineteenth century, science and technology were seen as motors of progress. As part of the Western outlook, it was broadly assumed that science and technology ensure welfare in all aspects of life, whether we are dealing with material production, economic resources, health care, or education. The organisation in 1851 of the Great Exhibition in the Crystal Palace in London, the erection in 1899 of the Eiffel Tower, and the presentation in 1900 of the Exposition Universelle in Paris symbolise the optimism that dominated the whole era. The very steel material used for the construction of the Eiffel Tower and the steel

©2024 Ole Skovsmose, CC BY-NC 4.0 https://doi.org/10.11647/OBP.0407.03

and glass used for the Crystal Palace announce the potentials of the coming century.

Naturally, this optimistic celebration of progress presupposed that a range of socio-political and economic factors were ignored. The horrible living conditions of the working class in industrialised countries came to be seen as unavoidable and, therefore, ignorable necessities for the modern world order. The broadly assumed racist outlook ensured that the brutality of colonialism was ignored as well. By the turn of the century, people, in particular those belonging to the well-protected layers of Western societies, could enjoy reading about world exhibitions—if not in fact going there—and be contented by living during a period of assumed ongoing progress.

Such visions of the future were shattered by the outbreak of the First World War. This catastrophe revealed a new dramatic connection between, on the one hand, science and technology, and on the other hand, war. While science and technology were supposed to constitute an integral part of peaceful and enlightened progress, they now appeared also as an integral part of the very machinery of war. The development of new and more powerful weapons was a science-based technological achievement. Submarines and airplanes became indispensable components of warfare. The application of poison gas likewise brought chemistry to the forefront of the battlefield. The First World War made evident that the image of science and technology as reliable motors of peaceful progress was an illusion.

A life

In 1940, as the Second World War was in dramatic development, Godfrey H. Hardy published the book *A Mathematician's Apology*. Rather than reading it as an immediate reaction to the outbreak of that war it could be seen as a profound, but delayed, reaction to the First World War. In the inaugural lecture that Hardy gave in Oxford in 1920, one finds an 'outline of an apology for mathematics' (Hardy, 1967, p. 74); so Hardy's first 'apology' was formulated long before *A Mathematician's Apology* was published. The First World War put the relationship between mathematics and war on the agenda, and certainly also on Hardy's agenda.

Hardy was born in 1877. In school, he was not particularly dedicated to mathematics, but from his early years demonstrated excellence in the subject. Hardy related that he primarily thought of mathematics in terms of competition, and found that there he could most decisively beat others.

In 1896, he entered Trinity College in Cambridge to study mathematics, and in 1900, he became a fellow. In 1898, he became a member of the Apostles, which was a closed elitist discussion group that also included George Moore (1873–1958), John Maynard Keynes (1883–1946), and Bertrand Russell (1872–1970). The Apostles was open only to brilliant scholars from the University of Cambridge, and at their meetings any topic could be addressed. The most famous non-member of the Apostles was Ludwig Wittgenstein (1889–1951), who was invited to join but did not find the group serious enough. Like Keynes and Russell, Hardy also joined the Bloomsbury Group, which focused on literature and art. Hardy was well located in the academic and intellectual circles at the time, and was aware of the current and controversial issues being discussed, in relation to politics, literature, or art. In 1906, he secured a position as lecturer in mathematics in Cambridge, and during the First World War he preoccupied himself with his teaching and research.

Russell was a declared pacifist, revolted by the English jingoism that accompanied the outbreak of the First World War. Hardy was not outspoken with respect to political issues, but well aware of Russell's sentiments. Russell held a position as lecturer at Trinity College, but in 1916 he was dismissed from this position as a consequence of his anti-war writings. In 1918, he was put in prison for five months, and during that time he wrote *An Introduction to the Philosophy of Mathematics* (Russell, 1919/1993). Hardy shared Russell's anti-war positions, and during the war he felt more and more uncomfortable staying in Cambridge, where jingoism was strongly articulated by some of his colleagues.

In 1919, Hardy took up a professorship in Oxford, and was received with enthusiasm by the younger mathematicians there. That he felt it important in his inaugural lecture to outline a defence of mathematics can come as no surprise. The atrocities of the First World War, and the roles played by mathematics, made such an apologia necessary. Its presentation made it possible for Hardy to concentrate completely on mathematical research, and the next ten years were very productive for him. In particular, his work with John Littlewood and Srinivasa Ramanujan became one of the outstanding collaborations in the history of mathematics.

A photo of Vladimir Lenin was displayed on the wall of Hardy's room in New College, Oxford. This information is noted by C. P. Snow, who wrote a biographical sketch of Hardy as preface to *A Mathematician's Apology*. I am not aware of any explanation of Hardy's choice of photo, but one should not conclude that Hardy was a communist. If he had leftist inclinations, they likely reflected a non-standard interpretation of the term. At that time, many intellectuals in England demonstrated an open curiosity for what was taking place in the Soviet Union.

In 1920, Russell visited the Soviet Union as a member of a British delegation and, during the visit, had the opportunity to meet Lenin in person. Russell became disillusioned, and back home he wrote a critique of what he saw: *The Practice and Theory of Bolshevism* (Russell, 1920/2017). However, Russell's critique was based on a profound political sympathy, and he states: 'The existing capitalist system is doomed. Its injustice is so glaring that only ignorance and tradition could lead wage-earners to tolerate it' (p. 2). I assume that Hardy had read Russell's book, and the picture of Lenin might represent some feeling of resonance.

Hardy was certainly in full accord with Russell's attacks on Christianity. In 1927, Russell gave the lecture 'Why I Am Not a Christian', wherein, among other things, he states that 'every single bit of progress in human feeling, every improvement in the criminal law, every step towards the diminution of war, every step toward better treatment of the coloured race, or every mitigation of slavery, every moral progress that there has been in the world, has been consistently opposed by the organized churches of the world' (pp. 20–21). The lecture was circulated as a pamphlet, and later included in several books as, for instance, Russell (1957). Hardy shared Russell's anti-Christian stance. He did not go to church, quite literally: he simply did not enter a church under any circumstances, not even when requested to do so for academic ceremonies.

When Hardy felt that his creative mathematical powers had declines, he experienced periods of post-creative depression. These moments provided the personal context that ultimately led him to write *A Mathematician's Apology*. It opens with the following statement:

'It is a melancholy experience for a professional mathematician to find himself writing about mathematics' (p. 61). Hardy considered this type of writing second-rate work. He thought of writing about literature, theatre, as inferior activities: 'Exposition, criticism, appreciation, is work for second-rate minds' (p. 61). Presumably, Hardy had postponed this activity until he had no better things to do.

In 1941, Hardy published the booklet *Bertrand Russell and Trinity* (Hardy, 1970), in which he provides an account of Russell's dismissal from Trinity College in 1916. In the booklet, Hardy also gives glimpses of his own position, and he mentions that he had been secretary of the Cambridge branch of the Union of Democratic Control, founded shortly after the outbreak of the First World War. This was an organisation that represented war-sceptic positions. Hardy's insider clarification of what took place in 1916 only appeared twenty-five years after the event. I have no doubt that Hardy maintained clear priorities in life: first things first, and mathematics was a clear number one. Only after his creative powers had left, and he had written his apology, did he find time for clarifying what he felt to be a grave injustice done to his friend Bertrand Russell. After publishing *Bertrand Russell and Trinity*, Hardy published nothing more. He died in 1947.

A mathematician

In mathematics, Hardy worked in close collaboration with others. During most of his career, he collaborated with John Littlewood (1887–1977), who had entered Trinity College in 1903. Together they published more than 100 papers. Hardy also established a collaboration with the Indian mathematician Srinivasa Ramanujan (1887–1920), and together they published several papers.

Much of Hardy and Littlewood's collaboration was in number theory, for instance about the distribution of prime numbers. It appears common sense to consider their density to be decreasing in the sense that one could expect the number of primes between, say, 18000 and 19000 to be smaller than the number of primes between 8000 and 9000. Since Antiquity, it has been known that the number of primes is infinite, so their decreasing density will never reach zero. The prime number theorem provided an estimation of how the density decreases, and this estimation was first proposed by Carl Friedrich Gauss (1777–1855).

One can also consider prime twins, pairs of primes like 11 and 13, 41 and 43, and 107 and 109 that differ by 2. However, as the density of primes is decreasing, one could expect that the space between primes will be ever-increasing with the possibility that there is a largest pair of prime twins. However, according to the prime twin conjecture, there exist infinitely many prime twins, with decreasing density. Hardy and Littlewood provided an estimation of how this density decreases, similar in nature to the one provided by Gauss for prime numbers.

With respect to his start as a mathematician at Trinity, which one can link to around the year 1900 when he became a fellow, Hardy (1967) states: 'I wrote a great deal during the next ten years, but very little of any importance; there are not more than four of five papers which I can still remember with some satisfaction' (p. 147). The important turns in Hardy's career came in 1911 when he started his collaboration with Littlewood, and in 1913 when he came to know Ramanujan. He wrote that 'All my best work since then has been bound up with theirs, and it is obvious that my association with them was the decisive event of my life' (p. 148). Then follows an emotional remark: 'I still say to myself when I am depressed, and find myself forced to listen to pompous and tiresome people, "Well, I have done one thing that *you* could never have done, and that is to have collaborated with both Littlewood and Ramanujan on something like equal terms"' (p. 148, italics in original).

I like very much his addition 'on something like equal terms'. Hardy fully recognises that Littlewood and Ramanujan, both ten years younger than him, are mathematical geniuses. He is certainly also aware of his own unique creative powers. With both honesty and satisfaction, he can claim that he has co-operated with them – not at equal terms – but on something like that.

An apology

In *A Mathematician's Apology*, Hardy (1967) presents a conception of mathematics which we can think of as Hardy's working philosophy of mathematics.[1] Throughout all his formulations, he expresses a clear Platonic outlook:

---

1 See Chapter 1 in this volume for an introduction to the notion of 'working philosophy of mathematics'.

> I believe that mathematical reality lies outside us, that our function is to discover or observe it, and that the theorems which we prove, and which we describe grandiloquently as our 'creations', are simply notes of our observations. This view has been held, in one form or another, by many philosophers of high reputation from Plato onwards, and I shall use the language which is natural to a man who holds it. (pp. 123–124).

In his research, Hardy sees himself as making discoveries, as for instance with respect to the distribution of prime twins. Many research mathematicians operate with Platonism as an implicitly assumed element of their conception of mathematics. This observation has been elegantly captured by Reuben Hersh (1997), when he states that mathematicians are formalist on weekends while Platonist during working hours. Hardy, however, was very aware of actual trends and positions in the philosophy of mathematics. In the article 'Mathematical Proof', Hardy (1929) refers to the ideas and positions of, among others, David Hilbert, L. E. J. Brouwer, Russell, Alfred Whitehead, and Wittgenstein. When Hardy assumes a Platonism, it is not as part of any implicit working philosophy of mathematics, but as a deliberate positioning.

In *A Mathematician's Apology*, one meets a deep concern about the possible roles of sciences as well as of mathematics, in particular in times of war. That science forms part of the war machinery was made evident by the First World War, and even more evident by the start of the Second World War. This is a deep preoccupation for Hardy. He sees war as an abominable phenomenon, and it is horrible for him of think of science as a resource for war technology. But what about mathematics? Should a mathematician feel responsible? Should a mathematician feel guilty? No doubt Hardy was troubled by such questions, but he states that 'a real mathematician has his conscience clear; there is nothing to be set against any value his work may have; mathematics is [...] a "harmless and innocent" occupation' (pp. 140–141).

This is the crucial claim in Hardy's conception of mathematics: we are dealing with a harmless and innocent occupation. However, Hardy is not talking about mathematics in general, but only about what he refers to as real mathematics.

Hardy's formulation could have been 'like a physicist, a chemist, and an applied mathematician, also a real mathematician has his conscience clean'. But Hardy does not want to say anything like this. Rather his claim is: 'in contrast to a physicist, a chemist, and many applied mathematicians, a real mathematician has his conscience clean'. He states:

> There is one comforting conclusion that is easy for a real mathematician. Real mathematics has no effects on war. No one has yet discovered any warlike purposes to be served by the theory of numbers or relativity, and it seems very unlikely that anyone will do so for many years. It is true that there are branches of applied mathematics, such as ballistics and aerodynamics, which have been developed deliberately for war and demand a quite elaborate technique: it is perhaps hard to call them 'trivial', but none of them has any claim to rank as 'real'. They are indeed repulsively ugly and intolerable dull: even Littlewood could not make ballistics respectable, and if he could not who can? (p. 140)

Then follows the conclusion as already quoted: 'A real mathematician has his conscience clear'.[2]

Rather than elaborating on the distinction between pure and applied mathematics, Hardy differentiates between real and trivial mathematics. According to Hardy, much mathematics is trivial, like school mathematics, calculus, and other such topics covered by introductory university textbooks, what can be referred to as engineering mathematics, and much applied mathematics. Contrary to real mathematics, such mathematics is 'trivial'. Hardy also finds it to be 'repulsively ugly' and 'intolerable dull'. These are very strong words that might reflect Hardy's profound aversion for the parts of mathematics, such as ballistics and aerodynamics, that are put into operation for purposes of warfare.

According to Hardy, mathematics developed as part of natural sciences can also be real, and he explicitly states: 'I count Maxwell and Einstein, Eddington and Dirac, among "real" mathematicians' (p. 131). He also states that he counts Isaac Newton as 'one of the world's three greatest mathematicians' (p. 71). As real mathematicians, they can

2 In a note, Hardy (1967, p. 152) makes some modifying observations with respect to §28 in the *Apology* (pp. 139–143) from where the quotations are taken. According to Hardy, the modifications are inspired by comments to the manuscript made by C. D. Broad and C. P. Snow. Hardy acknowledges that they might have some points and that he might have been too 'sentimental' in his formulations. However, he adds that he, anyway, decided not to make changes in this part of the manuscript. §28 is based on a short article that Hardy had published previously in 1940 in *Eureka*, the journal of the Cambridge Archimedean Society.

also have their consciences clear, while people contributing to trivial mathematics are not saved by Hardy.

I doubt that Hardy considered all elementary mathematics to be trivial. In the *Apology*, he refers to two classic mathematical proofs, one showing that the number $\sqrt{2}$ is irrational, and the other showing that there are infinitely many prime numbers.[3] We are dealing with elementary mathematical proofs, but I think that Hardy found them to be exemplars for making real mathematical discoveries. As mentioned, Hardy's use of the descriptor 'real' indicates that he embraces a Platonic view. Mathematical entities have a real existence, and properties of these entities become revealed through mathematical demonstrations. Thus, the two proofs that he refers to reveal the existence of non-rational numbers and the existence of infinitely many prime numbers. Together with Littlewood he tried to discover whether or not there exist infinitely many prime twins. When Hardy refers to real mathematics, he might well have in mind mathematics that contributes to revealing the properties of mathematical reality. He might think of trivial mathematics as not making such contributions, but operating within what already exists of mathematical entities. Trivial mathematics might combine techniques of huge complexities, it might provide a range of applications, but it does not contribute with mathematical discoveries.

One way of cleaning a mathematician's conscience could be to show that what is done through mathematics can be only 'good things'. Such a line of argumentation could take an almost religious format. For instance, Newton was a devoted believer in God. He revealed how mathematics captures the laws of nature, and therefore the way God had created the world. Mathematics could be thought of as an expression of the rationality of God, and as a consequence, one cannot say anything other than good things about mathematics. Versions of this line of thought have been repeated again and again. But not by Hardy. Any religious flavouring of an apology for mathematics was impossible to him.

Hardy cleans the real mathematicians' consciences by claiming that what they are doing is without any use. While trivial mathematics can be useful, there is no usefulness to be associated to real mathematics: 'I have never done anything "useful". No discovery of mine has made, or

3 In Chapter 7 of this volume, the proof for the infinity of prime numbers is presented and discussed with reference to intuitionism.

is likely to make, directly or indirectly, for good or ill, the least difference to the amenity of the world' (p. 150). Contrary to trivial mathematics, real mathematics does not make a difference, neither for the good nor for the bad. Real mathematics has no utilitarian value whatsoever, whether in times of peace or in times of war. Real mathematics is harmless and innocent.

Hardy uses different words to express that real mathematics is harmless and innocent, as, for instance, 'gentle' and 'clean'. After making some references to Gauss, he states the following:

> If the theory of numbers could be employed for any practically and obviously honourable purpose, if it could be turned directly to the furtherance of human happiness or the relief of human suffering, as physiology and even chemistry can, then surely neither Gauss nor any other mathematicians would have been so foolish as to decry or regret such applications. But science works for evil as well as for good (and particular, of course, in time of war); and both Gauss and lesser mathematicians may be justifying in rejoicing that there is one science at any rate, and that is their own, whose very remoteness from ordinary human activities should keep it gentle and clean (pp. 120–121).

Hardy does not propose any theory about the neutrality of science. In fact, he highlights the opposite, that 'science works for evil as well as for good', and that it does so, in particular, 'in times of war'. According to Hardy, mathematics is not any neutral science. The only thing he insists upon is that real mathematics operates beyond any evil-good controversies. Not because it contains any intrinsic goodness or any sublime ethical qualities, but because it is useless.

His utilitarian perspective on mathematics is consequential, as utilitarianism as an ethical position provides a non-religious perspective on ethical questions. Whether something is 'good' or 'bad' cannot be judged according to some sublime ethical or religious principles, but only with respect to its utilitarian implications. By stipulating that real mathematics has no such implications, Hardy saves this part of mathematics from being considered harmful in any way. It is simply innocent.

Why then work with real mathematics? As Hardy set aside any possibility for providing a utilitarian justification for such work, one needs to ask what kind of justification is then possible. Hardy is well aware that he leaves only a narrow space for himself to articulate justifications. His reaction, however, seems to be that he, in fact, does not need much such space. In some formulations in the *Apology*, he uses the notion of being 'serious'. Chess problems might be extremely challenging, but, according to Hardy, compared to real mathematical problems they are unimportant: 'The best mathematics is *serious* as well as beautiful – "important" if you like, but the word is very ambiguous, and "serious" expresses what I mean much better' (p. 89, italics in original).

Elaborating justifications for working with real mathematics does not appear necessary to Hardy. To ask for any such justification, utilitarian or not, is like asking Mozart to provide a justification, utilitarian or not, for making his compositions. Hardy would rather state that Mozart's work is serious (and innocent) like any other work of art, including real mathematics.

A doctrine

In *A Mathematician's Apology*, Hardy (1967) elaborates a thesis of innocence, which can be summarised in the following way: sciences might work for the evil as well as for the good. Within science, however, there exists a small domain that is not under suspicion for being harmful, and this is real mathematics. It is useless, and as such it is harmless and innocent.

Hardy's thesis of innocence can be related to, but also contrasted with, a doctrine of neutrality. While Hardy's thesis is well-articulated and refers to a particular domain within mathematics, the doctrine of neutrality often operates as a discursive pattern and includes any kind of mathematics. The doctrine is deployed whenever one wants to cut off a discussion of possible socio-political impacts of mathematics. It turns into an ideology by assuming that mathematics as such is harmless and innocent, and that one can conduct mathematics research without engaging in critical reflections about what might be done through mathematics. Contrary to Hardy's thesis of innocence, the doctrine of neutrality operates as a discursive given, and not as a claim in need of justification. The doctrine is part of an implicit working philosophy of mathematics. It is called into operation when socio-political issues are stipulated as irrelevant when doing mathematical research.

I see the doctrine of neutrality as a disproportionate and exaggerated shadow of Hardy's thesis of innocence. The doctrine concerns any kind of mathematics, so that mathematics as such becomes stipulated as being neutral. There is no need to make specifications with respect to mathematical topics: algebra, calculus, number theory – all such subjects are harmless and innocent. They are neutral. Nor is it necessary to make specifications with respect to levels of mathematics: elementary mathematics, advanced mathematics, research mathematics – all are neutral subjects.

The doctrine of neutrality is materialised in the organisation of university studies in mathematics. Naturally there is a variety of such study programmes, but what I have in mind here I refer to as the university mathematics tradition.[4] This tradition includes the following characteristics: (1) It defines the curriculum in well specified units such as Calculus 1, Calculus 2, Linear Algebra, Algebra 1, Algebra 2, Probability Theory, non-Euclidean Geometry, Projective Geometry, and so on. (2) Among these units there is no space for a philosophy of mathematics including ethical discussions related to the use of mathematics, and not much space for history of mathematics, which could include socio-political reflections. (3) Within the units, ethical and socio-political controversies that could be related to the particular mathematical subdiscipline are not addressed. (4) All tests and exams focus on mathematical competencies.

When we are dealing with a doctrine, the structure of its justification need not be explicit, nor even coherent. A doctrine is a general positioning, which can be articulated in different contexts and make part of a variety of discourses, insisting that mathematics is detached from socio-political issues. *A Mathematician's Apology* has turned into a most questionable publication, as it has enabled many to maintain a doctrine of neutrality as part of a working philosophy of mathematics. The doctrine leads to a conception of mathematics that fosters a banality of mathematical expertise (see Skovsmose, 2020). This banality embraces the ignorance of possible implications of what one is doing. It ignores the context within which mathematical research is conducted and where mathematics is brought in action.

---

4 One can find a characterisation of the school mathematics tradition in Skovsmose and Penteado (2016).

A sigh

Hardy (1967) did not elaborate his thesis of innocence starting from systematic philosophical observations, but rather from his experiences as a mathematician. As already referred to, his principal justification for the thesis built on observations such as: 'No one has yet discovered any warlike purposes to be served by the theory of numbers or relativity, and it seems very unlikely that anyone will do so for many years' (p. 140). The justification for his thesis is empirical, referring to what can be observed, or rather, to what has not (yet) been observed.

It is not surprising that Hardy refers to number theory, which is his paradigmatic case of real mathematics. That he also refers to relativity is a surprise to me. Relativity theory provides a mathematical conception of nature, and Hardy thinks of Albert Einstein, and other great physicists, as contributing to the domain of real mathematics. Hardy states: 'The great modern achievements of applied mathematics have been in relativity and quantum mechanics, and these subjects are, at present at any rate, almost as useless as the theory of numbers' (pp. 131–132). According to Hardy, such examples of applied mathematics are not trivial, but real.

However, already in 1940 when Hardy published this statement, it was possible, with developments in the theory of relativity, to conceptualise the possibility of an atomic bomb, as expressed dramatically in the equation $E = mc^2$. However, the route from this theoretical insight to the actual construction of a bomb only became identified in steps, many of which were kept as military secrets. In 1945, with the destruction of Hiroshima and Nagasaki, it was demonstrated to everybody that the theory of relativity was implicated as an integral part of modern war machinery. Hardy's justification of the thesis of innocence by referring to relativity is simply wrong.

What would Hardy make of this? He witnessed the conclusion of the Second World War and the destructions of Hiroshima and Nagasaki, but I am not aware he tried to make any revision of this formulation in his *Apology*. I also think that he did not really think of his remark about relativity as being that crucial for his justification. The remark appears as an aside from his principal argument referring to number theory.

In the article 'Formatting Power of "Mathematics in a Package": A Challenge for Social Theorising?', Keiko Yasukawa and I (2009) discuss modern cryptography.[5] Cryptography has a long history, and was applied already in Antiquity. The development of cryptography can be directly related to technological developments, and different mechanical machineries for coding and decoding have been invented, reaching an extreme sophistication during the Second World War.

Two years after the death of Hardy, Claude E. Shannon (1949) published the article 'Communication Theory of Secrecy Systems', which establishes the opening for an advanced mathematical approach. In *New Directions in Cryptography*, this approach was elaborated in detail by Whitfield Diffie and Martin Hellman (1976). As Yasukawa and I point out, the very identification of these new directions in cryptography is based on profound number theoretical insights. The idea is to construct a technique for encoding and decoding that can be handled automatically without compromising security measures. The computer makes such automatisation possible, so that huge amounts of data can be encoded and decoded. The whole process is complex, but the principal observation, which ensures the safety of the whole approach, is related to a simple observation: breaking the code turns out to be equivalent to being able to factorise a number that is the product of two huge prime numbers.

In *Number Theory in Science and Communication*, Manfred R. Schroeder (1997) states that if we are dealing with a 200-digit number that is the product of two prime numbers of more or less equal size, the factorisation cannot be completed within any conceivable time limit. He points out that 'not so long ago, the most efficient factorising algorithms on a very fast computer were estimated to take 40 trillion years, or 2000 times the present age of the universe' (p. 131). Certainly, this statement is time-dependent. The quotation here is from the third edition of the book, while in a previous edition from 1983, Schroeder makes the same statement, but referring to a 100-digit number. Newer editions of the book have been published, but I have not yet had the opportunity to check the possible reformulations of the quoted statement. Certainly new algorithms can be identified, and computers are becoming more and

5 See also Yasukawa, Skovsmose, and Ravn (2012).

more powerful. That we, independent of such development, continue to face a task that cannot be completed within any conceivable time limit is crucial for the whole cryptographic approach. To make such a claim about factorisation depends on a deep number theoretical insight.

How could it be that a factorisation explodes in complexity? Here comes a number n with 200 digits:

> 78335008712224185576663326829088442611090234337768177777325699
> 40096722618376645422511256670099959933385155736329922889009923
> 8238882812482500038888217888299994898921156667211390080900765334
> 87112387111

I just pressed number keys 200 times, so if this number n turns out to be a product of two prime numbers of more or less the same size, it would be a most unlikely coincidence. However, if so, we should expect that it would take trillions of years until we discover which two prime numbers we are dealing with. My intuition does not point towards such a conclusion. To me the number looks large, but that the equation $n = p^1 p^2$ represents such overwhelming computational complexities, I could never imagine. In order to reach such an insight, one needs to draw on profound number theoretical insights. Furthermore, new number theoretical insights concerning the distribution of prime numbers and efficient algorithms for factoring might lead to a modification, if not a direct falsification, of the claim.

As cryptography makes an indispensable part of modern war technology, number theory forfeits all claim to be harmless and innocent. This observation is devastating for Hardy's justification of the thesis of innocence. Number theory turns out to be extremely useful, in particular in times of war. It can be harmful in just the same way as ballistics and aerodynamics can be.

I imagine that Hardy is sitting in the same comfortable chair as shown at the cover of my edition of the *Apology*. He has an attentive look, his glasses are a bit down his nose, his hands are empty, and he seems just ready to grasp a book or a paper. I imagine that he gets an opportunity to look at the paper by Yasukawa and me. I have no doubt that, after a few moments, he will put aside the paper and ask for the original references. After looking through them, his expression will change. He is not really looking at anything or at anybody anymore.

His look turns inwards. I have no idea what he is going to say, but I can imagine his deep sigh.

While the maintenance of an articulated thesis might depend on the status of its justification, a doctrine could easily survive even the most downright falsification. Although Hardy's justification of the thesis of innocence has collapsed, its disproportionately exaggerated shadow, the doctrine of neutrality, continues to be seen everywhere. This shadow provides a cover for mathematical research and university studies in mathematics to maintain a profound silence with respect to ethical and socio-political issues.[6] Mathematics continues to be conceptualised as harmless and innocent. But it is not.

References

Diffie, W., & Hellman, M. E. (1976). New directions in cryptography. *IEEE Transactions on Information Theory, 22*(6), 644–654. https://doi.org/10.1109/TIT.1976.1055638

Hacking, I. (2014). *Why is there philosophy of mathematics at all?* Cambridge University Press. https://doi.org/10.1017/CBO9781107279346

Hardy, G. H. (1929). Mathematical proof. *Mind, 38*(149), 1–25. https://doi.org/10.1093/mind/XXXVIII.149.1

Hardy, G. H. (1967). *A mathematician's apology*. Cambridge University Press.

Hardy, G. H. (1970). *Bertrand Russell and Trinity*. Cambridge University Press.

Hersh, R. (1997). *What is mathematics, really?* Oxford University Press.

Russell, B. (1957). *Why I am not a Christian and other essays on religion and related subjects*. Simon and Schuster.

Russell, B. (1993). *Introduction to mathematical philosophy*. Routledge. (Original work published 1919)

Russell, B. (2017). *The practice and theory of Bolshevism*. Anodos. (Original work published 1920)

Schroeder, M. R. (1997). *Number theory in science and communication: With applications in cryptography, physics, digital information, computing and self-similarity*. Springer.

6 In line with such an observation, Hacking (2014) makes the following comment: 'G. H. Hardy's fantasy about pure mathematics, *A Mathematician's Apology*, did much harm to philosophers' (p. 185).

Shannon, C. E. (1949). Communication theory of secrecy systems. *The Bell System Technical Journal, 28*(4), 656–715. https://doi.org/10.1002/j.1538-7305.1949.tb00928.x

Skovsmose, O. (2020). Banality of mathematical experience. *ZDM Mathematics Education, 52*(6), 1187–1197. https://doi.org/10.1007/s11858-020-01168-4

Skovsmose, O., & Penteado, M. G. (2016). Mathematics education and democracy: An open landscape of tensions, uncertainties, and challenges. In L. D. English & D. Kirshner (Eds.), *Handbook of international research in mathematics education* (pp. 359–373). Routledge.

Skovsmose, O., & Yasukawa, K. (2009). Formatting power of 'Mathematics in a package': A challenge for social theorising? In P. Ernest, B. Greer, & B. Sriraman (Eds.), *Critical issues in mathematics education* (pp. 255-281). Information Age.

Yasukawa, K., Skovsmose, O., & Ravn, O. (2012). Mathematics as a technology of rationality: Exploring the significance of mathematics for social theorising. In O. Skovsmose & B. Greer (Eds.), *Opening the cage: Critique and politics of mathematics education* (pp. 265–284). Sense.

4. Formalism, structuralism, and the doctrine of neutrality

Ole Skovsmose

The doctrine of neutrality states that mathematics can be researched and developed without considering any ethical or socio-political issues. This doctrine became elaborated and argued in detail by the school of logical positivism. By the turn of the nineteenth century, a range of paradoxes and inexplicable mathematical phenomena appeared, a situation referred to as the foundational crisis of mathematics. To many, intuition was the scoundrel, and it had to be eliminated from mathematics. Formalism provided a principal approach by identifying mathematics with formal structures. This idea was embraced by logical positivists who claimed that mathematics as the language of science ensures the ethical neutrality of science. They considered mathematics not only as being neutral itself, but also as a guarantee for scientific neutrality in general. In this way, a most profound stipulation of the doctrine of neutrality was reached. Formalism developed into structuralism, which described mathematics as an architecture of pure formal structures. As part of the structuralist conception of mathematics, the doctrine of neutrality was expanded from being a conception of mathematical research to become also a doctrine shaping educational practices in mathematics. I am going to confront this conception. The doctrine of neutrality is a stipulation, which makes us ignore that a profound politicisation of both mathematics and mathematics education might be taking place.

Introduction

On the 22 June 1936, Moritz Schlick was murdered on the broad steps of the main entrance of the University of Vienna. He was shot from close range by a former student, Johann Nelböck. Schlick died immediately.

©2024 Ole Skovsmose, CC BY-NC 4.0 https://doi.org/10.11647/OBP.0407.04

In 1922, Schlick had been nominated as professor in *Naturphilosophie* (philosophy of nature) at the university, and Nelböck had studied there with Schlick as his advisor. In 1931, Nelböck graduated as a doctor in philosophy. During the trial, he gave different explanations for the killing, one referring to jealousy. He also claimed that Schlick's anti-metaphysical philosophy had troubled him. Nelböck was sentenced to ten years in jail.

The event became a controversial issue with much public attention. Although Schlick was a German Protestant, he became portrayed in the press as a key figure in suspicious academic Jewish circles, and Nelböck became celebrated by Nazi movements. In 1938, after the *Anschluss*, the German annexation of Austrian, Nelböck was released.

Soon after his nomination, Schlick organised a discussion group that later became known as the Vienna Circle (*Wiener Kreis*). This circle formulated a view on science and mathematics that came to be known as logical positivism. The circle was deeply engaged in recent developments in science, mathematics, and logic. They were conversant with developments in physics. They studied carefully the monumental work *Principia Mathematica* by Alfred Whitehead and Bertrand Russell, published in three volumes in 1910–1913. The *Tractatus Logico-Philosophicus* by Ludwig Wittgenstein, published in 1922, was also studied with extreme intensity.

The circle was also deeply engaged in the recent political developments in Austria and Germany. They launched a strong attack on any form of metaphysical thinking, including Nazi ideologies. No doubt, Schlick's anti-metaphysical philosophy troubled not only Nelböck, but many from the Nazi movement.

The anti-metaphysical philosophy initiated by the Vienna Circle ended up providing a broad platform for claiming that science and mathematics can be kept separated from socio-political and ethical issues, that they are neutral subjects. In the previous Chapter 3, we saw how Hardy's thesis of mathematics being innocent turned into a dogmatic claim of mathematics being neutral. In this chapter, we are investigating a much broader philosophical trend, which establishes a formidable manifestation of this dogmatism.[1]

---

1 This dogmatism was confronted by a critical conception of mathematics that I return to in Chapter 11 in this book.

As an initial step into this dogmatism, I refer to the *elimination of intuition* from mathematics, which represents a de-contextualisation of mathematical notions and reasonings. I refer to *formalism,* which emerges as the result of this elimination of intuition, and which identifies mathematics with formal structures. Then I provide an exposition of *logical positivism* that embraces formalism and claims that not only mathematics but science in general is neutral and detached from any socio-political stance. After that, I present *structuralism,* which represents a particularly elaborated version of formalism also embracing the dogma of neutrality. I show how the Modern Mathematics Movement emerged as an educational version of structuralism and manifests the de-policisation of mathematics education. As a kind of epilogue, I make a few comments about 'Poor Piaget!'.

Elimination of intuition

The elimination of intuition from mathematics includes three components: making *explicit* hidden axioms and assumptions that in fact are used in developing a mathematical theory; eliminating *ontological* issues referring to assumptions about the nature of mathematical objects; and *identifying* and formalising the logical patterns of deduction and reasoning used in mathematics.

Since Antiquity, Euclid's *Elements* has been taken as the paradigmatic case of mathematical deduction.[2] A deduction must start out from something, and this 'something' was presented by Euclid in terms of five axioms (which by Euclid were referred to as postulates). From these axioms, logical deduction leads to a range of theorems. If the axioms were true, all theorems would be true as well, since logical deduction provides a delivery of truth. As Euclid's axioms appeared simple (although the fifth axiom seemed less so), their truth could be grasped immediately by intuition, and due to the strict deductive organisation of the *Elements*, all theorems could be considered true as well. So it was generally assumed.

This perception of the *Elements* only became challenged during the nineteenth century, where different studies revealed that more than the

2 See Joyce (1998).

five axioms were involved in the deduction of theorems. Euclid had also applied intuition, for instance concerning space. This came as a shock: How could it be that this had been overlooked for more than 2000 years? In 1882, Moritz Pasch (1912) prepared the foundations for an extended axiomatics for Euclidean geometry. The process of making all applied axioms explicit was brought together in *Foundations of Geometry* (Hilbert, 1950), the first German version of which appeared in 1899. While the *Elements* includes five axioms, the *Foundations of Geometry* builds on twenty-one axioms (later it was showed that one of them was redundant). Besides the five included in the *Elements*, one also finds, for example, the axiom:

> Let A, B, C be three points not lying in the same line and let *a* be a line lying in the plane ABC and not passing through any of the points A, B, C. Then, if the line *a* passes through a point of the segment AB, it will also pass through either a point of the segment BC or a point of the segment AC. (Hilbert, 1950, p. 4)

This axiom, first made explicit by Pasch, was applied by Euclid, but just as an intuitive insight. It was taken as a given that, when a straight line cuts one of the sides of a triangle, it will also cut one of the other sides (except from the situation where the line passes through a vertex of the triangle). The axiom states that when a straight line cuts into a triangle, it will not disappear in the interior of the triangle, but cut out of the triangle as well. There are no Bermuda triangles to be found in Euclidean geometry.

In order to eliminate intuition from mathematics, any deduction should be based on explicitly stated axioms. This was exactly what was prepared for by Pasch and accomplished by David Hilbert with respect to Euclidean geometry.

Ontological issues have been an ongoing challenge for the philosophy of mathematics: What is a number, a point, a line, or any other 'mathematical object' for that matter? In Euclid's *Elements*, a point becomes defined as that which cannot be divided. This sounds clear enough, but it appears unclear what is the point of making this definition. It is not used later on in the deductive processes. Maybe the clarification of ontological issues is not crucial for mathematical proving. This point was clearly recognised by Hilbert (1950), who initiates the presentation in *Foundations of Geometry* by stating:

> Let us consider three distinct systems of things. The things composing the first system, we will call *points* and designate them by the letters *A*, *B*, *C* [...], those of the second, we will call *straight lines* and designate them by the letters *a*, *b*, *c*, [...], and those of the third system, we will call *planes* and designate them by the Greek letters α, β, γ, [...]. The points are called the *elements of linear geometry*; the points and straight lines, the *elements of plane geometry*; and the points, lines, and planes, the *elements of the geometry of space* or the *elements of space*. (p. 2, italics in original)

Hilbert's point is that, for presenting geometry in an axiomatic format, the very nature of its objects is irrelevant. They can be referred to as 'things'. In the quotation, he refers to 'points', 'lines', and 'planes', but, as he in 1891 had pointed out in a conversation with two other mathematicians at a train station, he could as well have referred to 'tables', 'chairs', and 'beer mugs'.[3] Hilbert did highlight this point in 1891. For a geometric theory, the intrinsic qualities of points, lines, and planes are irrelevant; only their relationships are relevant. Such relationships, and nothing else, become specified through the axioms of geometry. In this way, Hilbert tried to eliminate intuition from ontological considerations, simply by considering ontology superfluous.

If an elimination of intuition from mathematics reasoning is to be properly carried out, one needs a firm grasp of what logical deduction could mean. This was the principal idea of Gottlob Frege's life project. He wanted to provide an enumeration of all valid forms of logical deduction. But how to do this? It would become a long list, and in what order should it be organised? Frege had a clear approach in mind; he wanted to organise all valid logical deductions in an axiomatic system, and in the *Begriffsschrift* (Frege, 1967), the first German version of which was published in 1879, he presented how this could be done. His *Begriffsschrift* provides a start of the formulation of modern formal logic.

Frege presents a set of logical axioms, seven in total, and two specific rules of inference, claiming that this defines a system that has as theorems precisely all valid forms of logical deduction. The system concerns propositional logic, but on top of this Frege added predicate calculus. Here, however, let me concentrate on its propositional basis.

Frege used a particular formal terminology, which did not become common. However, his presentation in the *Begriffsschrift* has been

3 See Shapiro (2000, p. 151), and Hilbert (1935, p. 403).

carefully reworked in a symbolism that is now common in modern logic. A huge effort was made in *Principia Mathematica* by Alfred Whitehead and Bertrand Russell (1910–1913). The axioms that Whitehead and Russell used are a bit different from those suggested by Frege, but the scope of the axiomatic systems are quite the same. For the propositional calculus, Whitehead and Russell operated with the following five axioms:

1. $(p \vee p) \Rightarrow p$
2. $q \Rightarrow (p \vee q)$
3. $(p \vee q) \Rightarrow (q \vee p)$
4. $(p \vee (q \vee r)) \Rightarrow (q \vee (p \vee r))$
5. $(q \Rightarrow r) \Rightarrow ((p \vee q) \Rightarrow (p \vee r))$

Russell and Whitehead maintained the two rules of inference as formulated by Frege. The first states that if A⇒B *is a theorem or an axiom, and A is a theorem or an axiom, then B is a theorem.* The second states that one can substitute a symbol with another. If, for instance, one has proved the formula $p \Rightarrow (p \vee p)$, then one can also conclude that $q \Rightarrow (q \vee q)$. No other rules of inference than these two were applied.

In summary, the elimination of intuition from mathematics was carried out by making hidden axioms explicit, by eliminating ontological issues from mathematical theorising, and by capturing mathematical reasoning by formal axiomatic systems. This triple-strategy for eliminating intuition created a new way of looking at mathematics. The triple strategy became defining for the formalist outlook.

Formalism

Mathematics has been seen as a unique way of obtaining certainty. When a mathematical proof has been completed, we conclude that the proved theorem is true, and true with certainty. While doubt and uncertainly accompany many forms of human knowledge, it can apparently be eliminated from the domain of mathematics. Mathematics seems a fortification against any possible invasions of scepticism. Therefore, it came as a shock when the fortification seemed to be collapsing.[4]

4 In my discussion of the foundational crisis of mathematics I draw on Ravn and Skovsmose (2019).

By the turn of the nineteenth century, a range of paradoxes and inexplicable mathematical phenomena appeared, creating a situation referred to as the *foundational crisis of mathematics*. In 1901, one paradox was discovered by Bertrand Russell; it was also identified by Ernst Zermelo in 1899, but he communicated it only to a small circle of colleagues from Göttingen University, including Hilbert. The paradox has the following form: Let M denote the set of sets that are not members of themselves, thus $M = \{x \mid x \notin x\}$. Then let us ask: is M a member of itself? If the set M is a member of itself, it has the property $M \notin M$. If M is not a member of itself, it has the property $\neg(M \notin M)$. In other words, we can conclude $M \in M$. Thus, we have $M \in M$ if and only if $M \notin M$.

Not only did such explicit paradoxes emerge, but also strange phenomena were observed. Georg Cantor (1874) presented a new understanding of the notion of set, which until then had been taken as an uncomplicated intuitive notion. He showed that the infinity of real numbers is of a higher degree than the infinity of natural numbers. In fact, he revealed the existence of an infinity of degrees of infinities. Guiseppe Peano (1890) discovered a curve, commonly referred to as the Peano curve, which is a surjective and continuous function from the unit line the unit square. This curve, with the surprising property of being able of cover an area, has also been referred to as 'the bald man's hope'. If just one hair, long enough, is left, then the baldness can be properly covered.

How could it be that mathematics, which appeared so carefully elaborated through proofs and theorems, could run into logical contradictions? What did the occurrence of new strange mathematical objects signify? Something seemed to have gone wrong. But how?

Logicism, formalism, and intuitionism represent three main approaches for addressing the foundational crisis. To logicism and formalism, the scoundrel was intuition, and the elimination of intuition from mathematics forms part of these two approaches. Contrary to these positions, intuitionism claims that intuition is crucial to mathematics, and that the paradoxes emergence when mathematics procedures become led astray by formalist procedures.[5] In the following we concentrate on how Hilbert addressed the crisis.

---

5 In Chapter 7 in this volume, I discuss more carefully the intuitionist approach to mathematics.

Hilbert suggested a two-step metamathematical programme inviting a formalist outlook on mathematics. First, mathematical theories had to be formalised. This could be done by squeezing every juicy drop of intuition out of mathematics. Then only formal structures would remain. Second, these formal representations of mathematical theories had to be investigated, in particular with respect to completeness and consistency. If the completeness and consistency could be proved, then mathematical theories would be vaccinated against paradoxes.[6]

However, in 1931 Hilbert's programme suffered a knock-out, when Kurt Gödel (1962) published his famous incompleteness theorem. This theorem states that if a formal system of a certain complexity is consistent, it will be incomplete. The idea of representing mathematical theories by complete and simultaneously consistent formal systems was revealed as an illusion. Gödel's proof presupposes that the formal system in question is rich enough to include an axiomatisation of standard arithmetic, which was a minimal requirement for the whole metamathematical programme.[7]

The original idea of metamathematics was that mathematical theories could be *represented* by formal systems. Soon emerged the idea that mathematical theories could be *identified* with formal systems. This idea acquired much force, even after the metamathematical programme had stumbled over Gödel's incompleteness theorem. The claim of identity between mathematics and formal structures is defining for formalism as a philosophy of mathematics. Hilbert has often been referred to as the father of formalism, but I doubt if he thought of formal systems as being anything more than representations of mathematical theories.

Formalism appears a powerful position, as it provides straightforward answers to such classic philosophical questions as: What is mathematics? The question simply becomes identical to the question: What is a formal system? This later question can be answered in specific steps by clarifying the notions of *alphabet*, *formula*, *axiom*, *rule of inference*, *proof*, and *theorem*.

A formal system has to operate with an *alphabet*, which refers to the set of symbols that can be applied. Such an alphabet can include symbols such as p, q, r, (,), $\vee$, $\Rightarrow$, $\forall$, $\in$, and $\exists$. For any specific formal system, the

6 For a detailed presentation of metamathematics, see Kleene (1971).

7 For further discussions of Gödel's incompleteness theorem, see Budiansky (2021) and Goldstein (2005).

list of allowed symbols must be explicitly enumerated. It should be well-defined whether or not a symbol belongs to the alphabet or not. It needs to be specified which sequences of symbols count as formulas in the system. One can think of this definition as the grammar of the formal system. A grammar could, for instance, define the sequence $(p \vee q)$ as being correct, and the sequence $(\Rightarrow p \vee q$ as being incorrect. The whole grammar has to be formulated in such a way that it is well-defined whether or not a sequence of symbols is a formula or not.

Some formulas have to be enumerated as *axioms*. This set will serve as a departure for the deductions to be made. Naturally, there are many issues related to the selection of axioms, as, for instance, not selecting axioms that might lead to contradictions. The *rules of inference* that are going to be applied in the system have to be enumerated. Such rules specify how one, from one or more formulas, can derive other formulas. The basic idea is that *if* the original formulas (the premises) are true, *then* the derived formulas (the conclusions) will be true as well. No formal system demonstrates the actual truth of any theorems, but it shows what can be considered true if the axioms are considered true. In a formal system, the notion of truth is of a hypothetical if-then nature.

A *proof* can be defined as a sequence of the formulas $F_1, \ldots, F_n$, where any formula F_i (where $1 \leq i \leq n$) is either an axiom or can be derived from one or more of the formulas in the sequence $F_1, \ldots, F_{i-1}$ in accordance with the rules of inference. This definition of proof brings us to the definition of a *theorem* as a formula which occurs as the last formula F_n in a sequence of formulas $F_1, \ldots, F_n$, that composes a proof.

By such a clarification of alphabet, formula, axiom, rule of inference, proof, and theorem, one gets a definition of formals system, and, as a consequence, a definition of mathematics according to formalism.

Logical positivism

The formalist interpretation of mathematics had a huge impact on the formulation of logical positivism as a philosophy of science in general. According to logical positivism, mathematics and scientific theories have to be kept strictly detached, not only from intuition, but from any form of contextualisation. They have to be kept separated from subjective preferences, religious convictions, ethical principles, cultural traditions, political priorities, and from any form of metaphysical thinking.

In *A Mathematician's Apology*, Hardy formulated a thesis of innocence, with respect to what he referred to as 'real' mathematics. This thesis, however, invoked the much broader dogma of neutrality, according to which any form of mathematics can be researched and developed separately from ethical and political considerations. Logical positivism establishes an even much broader dogma of neutrality according to which not only mathematics, but also science in general, can be kept, and must be kept, ethically and politically neutral. This dogma came to dominate the perspective on mathematics and science, and was rarely questioned until the late 1960s, when critical conceptions of mathematics and of sciences were formulated.[8]

The Vienna Circle, as organised by Moritz Schlick (1886–1936), included philosophers, scientists, and mathematicians. Rudolf Carnap (1891–1970), Herbert Feigl (1902–1988), Kurt Gödel (1906–1978), Hans Hahn (1879–1934), Otto Neurath (1882–1945), and Friedrich Waismann (1896–1959) were among them.[9] The Circle was deeply engaged in actual developments in science and mathematics. They studied Albert Einstein's formulation of the theory of relativity, and the principles of quantum mechanics. Hilbert's metamathematical programme was carefully discussed, and recent developments in formal logic were investigated. In 1929, Gödel completed his PhD studies in formal logic with Hahn as his supervision,[10] and two years later he presented his famous incompleteness theorem. Wittgenstein's *Tractatus*, published in 1922 in a German-English parallel edition, was studied carefully by the Circle. It provided a principal inspiration for formulating the

---

8 See Chapter 11 in this volume for the formulation of a critical conception of mathematics.

9 Several other people became associated with the Vienna Circle, for instance Hans Reichenbach, who worked in Berlin. Together with Carnap, he edited the journal *Erkentnis* (*Knowledge*) that expressed the outlook of logical positivism. Carl Hempel also worked in Berlin. Karl Popper was around, but even though he was actively contributing to the discussion of science and shared many of the concerns of the Vienna Circle, he was never invited by Schlick to join.

10 The result of this study is found in Gödel (1967). By proving the completeness theorem of predicate logic, Gödel demonstrated that Frege's intuition was sound; the axiom system that he presented in the *Begriffsschrift* as the foundation of predicate logic was in fact complete.

overall position of logical positivism including the dogma of neutrality. Wittgenstein was also invited by Schlick to join meetings of the Circle.[11]

Members of the Vienna Circle were deeply concerned about political developments including the growing anti-Semitism. For them, Nazi conceptions such as 'Arian Physics' or 'degenerate Jewish physics' had nothing to do with science; as meaningless metaphysical notions, they had to be eliminated from any scientific outlook. Looking more carefully at scientific theories, one might find a broader range of metaphysical assumptions and preconceptions, not only of political but also of philosophical, religious, and psychological nature. According to the Vienna Circle, all such features of metaphysics had to be eliminated from science. They found that they were facing a huge task in a most difficult period of time, namely to clean up science and to ensure that it got its proper neutral format.[12]

In an attempt to eliminate metaphysics from the domain of science, the Vienna Circle formulated the principle of verification. According to this principle, a statement has a meaning if and only if it is possible to specify some observations that can serve as empirical evidence for that statement. If such a specification is not possible, the statement is meaningless. As an example, we can take the statement 'God is almighty.' As one cannot point out any possible empirical observations that could support this statement, it is meaningless. In general, religious claims end up in the waste bin together with any other forms of supposed nonsense. So do many statements from psychology and psychoanalysis. The waste bin also becomes stuffed with ethical statements, as no empirical evidence for such statements can be identified. Furthermore, established disciplines such as physics need critical investigations since, for instance, the concept of force might include metaphysical features.[13]

A variety of specific formulations of the principle of verification was carefully investigated by the Vienna Circle. However, it turned out that whatever formulation one gave the principle, one could not escape the dilemma that either the formulation would be too loose, meaning

11 For a careful study of the Vienna Circle, see Stadler (2015). For captivating presentations of the Vienna Circle, see Edmonds (2020), and Sigmund (2017).

12 Carnap (1959, first published in *Erkenntnis* in 1932) made a powerful presentation of this cleaning programme.

13 See Jammer (1957).

that obvious metaphysical statements came to count as meaningful, or it would be too tight, meaning that general natural laws of physics became relegated as meaningless. Furthermore, what about the very principle itself? How could you verify the principle of verification? As it appears impossible to specify what empirical evidence might support the principle, it seems itself to become meaningless.[14]

The approach to eliminate metaphysics, however, also followed another much more powerful departure. Logical positivism expressed a huge doubt with respect to natural language. In doing so, it drew directly on the formalist conception of mathematics. The grammar of natural language was all too loose, making ample space for formulating any kind of statement, also with a profound metaphysical content. Natural language opens an extensive space for expressing nonsense in grammatically correct ways. When used as the language of science, natural language must be under suspicion.[15]

In the *Tractatus*, Wittgenstein (1992) assigned a particular role to formal language. Here he consistently talks about language as singular, and it really has to be read as *the* language. This is the formal language from *Principia Mathematica*, and the language that formalism had cultivated. This language Wittgenstein sees as *the language of science*, emphasised throughout the *Tractatus* and brought together in §6 and §7. In the concluding paragraph of the *Tractatus*, one can read:[16]

> §7 Whereof we cannot speak about, thereof one must be silent.

Many times, this paragraph has been read as an elegant and artistic conclusion of the book, but it is much more than that. It condenses Wittgenstein's whole conception of science and ethics. §7 has to be read together with §6. While §7 states what *cannot* be said, §6 states what *can* be said:

> §6 The general form of a truth-function is: $[\bar{p}, \bar{\xi}, N(\bar{\xi})]$. This is the general form of proposition.

14 An overview of the discussion of the principle of verification is presented by Hempel (1959), first published in 1950.

15 One important contribution to the critique of natural language was previously formulated by Russell (1905) in the article 'On Denoting'.

16 There exist several translations of the *Tractatus* into English. Here I cite the first translation by C. K. Ogden from the original German-English edition of 1922.

By a truth-function, Wittgenstein refers to a property of a composed proposition, namely that its truth value is determined by the truth values of the propositions of which it is composed. The expression $[\bar{p}, \bar{\xi}, N(\bar{\xi})]$ is Wittgenstein's shorthand for an arbitrary proposition built up by logical connectives.[17] Wittgenstein claims that any proposition has this form. If a linguistic formulation does not have the property of being a truth-function, it must, according to §7, be passed over in silence. Wittgenstein's claim is that the language of science is truth-functional.[18]

This claim was carefully discussed by the Vienna Circle. In the *Logical Syntax of Language*, the original German version of which was published in 1934, Carnap (1937) elaborated on the claim that a formal language can serve as the language of science. The discussion is rich in details, and Carnap recognised that one needs to apply a version of formal logic with a higher degree of complexity than the one Wittgenstein referred to in the *Tractatus*. I see Carnap's discussion in *Logical Syntax of Language* as a careful elaboration of the clue provided by Wittgenstein in §6 of *Tractatus*.

What now to think of that which cannot be expressed in formal language? As mentioned, Wittgenstein's answer comes in §7: Remain silent! Logical positivism agrees: Science has to concentrate on what can be expressed in formal language, and to leave the rest aside. Thus, §7 is a condensed expression of the claim that no metaphysical elements can be part of science, whether they take the form of religious convictions, political positions, or ethical principles. Together §6 and §7 provide as condensed formulation of the total separation between science and any value statements.

---

17 The connectives could be and , as used by Whitehead and Russell in *Principia Mathematica*. In 1913, Henry Sheffer showed that it is possible to define the other connectives from just one connective, now referred to as the Sheffer stroke. Informally, the Sheffer stroke can be defined as 'not both', meaning that it occurs as a negation of a conjunction. In Wittgenstein's symbolism, the Sheffer stroke is referred by be the symbol . Thus by Wittgenstein refers to an arbitrary truth-function expressed by means of the Sheffer stroke.

18 The formulation seems to indicate that Wittgenstein thinks of the language of science as being that of proposition logic. That is certainly a simplistic claim. As a minimum, one should think of the language of predicate calculus as a prerequisite for the language of science. Any formulation of laws of nature would presuppose such a language.

Acknowledging the principal idea of logicism that mathematics and logic are the same, logical positivism claims that mathematics as the language of science ensures that science does not include metaphysical elements and that science turns ethically neutral. In this way we have reached a most profound legitimation of the doctrine of neutrality.

Due to their explicit anti-Nazi positions and, in several cases, also to their Jewish origins, many members of the Vienna Circle left Austria after Adolf Hitler came to power. Many escaped to the United States of America.[19] As part of the transplantation into an English-speaking context, logical positivism did change.[20] From being a critical stance with respect to the present state of science, including a profound critique of Nazi ideologies, it turned into a devise for legitimising science as, in fact, being detached from socio-political issues. In this way, the transplanted version of logical positivism came to operate as a legitimation of what was taking place in most university studies in sciences and mathematics, not only in the USA but the world over. Logical positivism turned into a legitimation of not engaging in socio-political issues as an integral part of any such study programmes. From providing a departure for a critique of science, logical positivism turned into a broadly assumed convenient dogma about neutrality.

Structuralism

Structuralism can be interpreted as an elaborated version of formalism, and structuralism had a profound impact on mathematical research. It

19 In 1935, Carnap emigrated to the USA. Feigl's parents were not religious, but they were Jewish, and in 1931 he left for the USA. In 1939, Gödel got a position at the Institute of Advanced Studies in Princeton, with which also Einstein was associated. In 1934, Neurath fled to the Netherlands and later on to England. Waismann was a Jew, and in 1938 he emigrated to the USA. In 1933, Reichenbach was dismissed from his work due to his Jewish background, and in 1938 he moved to the USA. Hempel's wife was of Jewish origin, and in 1937 they emigrated to the USA. Popper had Jewish origins as well, and in 1937 he emigrated to New Zealand.

20 The first English introduction to the ideas of the Vienna Circle was presented by Ayer (1970), when he published *Language, Truth and Logic* in 1936. Other presentations in English are found in Ayer (1959), Carnap (1962), Hempel (1970), and Reichenbach (1966). In 1961, Newman (1979) published *The Structure of Science*, which is a textbook-like presentation of how to do science.

resulted in a restructuration of mathematical theories, which included the formation of new mathematical notions and structures.

Structuralism acknowledges the importance of outlining the alphabet, defining the formulas, and enumerating the axioms for developing a mathematical theory. Structuralism also emphasises the importance of specifying the nature of proof, although without operating with an explicit enumeration of rules of inference. With respect to proving, structuralism sticks to the practice of mathematics, according to which proving must be strictly logical and transparent. There is no application of any form of intuition in mathematical proving; no figures or diagrams are necessary, not even in geometry. In this sense, structuralism assumes the whole approach of eliminating intuition from mathematics.[21]

Nicolas Bourbaki was an important exponent of structuralism. In some places, one can read that he worked at the Royal Academy of Poldavia, in other places that he was associated with the University of Nancago. However, behind the collective pseudonym one finds mathematicians including André Weil (1906–1998), Henri Cartan (1904–2008), Claude Chevalley (1909–1984), and Jean Dieudonné (1906–1992). Over time, many more people have contributed to the collected works of Bourbaki.[22]

The Bourbaki working group was established in the mid-1930s. The original idea was to write a university textbook in mathematical analysis covering recent developments in mathematics. Soon, however, the work became much more ambitious and turned into a project of providing a systematic presentation of major parts of mathematics. The first volume of *Elements of Mathematics* (*Éléments de Mathématique*) was published in 1939 (Bourbaki, 2004). It provides a presentation of set theory, which

21 The formal logical systems, as presented in the *Begriffsschrift* or *Principia Mathematica*, operate with two rules of inference that easily can be stated explicitly. But if we are dealing with a mathematical formal system, such as Peano's axiomatics for the natural numbers, many more rules of inference are going to be applied. But which? One could stipulate that the set of possible inferences for a Peano axiomatics correspond to the theorems in, say, *Principia Mathematica*. This seems consequential, as *Principia Mathematica* presents a system of valid inferences. However, the situation is more complex than that. There are forms of mathematical reasoning which are not captured by any theorem in *Principia Mathematica*, but which are broadly applied in making mathematical deduction. For mathematical research practice, also as shaped by structuralism, the implication is that the rules of inferences are not enumerated, but inferences are kept as transparent as possible.

22 See, for instance, Bourbaki (1950) and Dieudonné (1970).

was considered the basis of mathematics; this idea Bourbaki shares with Frege, Whitehead, and Russell, and logicism in general. Many more volumes of *Elements of Mathematics* followed covering topics like algebra, topology, and topological vector space.

Historically, there is a connection between formalism and Bourbaki's structuralism via Emmy Noether (1882–1935) who, for a period, worked at the mathematical department at Göttingen University, directed by Hilbert. Bartel van der Waerden (1903–1996) was one of her students, and his book *Modern Algebra*, first published in two volumes in 1930 and 1931, was deeply inspired by Noether's lectures. The book is referred to by Dieudonné as an important resource for the Bourbaki group, preparing as it did for the definition of several of the formal structures to which they referred.

The Bourbaki group met a few times per year. At such meetings, manuscripts were presented and discussed carefully, sometimes being read aloud and criticised sentence by sentence. Alternative suggestions for completing a proof were suggested as well as alternative definitions of concepts. The meetings had no chair, and the discussion could be heated. When a manuscript had been worked through, a different member of the group got the task of presenting a revised version of the manuscript at the following meeting. This procedure continued meeting after meeting until consensus was reached.

Only one formal rule guided the work in the group, namely that, on turning fifty years old, the member had to leave the group. New members had to be recruited, and if members became aware of particular gifted students, they could be invited to join a meeting. Any newcomers who did not make significant contributions were dropped, though a second invitation could be considered.

It was presupposed that the members of the group had broad interests in mathematics since the work in the group was not for narrow specialists, the principal aim being to identify relationships between different areas of mathematics. Bourbaki tried to identify how structures and proofs in one area appeared similar to structures and proofs in other areas. When such similarities were identified, the challenge was to make them explicit, and Bourbaki identified a range of such overlapping structures.

Mathematics is in rapid development, new theories and new concepts are constantly emerging. How can we effectively integrate and update all these developments? Bourbaki provided a suggestion. The *Elements of Mathematics* can be read as a kind of mathematical encyclopaedia, organised not in alphabetic order, but structurally.

Chapter 1 in the first volume of *Elements of Mathematics* makes a presentation of what is to be understood by formal mathematics. By making this start, Bourbaki explicitly takes a formalist departure. Chapter 2 presents set theory, defining notions like order pair, function, and correspondence. Chapter 3 addresses ordered sets, cardinals, and integers, while the final Chapter presents the notion of structure.

The crucial notion is *structures*, which became the building blocks of Bourbaki's architecture of mathematics. In order to describe a structure, the properties of its elements are without significance. Bourbaki agreed completely with Hilbert's formulation in the *Foundations of Geometry*, when he enumerated objects like 'point', 'line', and 'plane' without specifying anything about these objects. The only thing relevant is to specify how they relate to each other, and this is done in terms of the axioms defining the structure.

Through Bourbaki's profound studies of a variety of mathematical theories, three *mother structures* were identified: (1) a set organised by an operation; (2) a set organised by a relation; and (3) a set organised by a topology. The group $(G,\circ)$ is an example of a set G organised by an operation $\circ$ which is a function of two variables from $G\times G$ to G. The group $(G,\circ)$ fulfils the axioms:

1. $\forall a,b,c \in G\colon (a\circ b)\circ c = a\circ(b\circ c)$
2. $\exists n \in G\colon \forall a \in G\colon a\circ n = n\circ a = a$
3. $\forall a \in G\colon \exists a^{-1} \in G\colon a\circ a^{-1} = a^{-1}\circ a = n$

One theorem in group theory states that there is only one neutral element. The proof runs like this: Assume that there exists two neutral elements n_1 and n_2. According to the definition, we would have $n_1 = a\circ a^{-1}$ as well as $n_2 = a\circ a^{-1}$. From this we can conclude that $n_1 = n_2$. Group theory developed further along such lines. The departure is the axioms, and nothing but axioms, and the proving needs to be logically straightforward and transparent. The group structures can be recognised in a variety of mathematical disciplines: number theory, geometry, vector calculus, etc.

By means of the mother structures, a huge amount of other mathematical structures can be defined. Notions like ring, field, ordered field, vector space, and Hilbert space can be defined, and the many different classic disciplines start growing together in the same architecture.

By emphasising the importance of mother structures, Bourbaki diverged from a traditional formalist outlook as, for instance, summarised by Haskell Curry (1970) when he states that the 'essence of mathematics lies [...] not in any particular kind of formal system, but in formal structure as such' (p. 56). Bourbaki does not assume any such relativism, but finds that some structures are more important than others to the extent that they express fundamental similarities between apparently different mathematical disciplines. That we are dealing with three mother structures is not any *a prior* given. It is an insight that emerged from the discussions in the Bourbaki group. More mother structures could be identified as mathematics develops. What we are dealing with is just a summing-up of structures identified by a certain group of mathematicians at a certain moment in the history of mathematics.

The Bourbaki group took as its point of departure the current state of mathematics. For identifying structures, they did not consider any historical developments that have brought forward the mathematical ideas and theories. Nor did they pay attention to possible applications of mathematics. Applications were not considered relevant for identifying mathematical structures.

Through a profound de-contextualisation of mathematics, Bourbaki's structuralism repeats the separation between mathematics and socio-political issues as advocated by logical positivism. I interpret structuralism as a principal example of how the dogma of neutrality can be acted out within mathematics research. Most ironically, however, structuralism gained a profound social impact through a widespread reformation of mathematics education.

The Modern Mathematics Movement

The seminar *New Thinking in School Mathematics* took place over twelve days in 1959 at Cercle Culturel de Royaumont, a more than

700-hundred-year-old abbey located north of Paris. The seminar was organised and financed by the Organisation for European Economic Co-operation (OEEC), later to become the Organisation for Economic Co-operation and Development (OECD).

In the peaceful environment provided by the Royaumont Abbey, an important feature of the Cold War was addressed. The tension between the East and the West had been steadily growing, and the military potentials were a crucial factor. The assumption had been, at least in the West, that the USA was well ahead of the Soviet Union with respect to technology in general, and military technology in particular.

One important element of military technology was the capacity for deploying rockets, and it came as a major shock to the West when in 1957 the Soviet Union launched their first Sputnik.

The seminar *New Thinking in School Mathematics* was provoked by the Sputnik shock. It became accepted that in order to advance technology, recognised as an urgent matter, radical improvements in mathematics education were necessary. At the seminar, the mathematician Marshall H. Stone gave the introductory lecture and highlighted that the 'teaching of mathematics is coming to be more and more clearly recognized as the true foundation of the technological society which it is the destiny of our time to create' (p. 18). This and others of his formulations resonated nicely with the overall OEEC rationales for organising the seminar. However, right after the opening lecture, the seminar took an abrupt turn and references to social and technological issues were forgotten.

In his lectures, which turned out to become the principal reference for the whole seminar, Dieudonné presented drastically new ideas about the content of secondary school mathematics. (As mentioned, Dieudonné was born 1906, meaning that he had turned fifty years old and therefore had to leave the Bourbaki group. This might have created space for him to engage in other activities.) He started his lecture this way:

> My specific task today is to examine, from the point of view of present curriculum in mathematics in universities and engineering schools: (a) What mathematical background professors in these institutions would like to find in the students at the end of their secondary school years. (b) What they actually get. (c) How it would be possible to improve the existing situation. (OEEC, 1961, p. 31)

Dieudonné's perspective is clear: a reform of the mathematical curriculum at secondary schools has to be guided by the actual curriculum at the university level. He asks for a radical updating of the curriculum:

> The curriculum of the secondary schools has to be reorganised in order to eliminate any undue waste of time and to absorb as much as possible of the burden now resting entirely of the university as is compatible with the intellectual capacities of the children. (p. 34)

What reorganisation, then?

> In the last 50 years, mathematicians have been led to introduce not only new concepts but a new language, a language which grew empirically from the needs of mathematical research and whose ability to express mathematical statements concisely and precisely has repeatedly been tested and has won universal approval. But until now the introduction of this new terminology has (at least in France) been steadfastly resisted by the secondary schools, which desperately cling to an obsolete and inadequate language. And so when a student enters the university, he will most probably never had heard such common mathematical words as, set, mapping, group, vector space, etc. (p. 34)

Dieudonné wants a conceptual updating of secondary school mathematics. He is not referring explicitly to the work of Bourbaki, but it is clear that his suggestion reflects his structuralist outlook. The curriculum of secondary school mathematics has to be developed around the basic mathematical structures.

This demand, Dieudonné turned into a slogan: 'Euclid must go!' For centuries, Euclid's *Elements* had existed as a principal departure for mathematics education. The *Elements* provided a path whereby proofs led to one theorem after the other. This path has been assumed to reveal the genuine nature of mathematics. But according to Dieudonné, this approach belongs to the museum of mathematics. Euclid must go in order to make space for a relevant updating of the whole discipline.[23]

'Euclid must go!' condenses clearly the structuralist concern with respect to intuition. Intuitions had been incorporated in the whole Euclidean presentation of geometry. This became obvious when Pasch and Hilbert made explicit the many 'hidden axioms' in Euclid's *Elements*. That intuition had brought the deductive processes forward had been

23 See also Dieudonné (1973).

hidden by the presence of diagrams. Diagrams should not have any role to play in mathematics, but in Euclid's *Elements* they did. According to structuralism, this diagram-based intuition had to be eliminated, and in particular structuralist presentations of geometry could be completed without any use of diagrams.

Mathematicians from around the world with an interest in mathematics education joined the seminar. They listened to Dieudonné's presentation, discussed over the twelve days, and gained much inspiration. From Denmark participated Svend Bundgaard, a mathematician from Århus University, and Ole Rindung, particularly interested in secondary school mathematics. At the seminar, it was decided that an expert group should be brought together in order to provide a synopsis for the new curriculum for secondary school mathematics. The group had sixteen members, including Erik Kristensen, also from Århus University. In August–September 1961, the group met in Dubrovnik, and in 1961 their report *Synopsis of Modern Secondary School Mathematics* (OECD, 1961) was published.

A few years later, Rindung and Kristensen published the first volume of a mathematical textbook for the Danish *Gymnasium* for sixteen- to nineteen-year-old students. This textbook was radically different from what had been seen until then. It started with set theory, and right from the beginning the symbolic language of formal logic was brought into operation. The principal mathematical structures were presented, and a new path into the whole landscape of mathematics was defined.

Soon there appeared textbooks for fourteen-year-old students at the Danish *Folkeskole* for six- to sixteen-year-old students, starting with set theory. Simultaneously, textbooks for teacher education and for in-service training of teachers became published, all reflecting the idea that set theory provided the start of learning mathematics. Bent Christiansen from the Royal Danish School of Educational Studies was deeply engaged in implementing the reform by developing material for teachers as well as for students. Soon appeared textbooks for six- to seven-year-old children starting with set theory. In the end, the structuralist approach came to dominate mathematics education in Denmark, at least for a while.

My reference to the development in Denmark serves as an illustration, as what took place in Denmark took place, *mutatis mutandis*, in many other countries as well. We are dealing with a most powerful reform movement. I am not aware of any other educational reforms with such an immediate impact.

The Modern Mathematics Movement covered mathematics education through new structures together with an implicit claim about neutrality, totally distancing it from socio-political issues. Mathematical structures were the focus, not what could be done by means of mathematics. Although the rationale for the Royaumont Seminar was both economic and political, the structuralist outlook annihilated all such 'externalities'. Structuralism focused on intrinsic features of mathematics, and it represented the ultimate de-contextualisation of both mathematics research and of mathematics education. It provided the final step of the ambition of logical positivism of characterising mathematics as neutral, establishing the dogma of neutrality.[24]

Poor Piaget!

In the middle of the 1970s, when the Modern Mathematics Movement was in full swing, and when I started studying mathematics education, one found references to the work by Jean Piaget everywhere. There appeared to exist a clear connection between his formulations of a genetic epistemology and the Modern Mathematics Movement.

When I first looked through the report from the Royaumont Seminar, I was surprised not to find any references to Piaget. It appeared to me that the implementation of the Modern Mathematics Movement came before its epistemological justification. I became interested in clarifying better the nature of Piaget's genetic epistemology. An important resource for me was the book *Mathematical Epistemology and Psychology*, written by Ewert Beth and Jean Piaget (1966), which first appeared in French in 1961. The book is divided into two parts, the first written by Beth, the second by Piaget.[25]

24 In this way, structuralism cemented the ground-zero from which critical mathematics education was to sprout, to which I return in Chapter 11 of this volume.

25 Beth was deeply interested in the foundation of mathematics. His book of more than 700 pages, *The Foundations of Mathematics* (Beth, 1968), first published in 1959, provides a most elaborated discussion of foundational issues. See also Piaget (1970).

In his part, Piaget refers to a seminar that took place in 1952, in which both Dieudonné and he participated. Dieudonné presented the structuralist view on mathematics as formulated by Bourbaki, and he outlined the nature of the three mother structures. Piaget presented how he had studied children's operations with objects, and how he had condensed his observations by means of three operational structures. Piaget tells that the high degree of correspondence between the three mother structures and the three operational structures appeared as a surprise to those participating in the seminar, and also to Piaget himself.

What can be concluded from such an observation of similarity? One can make a step further than just acknowledging similarities by claiming that there exists an intrinsic connection between the two types of structures. As the mother structures are the basic building blocks in Bourbaki's architecture of mathematics, one can be tempted to stipulate children's operational structures as being the genetic roots of mathematics. To me, this stipulation constitutes the departure for Piaget's genetic epistemology. The seminar in 1952 might be the occasion where this idea emerged.

The idea of a genetic epistemology is original. A classical empirical interpretation of the roots of mathematical knowledge has highlighted that mathematical concepts and insights emerge from *observations of properties* of physical objects. One experiences a very smooth surface, and one gets to the concept of a plain. One makes addition of different objects, and one gets to the basic laws of arithmetic. Piaget's idea is different. He sees *reflections on operations* with objects as being the root of mathematics.

On various occasions, Hans Freudenthal pointed out that Piaget completely misunderstood the nature of Bourbaki's work.[26] According to Freudenthal, Bourbaki's suggestion for an architecture of mathematics just represents a particular event in the history of mathematics. The identified mother structures could have been different; their identification depended on the heated discussions in the Bourbaki group. What ended as the architecture was just a historical coincidence.

To Freudenthal it appears arbitrary, if not simply misunderstood, to conclude that – due to similarities between children's operational

26 In Chapter 7 of this volume, we will look more carefully into Freudenthal's view on mathematics; here I restrict myself to mentioning his critique of Piaget.

structures and some structures identified during the late 1930s by a group of French mathematicians – one had identified the genetic roots of mathematics. To me as well, it appears arbitrary, if not misunderstood. I find that the references to Piaget accompanying the Modern Mathematics Movement first of all served as a questionable legitimisation of what was taking place. Freudenthal (1973) points out the following:

> Poor Piaget! He did not fare much better than Kant, who had barely consecrated Euclidean space as 'a pure intuition' when non-Euclidean geometry was discovered! Piaget is not a mathematician, so he could not know how unreliable mathematical system builders are. Bourbaki's system of mathematics was not yet accomplished when the importance of categories was discovered. There can be little doubt that categories will be a new organizing principle and that rebuilding of Bourbaki's structure in categorical style will leave no stone left on top of another. If a leading development psychologist could then convince us of the categorizing genesis of all mathematical concepts – which will certainly eventually happen – then it will just be in time to see the categorical style mathematics, before it is ready, being pulled down in favour of some new principle, which will certainly have its day. Mathematics is never finished – anyone who worships a certain system of mathematics should take heed of this advice. (pp. 45–46)

Piaget's genetic epistemology recapitulates the complete separation between the learning of mathematics and socio-political issues. His theory is about patterns of 'natural growth', and not about social and critical reflections.[27]

The dogma of neutrality was established as integral part of the formalist outlook on mathematics. From there is became articulated by logical positivism as a much broader dogma of neutrality, not only with respect to mathematics but with respect to science in general. Structuralism represents a further development of the formalist outlook with a profound impact on the mathematical research practice and the formation of mathematical theories. Structuralism embraces the dogma

---

27 This separation is repeated by Ernst von Glasersfeld's radical constructivism, which represents a further elaboration of Piaget's genetic epistemology. In a conversation with me, Christine Keitel told that once she had the opportunity to ask Glasersfeld how he saw social and political issues related to mathematics education. Glasersfeld found the question interesting, but admitted that he had never thought about it.

of neutrality, and via the Modern Mathematics Movement this dogma became propagated in mathematics education.

References

Ayer, A. J. (Ed.). (1959). *Logical positivism*. Free Press.

Ayer, A. J. (1970). *Language, truth and logic*. Victor Gollancz.

Beth, E. W., & Piaget, J. (1966). *Mathematical epistemology and psychology*. Reidel.

Bourbaki, N. (1950). The architecture of mathematics. *The American Mathematical Monthly, 57*, 7–30. https://doi.org/10.1080/00029890.1950.11999523

Bourbaki, N. (2004). *Elements of mathematics: Theory of sets*. Springer.

Budiansky, S. (2021). *Journey to the edge of reason: The life of Kurt Gödel*. New Norton.

Carnap, R. (1937). *The logical syntax of language*. Routledge and Kegan Paul.

Carnap, R. (1959). The elimination of metaphysics through logical analysis of language. In A. J. Ayer (Ed.), *Logical positivism* (pp. 60–81). Free Press.

Carnap, R. (1962). *Logical foundations of probability*. University of Chicago Press.

Curry, H. B. (1970). *Outlines of a formalist philosophy of mathematics*. North-Holland.

Dieudonné, J. (1970). The work of Nicolas Bourbaki. *The American Mathematical Monthly, 77*, 134–144. https://doi.org/10.2307/2317325

Dieudonné, J. (1973). Should we teach 'modern' mathematics? *American Scientist, 61*, 16–19.

Edmonds, D. (2020). *The murder of professor Schlick: The rise and fall of the Vienna Circle*. Princeton University Press.

Frege, G. (1967). *Begriffsschrift: A formula language, modelled upon that of arithmetic, for pure thought*. In J. van Hiejenoort (Ed.), *From Frege to Gödel: A source book in mathematical logic, 1879–1931* (pp. 1–82). Harvard University Press.

Goldstein, R. (2005). *Incompleteness: The proof and paradox of Kurt Gödel*. New Norton.

Gödel, K. (1962). *On formally undecidable propositions of Principia Mathematica and related systems*. Basic Books.

Gödel, K. (1967). The completeness of the axioms of the functional calculus of logic. In J. van Hiejenoort (Ed.), *From Frege to Gödel: A source book in mathematical logic, 1879–1931* (pp. 582–591). Harvard University Press.

Hardy, G. H. (1967). *A mathematician's apology*. Cambridge University Press.

Hempel, C. G. (1959). The empiricist criterion of meaning. In A. J. Ayer (Ed.), *Logical positivism* (pp. 108–129). Free Press.

Hempel, C. (1970). *Aspects of scientific explanation and other essays in the philosophy of science.* Free Press.

Hilbert, D. (1935), *Gesammelte Abhandlungen, Dritter Band*. Springer.

Hilbert, D. (1950). *Foundations of geometry*. Open Court.

Jammer, M. (1957). *Concepts of force: A study in the foundations of dynamics.* Harvard University Press.

Joyce, D. E. (1998). *Euclid's elements.* http://aleph0.clarku.edu/~djoyce/java/elements/elements.html

Kleene, S. C. (1971). *Introduction to metamathematics*. Wolters-Noordhoff.

Lakatos, I. (1976). *Proofs and refutations*. Cambridge University Press.

Newman, E. (1979). *The structure of science: Problems in the logic of science.* Routledge and Kegan Paul.

OECD (1961). *Synopsis for modern secondary school mathematics*. Author.

OEEC (1961). *New thinking in school mathematics*. Author.

Pasch, M. (1912). *Vorlesungen* über *neuere Geometrie* [Lectures on more recent geometry]. Teubner.

Piaget, J. (1970). *Genetic epistemology*. Columbia University Press.

Popper, K. R. (1965). *The logic of scientific discovery*. Harper and Row.

Reichenbach, H. (1966). *The rise of scientific philosophy*. University of California Press.

Russell, B. (1905). On denoting. *Mind, 14*(56), 479–493.

Russell, B. (1993). *Introduction to mathematical philosophy*. Routledge.

Shapiro, S. (2000). *Thinking about mathematics: The philosophy of mathematics.* Oxford University Press.

Sigmund, K. (2017). *Exact thinking in demented times: The Vienna Circle and the epic quest for the foundations of science*. Basic Books.

Stadler, F. (2015). *The Vienna Circle: Studies in the origins, development, and influence of logical empiricism.* Springer.

Whitehead, A., & Russell, B. (1910). *Principia mathematica I.* Cambridge University Press.

Wittgenstein, L. (1974). *Tractatus logico-philosophicus*. Routledge and Kegan Paul.

5. Dehumanisation through mathematics

David Kollosche

Mathematics can be seen as a project of dehumanisation in the sense that it allows us to work with a disregard for personal uniqueness. While the word 'dehumanisation' has a negative connotation and invites us to study this property of mathematics carefully, dehumanisation is also one of the great strengths of mathematics and has proved invaluable for modern society. We will trace this field of tension along two indisputable ingredients of mathematics activity: calculation and logic. As there is enough literature praising mathematics, we allow ourselves to take a more critical stance towards dehumanisation through mathematics. We end with a sceptical discussion on whether mathematics can be rehumanised.

Dehumanisation

I will argue that mathematics is a project of *dehumanisation*.[1] What do I mean with that term? Dehumanisation has become a present concept for the analysis of the psychological and sociological phenomenon of denying people their full humanness, usually in order to justify practices of injustice, violence, and silencing. This understanding of the term dates back to Herbert C. Kelman's (1973) analysis of the mental configuration which allowed people to be well-educated and family-loving while, at the same time, committing some of the most devastating crimes in human history, especially in the Holocaust. Nick

1 For a discussion of dehumanisation through mathematics *education*, see Bishop (1988, especially pp. 12–13).

©2024 David Kollosche, CC BY-NC 4.0 https://doi.org/10.11647/OBP.0407.05

Haslam (2015) provided a good overview of the research field that has developed on this basis, as does the freshly published *Routledge Handbook of Dehumanization* (Kronfeldner, 2021).

An older use of the term dates back to the work of Max Weber, one of the pioneers of sociology. In his monumental work *Economy and Society*, Weber (1978) argued that the bureaucratic organisation of administration is technically superior to any other form of administration because it is 'dehumanised' (*entmenschlicht*). Allow me to present the broader context of Weber's thoughts:

> Bureaucratization offers above all the optimum possibility for carrying through the principle of specializing administrative functions according to purely objective considerations. Individual performances are allocated to functionaries who have specialized training and who by constant practice increase their expertise. 'Objective' discharge of business primarily means a discharge of business according to *calculabe rules* and 'without regard for persons.' [...] When fully developed, bureaucracy also stands, in a specific sense, under the principle of *sine ira ac studio* [without anger and passion]. Bureaucracy develops the more perfectly, the more it is 'dehumanized,' the more completely it succeeds in eliminating from official business love, hatred, and all purely personal, irrational, and emotional elements which escape calculation. (p. 975)

Note that, first of all, Weber was not writing about people who become dehumanised. Instead, it is a form of administration, comprising a certain body of knowledge, specific practices and particular perspectives on social affairs, which are 'dehumanised'. What is this supposed to mean? Clearly, bureaucracy has never been a human being, which could suddenly come to be denied this status. However, administrative action is executed by people and affects people. The idea of bureaucracy is that the practice of administration is organised in ways which ensure that the individuality of both administrators and administered is denied. All the administrator is supposed to do is 'calculation', a mechanical processing of official affairs, while the administered is relevant only in terms of the data retrieved for the processing of a specific administrative act. Dehumanisation, here, is understood as a social practice which contributes to the denial of somebody's humanity.[2]

2 There is an interesting debate concerning the differences between the concepts of 'dehumanisation' and 'objectification' (Mikkola, 2021), but this is not the place to continue this debate.

Turning the focus to the people performing bureaucratic work opens up another perspective on dehumanisation. Weber (1978) noted that 'the spirit in which the ideal official conducts his office' is dominated by 'a spirit of formalistic impersonality'. Administrators have to work 'without hatred or passion, and hence without affection or enthusiasm' for their work. 'The dominant norms are concepts of straightforward duty without regard to personal considerations' and everybody 'is subject to formal equality of treatment' (p. 225). In consequence, the administrator has to work like a machine, has to behave in a way that could be called 'dehumanised'. That means that the *dehumanised practice* of bureaucratic administration does not only lead to the *dehumanisation of others*, first of all it requires a *dehumanisation of the self*. It appears that these dimensions of dehumanisation necessarily go together.

Note that Weber (1978) wrote that bureaucracy would be *technically* superior to other forms of administration (p. 973)! Apparently, he was careful enough to reserve some scepticism about the good of a fully dehumanised practice. This scepticism was well placed. To address an extreme example, mind Bauman's (1989) analysis that the highly demanding administrative organisation of the Holocaust was possible only because administrators worked in a demoralised, mechanical way. Despite his caution, Weber was a great admirer of bureaucracy which becomes clear when considering his historical situation. At his time, Weber witnessed a transformation from a poorly organised society, which suffered from poverty, starvation and extreme inequalities and in which support and rights depended largely on birth right, to a highly organised society, in which support and rights were allegedly equally distributed. Bureaucracy, then, was perceived as a tremendous step towards efficiency and equality. Obviously, it would be naïve to simply consider dehumanisation something good or bad. However, both the positive and the negative potential call for a closer analysis.

Mathematics

Alan Bishop (1988) brought up the concept of dehumanisation when discussing rationalism and objectism (the study of objects instead of actions) as cultural values of mathematics:

> So, once again we see, with objectism as with rationalism, an ideology which is in some sense dehumanised. Rationalism is about certain criteria of theories, divorced from their human creators, while objectism is based on inanimate objects and not on animate phenomena, such as humans. Mathematics favours an objective, rather than a subjective, view of reality. (p. 66)

Bishop's remark suggests that dehumanisation is a central ingredient of mathematics, and that dehumanisation through mathematics is a concern for understanding our culture. I want to depart from Bishop's remark and attempt to provide a more elaborated discussion of dehumanisation through mathematics.

My initial statement that mathematics is a project of dehumanisation could now be rephrased to claim that mathematics is a social practice which seeks to work with a disregard for personal uniqueness, with an emphasis on mechanical predictiveness. In this reading, my initial statement appears to be a negatively connotated account of something that is all too obvious: of course, there is an ideal that practices such as proving or calculating are independent of the individual. The belief that such practices are possible explains, to a large extent, the fascination for and use of mathematics. The technical possibility of this independence has been documented in the execution of such practices by electronic machinery. The ontological and epistemological status of this independence has been discussed under terms such as truth and objectivity.

Independent of the legitimacy of the perspective outlined in the last paragraph, there are good reasons to study mathematics as a dehumanised practice. Firstly, the possibility and reality of mathematics as a dehumanised practice is a psychological and sociological phenomenon which deserved attention. In what ways does mathematics achieve dehumanisation? How is such a practice even possible? Why would humans want to engage in it? All these questions focus on *processes* of dehumanisation. Secondly, the dehumanised practices of mathematics result in a dehumanised handling of the issues mathematics is applied to. This is how the prosperity of a dehumanised practice such as mathematics effects culture as a whole. This perspective proposes to focus our discussion also on *consequences* of dehumanisation. All these discussions have the potential to deepen our understanding of mathematics from a sociological, psychological, and philosophical perspective.

Admittedly, one might want to add that there should also be a focus on the *ideology* of dehumanisation, which plays a central role when the values of objectivity and truth are wrongfully projected from 'pure' mathematics to applications of mathematics. Especially the belief that the choice of mathematical models in application is not arbitrary but objectively necessary turns mathematics into a questionable tool of power. However, as intriguing as such a perspective is, much has already been written about it (Davis & Hersh, 1980; Desrosières, 1993; Dowling, 1998; Porter, 1996; Skovsmose, 1994; Ullmann, 2008), and it is not the focus of this chapter.

Some readers might be uneasy with my description of mathematics as a dehumanising practice. Has the turn from mathematics-as-a-product to mathematics-as-a-practice (stimulated, e.g., by Pólya, 1945) not been a major step forward in the philosophy of mathematics, allowing for sociological perspectives on the human side of doing mathematics? Is the description of mathematics as a dehumanising practice not a step back to rightfully outmoded perspectives on mathematics? I argue otherwise, and for that I want to return to Weber's discussion of bureaucracy once more. What did Weber mean when he argued that bureaucracy works the better 'the more completely it succeeds in eliminating from official business love, hatred, and all purely personal, irrational, and emotional elements which escape calculation'? Note that Weber did not say that bureaucracy *was* the practice that realised these attributes! Of course, even bureaucratic administration leaves open doors for some degree of personal variation based, for example, on annoyance or compassion.[3] Dehumanisation can instead be understood as an *ideal* of bureaucratic practice. Bureaucratic administration may never be fully dehumanised, but it is conceived the better, the more it reaches this ideal. Nevertheless, it is clear that human beings execute administration, and research could come to analyse the rather diverse practices that exist within bureaucratic institutions. In the same sense, dehumanisation can be understood as an ideal of research and school mathematics today without saying that

3 Examples might be: receiving petitioners after closing time, allowing petitioners to hand in attachments for applications after their deadlines, offering superficial or profound consulting, providing tips how to get the most out of a specific situation.

research and school mathematics is not still human-made or that the mathematical practices we engage in are no worthy part of the study of what we call mathematics.

What then is this 'mathematics' that is asserted to be a project of dehumanisation? I do not wish to attempt to answer the highly controversial question what mathematics *is*. Instead, I propose to further discuss two very general mathematical practices which have, beyond any doubt, become paradigmatic for the question what mathematics might be, namely *calculation* and *proof*. Calculation refers to a more hands-on characteristic of mathematics. It is closely connected to applications of mathematics and can be traced back to the very beginnings of human civilisation. Proof, then, at least in the Western tradition, refers to a more philosophical approach towards mathematics with only indirect connections to applications. It is closely connected to the manifestation of mathematics as part of academics and can be traced back to Ancient Greece.[4] The following parts of the chapter will therefore be dedicated to the discussion in how far calculation and proof can be understood as a project of dehumanisation.[5]

Dehumanisation through calculation

Calculation is probably the central driving force of dehumanisation through mathematics. First, calculation owes its efficiency to a certain disregard of the objects of investigation, thus, when these objects are people, opening a space for the dehumanisation of others. Second, calculation demands from its applier a certain mindset that is not unlike the dehumanised mindset of the bureaucrat. Consequently, calculation appears to be a worthy start for a discussion of dehumanisation through mathematics.

Calculation, here, should be understood in a wide sense as any manipulative practice within a calculus and any application of such

4 Note that there are traditions of the justification of mathematical knowledge that are closely connected to calculation and application, for example in Ancient Chinese mathematics (Chemla, 2012).

5 Some readers might notice that I have already discussed these issues elsewhere (e.g., Kollosche, 2014), but while my earlier approach towards that topic had been guided by socio-economic and didactical perspectives, I want to dare a closer look at epistemological aspects of mathematics here.

calculi. 'Calculus', in the English language often closely associated with the infinitesimal calculus, refers to a system comprising a set of allowed signs, rules for their combination to statements, and rules for the manipulation of such statements (Krämer, 1998, p. 29). Besides infinitesimal calculus and among others, we have arithmetic, algebraic, and propositional calculi. The following discussions will circle around arithmetic and elementary algebra, but the points made are meant to be valid for any form of calculation.

Calculation as a shared practice

For reasons of administration, every highly organised state appears to develop some proficiency in calculation. In the Rhind Papyrus, one of the oldest surviving textbooks for mathematics, dating back to Ancient Egypt around 1550 BC, we find the following mathematical problem:

> A quantity, ¼ added to it, becomes it: 15.
> Operate on 4; make thou ¼ of them, namely, 1; the total is 5.
> Operate on 5 for finding of 15. There become 3.
> Multiply: 3 times 4. There become 12.
> The quantity is 12, ¼ of it is 3, the total is 15. (Chace et al., 1929, Vol. 2, Plate 48)[6]

We can see that the problem is posed without any contextualisation. Later examples in that script for solving linear equations feature measures for volumes without any change in the calculative techniques. It can be assumed that those who worked with this textbook were ready to apply this kind of calculative practice to a variety of situations. Or, to put a different emphasis on that last statement: a variety of situations came to be dealt with using identical mathematical techniques. The papyrus also includes many distribution problems, in which usually bread loaves are distributed among men, whereas in one problem measures of beer are distributed without changing the calculative techniques, except for measure conversions in some cases. By contrast, it is fair to doubt that bread loaves and measures of beer were the most pressing problems of distribution for the Egyptian administration. Calculation seems not to depend on what the numbers stand for, be it abstract entities such

6 Chace et al. (1929) provide literal translations, hence the bumpy expression.

as measured quantities and distribution shares or rather real and alive objects such as flocks of animals. For the practice of calculation, context is irrelevant.

Now, imagine the problem presented above had a context, featuring, for example, a shepherd who lent his flock for ¼ interest and was paid back a flock of 15 animals. The question how many animals he lent in the first place could perfectly well be computed by the Egyptian solution. But what would '5' in the second line or '3' in the third line of the cited problem actually mean in our context? We cannot tell and we do not need to care. Calculative techniques work in a mechanical way, irrespective of context, which means that they can be applied also to human affairs without any regard for individual concerns of the objects they are applied to. They bear the possibility of dehumanisation.

In his *Remarks on the Foundations of Mathematics*, Ludwig Wittgenstein (1978) reflected on the nature of rule-following in mathematics. He sees this phenomenon based not only on the experience of shared perception, but also on our experience that imitation (in the sense of copying somebody's actions and obtaining the same result) is possible. For Wittgenstein, the possibility of this kind of conformity is a basic truth of our experience and the beginning of any explanation. Following rules is only possible in the areas of our experience which allow for such conformity. The very idea of a 'rule' results from this experience of repetition of perception and act. Verbalised rules then are first and foremost descriptions of repeating perceptions and actions. Only on this basis can they be understood as prescriptive (in the sense of prescribing perceptions and actions whose description would be that very rule). This line of thought shows, as Wittgenstein stressed, that following rules is a cultural achievement and specifically human. Nevertheless, the very nature of following rules is not to be oneself but to follow the other, to universalise perception and action, to ignore the peculiarities one might experience, to eventually dehumanise the processing of our affairs.

The annulment of meaning

Already the prehistorical example from Ancient Egypt teaches us that calculation is a technique whose internal working is ignorant to what it processes. This ignorance is one source for dehumanisation through calculation, for calculation serves all its objects equally – be they pebbles, bread loaves, sheep, or human beings. Already here, calculation means following specific rules for perceiving and acting with numbers.

However, the history of formalisation teaches us that ignorance can have limitations. For example, the disputes accompanying the introduction of the number zero and of infinitesimals were primarily based on the question what these entities ought to be (Kleiner, 2001; Krämer, 1988a). We might not be able to say what '5' means in the second line of the problem discussed in the last section, but at least we can refer to perceptions where we see five of something, which is a strategy for manifestation that we cannot use for zero and infinitesimals. Calculating with zero and infinitesimals actually *requires* to ignore the question what they might mean in reality. Apparently, not everybody was willing to make this sacrifice. Indeed, we can see that such instances of the ignorance of the question where concepts of mathematics are to be found in our world is a matter of modern times. When trying to locate where this attitude towards meaning has changed, we suddenly encounter developments that go far beyond mathematics and will prove relevant in a variety of ways.

In *The Order of Things*, subtitled *An Archaeology of the Human Sciences*, Michel Foucault (2007) tracked changes in 'the epistemological field [...] in which knowledge, envisaged apart from all criteria having reference to its rational value or to its objective forms, grounds its positivity' (p. xxiii). With central importance to this project, he discusses changes in the use of signs in different cultural arenas. Foucault's central finding is that at the beginning of the seventeenth century, signs were no longer assumed to be inseparably connected to the signified as they had been conceived ever before. Suddenly, they became to be considered arbitrary human constructions. In the arena of literature, he analyses the monumental novel *Don Quixote* by Miguel de Cervantes (1605), in which the protagonist imagines adventures as a knight only to be cast back to his profane reality. Foucault (2007) concluded that

> writing has ceased to be the prose of the world; resemblances and signs have dissolved their former alliance; similitudes have become deceptive and verge upon the visionary or madness; things still remain stubbornly within their ironic identity: they are no longer anything but what they are; words wander off on their own, without content, without resemblance

> to fill their emptiness; they are no longer the marks of things [...]. The written word and things no longer resemble one another. (p. 53)

In the arena of the price for material goods, prices were no longer inseparably linked to the expenses of production, storage, and distribution but were legitimate to take any value. In the arena of medicine, the idea that plants sensuously resembled the body parts or infestations they were able to cure was replaced by a logic of empirical enquiry guided by measurement and logical order. In general, the idea of resemblance was replaced by the idea of an ordered choice of signs. What such an order could and should look like is not an essentialist question; it is a question of logic. In consequence, the seventeenth century sees the rise of a vivid academic discourse on how concepts should be framed and related.

Algebra is no arena of Foucault's (2007) study, but it could just as well have been. Variables are a special sort of mathematical signs, and they change their nature within the time frame discussed by Foucault. Sybille Krämer presented an intriguing study of the change of the use of operative signs in mathematics.[7] Already in the Ancient Egyptian problem discussed above, we see the appearance of variables, there translated as 'a quantity' (elsewhere as 'a heap'), but noted as only one sign in the original hieroglyph script. Krämer (1988a) noted that this use of variables follows the idea that it stands *in place of* a well-defined and yet-unknown number, which eventually can be computed. Variables, here, have a very specific meaning.[8] They represent distinct numbers. Euclid's geometry of Ancient Greece did not include variables for numbers, but it used line segments as general entities irrespective of their actual lengths. In this sense, the abstract line segment can be understood as a variable for lines segments with specific lengths. In some cases, the specific length will follow with necessity from other data in a geometrical construction, while in other cases, the general lines segment is allowed to assume any length. In the latter sense, the line segment can be understood as a geometric variable, as it no longer

7 Krämer (1988a) and Krämer (1991) are two rich and original studies in German. Where possible, I will refer to Krämer (1988b), which is an early summary in English.

8 The epistemology of the term 'variable' from Lat. *variabilis*, meaning 'changeable', is misleading in this case. Some scholars speak of 'apparent variable' or 'bound variable' when they refer to variables in the function of a placeholder.

stands in place of a well-defined and yet-unknown length. It took a long international development to introduce the idea of the variable to algebra as well, including a lot of work by scholars from the Islamic world. In Europe, the popularisation of use of variables in mathematics is often attributed to François Viète's *Isagoge in Artem Analyticam* from 1591, where variables are used to represent a wide range of numbers as in many of today's equations such as $a + b = b + a$ or $y = 7x - 2$. Krämer (1988b) analysed:

> Thus it becomes possible to formulate rules of algebra with universal validity. Unlike the ciphers, the letters are no longer signs for single numbers, but rather signs for the whole class of numbers satisfying a given equation by substitution. The rules which refer to the transformation of equations can in this way be written down in a formal language. This means that their validity is independent of the numerical values entering into the calculation. Algebra becomes the transformation of series of signs according to rules which have no relation to the meaning of the signs. (p. 182)

We could also say that the meaning of Viète's signs is defined by their use in the calculus, the formal language of mathematics, alone, and not by any reference to a meaning beyond them. This breakthrough laid the foundations for many influential developments to come, all depending on the use of signs as ontologically independent entities. René Descartes established an analogy between algebra and geometry and thus opened geometry up for calculation as a tool for solving problems. Isaac Newton and Gottfried Wilhelm Leibniz developed an infinitesimal calculus, in which infinitesimals as well as functions become entities of calculation. Leibniz already worked on a logical calculus and developed the idea that all the truths of the world could be computed on the basis of a sufficiently developed formal language. Here, truth becomes a question of the logic of signs, which no longer represent anything.

The cultural impact of calculation

Now, if thought indeed changed from a logic of resemblance towards a logic in which signs were set loose, would mathematics be a leader or a follower in this process? Viète published his algebra in 1591, whereas *Don Quixote* was published in 1605. But this difference might be misleading, as it can be assumed that in all social arenas, the change came slowly and was only catalysed by intellectual pioneers, whose appearance in time is somewhat random. Foucault (2007) stated that the change occurred 'roughly half-way through the seventeenth century' (p. xxiv) and that the logic of resemblance was in use still 'at the end of the sixteenth century, and even in the early seventeenth century' (p. 19). Note that Descartes's analytical geometry, the next big step in the history of mathematics and in the use of signs as independent entities, was not published before 1637 and was received with astonishment even then!

Still, there are other reasons to assume that mathematics was not a mere follower in this transition in the use of signs. Calculation techniques as in the Egyptian problem above had already illustrated that the manipulation of signs is possible without any reference to their meaning. The signs themselves were required to resemble something real, but their handling was not. The geometry of Euclid had already made implicit use of the idea of variables, albeit restricted to geometrical contexts. In the third century AD, Diophantus of Alexandria had introduced ancient Asian techniques for adding lengths and areas without the scruples of the earlier Ancient Greek tradition, as did many Persian and Arab scholars in the middle ages. Fifteenth-century Europe also saw the introduction of the Indian positional notation system (popularised through economic applications), which brought with it a further appreciation for the efficiency of sign manipulation as in the algorithms of written calculation and a raising tolerance for meaningless signs such as the zero. These preconditions and developments have made it easier for Viète and those who followed to take the next step. So, mathematics seems to have been *a*, if not *the*, protagonist in the culture-wide change of the understanding of signs (Krämer, 1991).

Problematising dehumanisation through calculation

While calculation practices which are ignorant of the meaning of its manipulative steps have flourished for more than 3000 years, modern mathematics refuses to ask for the meaning of the values and expressions of calculation altogether. Roland Fischer (2006) pointed out that this ignorance is a virtue: mathematics would not be useful for practical affairs, if it was compelled to explain the meaning of every concept and manipulation. It is useful exactly because it can rely on calculation alone. All that may be so, but this potential of calculation comes with a price. The disregard for meaning requires those performing calculations to deny their individual thoughts in calculation and can operate on human beings only after reducing them to calculable magnitudes.

Formalism is the elaboration of the denial of meaning as an attitude in the philosophy of mathematics. It assumes mathematics to be nothing but a rule-based game with signs. The signs, representing and constituting mathematics, are held to have a meaning only within the game of mathematics. Reuben Hersh (1997) stressed that, from such a perspective, the applicability of mathematics cannot be explained; it must appear as an astonishing coincidence.[9] However, applicability requires explanation, for, as Gottlob Frege (1960) argued, 'it is applicability alone which elevates arithmetic from a game to the rank of a science' (p. 187).

By abandoning the logic of resemblance, mathematics was able to set loose the power of its formalistic apparatus, but it lost a dimension of self-reflection. No longer asking for more than formal explanations of what a mathematical concept stands for, what a mathematical proposition says, what a mathematical procedure does, means losing the ability to critically reflect on our use of mathematics. Of course, mathematics is still widely applied in our societies, but the question if these applications are justified, the question in how far the mathematical model actually resembles our worldly problem, is no longer a matter of mathematics.

The dialectics of the use of calculation for the processing of social affairs were best described by the Frankfurt School in philosophy. Max Horkheimer (2004) argued:

> As soon as a thought or a word becomes a tool, one can dispense with actually 'thinking' it, that is, with going through the logical acts involved in verbal formulation of it. As has been pointed out, often and correctly, the advantage of mathematics – the model of all neo-positivistic thinking – lies in just this 'intellectual economy.' Complicated logical operations are carried out without actual performance of all the intellectual acts upon which the mathematical and logical symbols

9 And mathematicians *are* astonished: Check, for example, Eugene Wigner's (1960) infamous paper on 'The Unreasonable Effectiveness of Mathematics in the Natural Sciences'.

> are based. Such mechanization is indeed essential to the expansion of industry; but if it becomes the characteristic feature of minds, if reason itself is instrumentalized, it takes on a kind of materiality and blindness, becomes a fetish, a magic entity that is accepted rather than intellectually experienced. (p. 16)

A central line of critique of calculation as a social practice questions the legitimacy and effects of a practice which has to atomise its perception of the world into countable entities. Horkheimer and Adorno (2002) explained:

> Bourgeois society is ruled by equivalence. It makes dissimilar things comparable by reducing them to abstract quantities. For the Enlightenment, anything which cannot be resolved into numbers, and ultimately into one, is illusion; modern positivism consigns it to poetry. Unity remains the watchword from Parmenides to Russell. All gods and qualities must be destroyed. (pp. 4–5)

They go further in proposing that the mathematical handling of affairs is actually rendering our perception of reality and blocking aspects which cannot be dissolved into patterns of sameness and repetition:

> When in mathematics the unknown becomes the unknown quantity in an equation, it is made into something long familiar before any value has been assigned. Nature, before and after quantum theory, is what can be registered mathematically; even what cannot be assimilated, the insoluble and irrational, is fenced in by mathematical theorems. (p. 18)

These practices do of course leave an imprint both on people who are subjected to them and on people who are performing them:

> Not only is domination paid for with the estrangement of human beings from the dominated objects, but the relationships of human beings, including the relationship of individuals to themselves, have themselves been bewitched by the objectification of mind. Individuals shrink to the nodal points of conventional reactions and the modes of operation objectively expected of them. (p. 21)

Concerning the conduct of the self when calculating, Wittgenstein (1978) demonstrated that calculation rests on rules. For Wittgenstein, rules do not hold any inner truth, they are nothing but patterns of repeated action. The whole sense of rules is securing that people can agree on procedures that yield the same results irrelevant of who is performing them. Such predictability lies at the basis of games such as chess, and we would just as well expect it from calculation. It follows that learning to calculate includes learning to follow pre-given rules, to perform acts that are not original to the self but copied from others. This is how calculation dehumanises the self.

In mathematics education research, there remains a strange lack of reflection on how learners cope with the dehumanising side of calculation. Renate Voswinkel (1998), a pastor reflecting on her troubles with mathematics in school, reported:

> We learned our times tables. I memorized the rows and only then checked if it was right that 7 times 3 is 21. 7 times 8 is 56, that is what I told myself in the morning when washing before we would write a test on the tables of eight. I did not keep this in mind because I kept thinking about further-reaching things that I cannot remember today. The result was an increasing quiet devaluation of my own thoughts. I forbade myself to think, because it confused me, although I made my own connections in all other subjects, had ideas, developed a lot of imagination [...]. (p. 18, my translation)

Concerning the discourse on others, there are plenty of examples of how calculation is instrumental in dehumanising people. Andreas Bell (2016) discussed how to allocate donor organs, where the usual practice in Germany relies on a mathematical model that calculates the individual claim on a donor organ on the basis of a few personal variables. Bell argues that although a society has an interest in installing a mathematical mechanism for the transparent allocation of donor organs on the basis of principles such as the social optimisation of this allocation, no mechanism can satisfy all possible expectations concerning optimisation, equity, and compassion. Here, reducing individual cases to a pre-defined set of variables is necessary in order to apply any systematic form of decision-making at all, and yet this process necessarily leads to a dehumanisation of those waiting for a donor organ.

Let me also cite an example from Philip Davis' (2012) epilogue to a recent study edition of *The Mathematical Experience*:

> We are indeed living in an increasingly techno-mathematized world. A recent hospitalization for a minor complaint drove this home to me. I was subjected to a battery of tests carried out on a variety of devices each of

which produced either numbers or a waveform. The medical attendant marked down all the numbers and perhaps a fast Fourier transform was applied to the waveform to obtain more numbers. As a patient, I was transfigured – some might say dehumanized – into a multicomponented vector. (p. 491)

My last example is an extreme case but rather revealing in its simplicity. Look at the following problem from a mathematics textbook in Nazi Germany:

In 1936, the annual expenditure for

1)	33 770	welfare children	19 881 000 RM[10]
2)	131 942	insane and mentally deficient	94 636 600 RM
3)	238 094	hereditary defective (deaf-mutes etc.)	166 000 000 RM

Calculate the cost per head [...].

How many single-family houses at 5000 RM could be built with the sum required for the insane (or the hereditary defective)?

How many families could make their living from these sums (1500 RM per year)?

(Frank, 1939, p. 38, cited in Kütting, 2012, p. 11, my translation)

The scandal here is that the legitimacy of care for people in need is reduced to only one variable of their existence, namely to what they cost society. The comparisons that the demanded calculations suggest, though they are economic nonsense, were meant to raise the acceptance for the euthanasia policy of Nazi Germany. While this example is extreme, we will find similarly ambiguous uses of calculatory practices as legitimisations throughout today's public life (Porter, 1996). The dehumanisation of human beings through mathematics is not a sporadic accident of the application of calculation, it is its predominant mode of operation, as has long been proposed by Davis and Hersh (1986):

The final intent of the application of mathematics to people is to be able to compare two individuals or groups of individuals, to be able to arrive at a precise and definitive opinion as to which is taller, smarter, richer, healthier, happier, more prolific, which is entitled to more goods and more prestige, and ultimately, when this weapon of thought is pushed to its logical limits and cruelly turned around, which is the most useless and hence the most disposable. Whenever anyone writes down an

10 RM stands for *Reichsmark*, the official currency of Germany at that time.

> equation that explicitly or implicitly alludes to an individual or a group of individuals, whether this be in economics, sociology, psychology, medicine, politics, demography, or military affairs, the possibility of dehumanization exits. [...] What is not often pointed out is that this dehumanization is intrinsic to the fundamental intellectual processes that are inherent in mathematics. (p. 283)

Dehumanisation through logic

The relationship between mathematics and logic can be heavily debated. There had been the ambitious but failed attempt by Whitehead and Russell to ground all mathematics on formal logic (George & Velleman, 2002). Others might argue that mathematical work can be disturbingly illogical, only to return to logical forms after a rather wild process of exploring and conjecturing. One way or the other, the *product* of mathematical work will be a theory, which is expected to follow certain criteria of logic, for example that it does not allow to deduce within it two mutually contradictory statements. In the *process* of creating such products, mathematicians will, to some extent or the other, use logical thinking. Eventually, even school mathematics, usually mirroring more elaborated mathematical theories in simpler forms, is a logically organised product.

A common assumption is that logical thinking is an innate capacity of human beings and that self-discipline and good education allow the individual to exploit this capacity to the fullest. From this point of view, logical thinking could be said to be a central part of evolving one's humanity. Psychologically, that may be a way to see it, but sociology casts doubt. Is logical thinking really an innate capacity? Valerie Walkerdine (1988) radically criticised traditional psychology and showed in many experiments that what we call rationality is actually a form of the conduct of the self that is learnt in social interaction. But if we, following this insight, begin to understand logical thinking as a cultural phenomenon, it appears to be astonishing that, using logical thinking, different people come to the same conclusions, find the same arguments compelling, see the same contradictions. Here, I will explain this particularity by demonstrating that logic is a dehumanised practice in that it offers a mechanism of thought which negates individual concerns.

Fundamentals, and how (not) to read them

There is no single answer to the question what logic is. *Formal logic* is highly mathematical, providing different calculi in which statements can be noted and manipulated through computation. While formal logic is a modern phenomenon, we find an abstract and, to some extent, already formalised approach in Aristotle's discussion of certain and uncertain *forms of inference* (Aristotle, trans. 1989). Such descriptions of logical thinking are, in modernised form, of interest for psychology, which studies individual capacities to perform such forms of thought. However, even Aristotle's approach can be argued to rest on some epistemological assumptions (as argued, e.g., by Leibniz, 1765/1896, pp. 404–410), which were already, though not systematically, mentioned by Aristotle, and systematically discussed by scholars such as Arthur Schopenhauer (1903). While categorising forms of inference and discussing logical calculi is rather technical work, the underlying assumptions are very far-reaching decisions of how to think about our world. My analysis of dehumanisation through logic will begin here.

This perspective will follow a somewhat Eurocentric interpretation of what logic might be. Notwithstanding the fact that other cultures developed reflections on logic or even described other forms of reasoning as logical, there are good reasons for the focus on Ancient Greek philosophy: first, it provides us with very early sources on the philosophy of logic, which allows a far-reaching look into the history of such reflections. Second, Ancient Greek logic has been studied by many scholars, upon which we can rely here. Third, Ancient Greek logic has been highly influential for European and modern philosophy and mathematics. However, it should be noted that there was no monolithic 'Ancient Greek logic', that the subject itself was much debated at that time, and that the philosophical worship of Ancient Greek logic as it has been perceived and retold by philosophical tradition may have clouded our view on epistemological alternatives. In any case, it should be noted that when I write 'logic' I refer to the Eurocentric reception of Ancient Greek logic. This use of the word is not meant to deny the existence and legitimacy of other forms of logic.

The formulation and meaning of these foundational assumptions of logic are a matter of ongoing debate, so that any presentation is already biased by a specific interpretation. Allow me to present the assumptions, which are canonically called 'principles' or 'laws', in an interpretation following Klaus Heinrich (1981), only to provide more diverse context later:

1. Things stay the same; they do not change. (Law of identity)
2. Everything is or is not; there is no other way. (Law of excluded middle)
3. Nothing both is and is not. (Law of excluded contradiction)
4. Everything has a reason and is defined by it. (Law of sufficient reason)

Even though these words may provoke many associations, their meaning appears not to be straightforwardly clear. I do not wish to summarise the vast landscape of interpretational controversies here, but let me give some short examples for the interpretation of the law of identity. In the years around 1700, Leibniz (1896) held the law of identity to say that 'everything is what it is' (p. 404), or, more formally, that 'A is A' (p. 405), thus reducing the law of identity to a mere tautology. Leibniz assumes that such 'primate truths of reason [...] seem only to repeat the same thing without giving us any information' (p. 404). He cautiously added the word 'seem' because he saw a function of the law of identity for the manipulation of formal logical statements, but he did not see in the law of identity anything more than a self-evident statement. But would the law of identity have fascinated philosophers over centuries if it was a mere tautology, if it was not 'giving us any information'?

Foucault sets the scene for a different perspective. In his study on insanity, Foucault (1954) showed that the idea of insanity came into being only in modern times, perceived as a threat to reasonable thinking and accompanied by asylums as new institutions and psychology as a new academic discipline. Apparently, as natural and indispensable as the idea of insanity may seem to us today, there had been a kind of thinking in which this idea played no role at all for understanding our world. Based on this insight, Foucault (1966) studied more general patterns of thinking and reasoning over time, and showed that they change severely, including the role of logico-mathematical perspectives of understanding. He called his approach *genealogy*. Foucault (1984) wanted to historically trace ideas not in order to show an inevitable way to any presumably necessary understanding we might have today, but in order to reveal the implicit meaning of the ideas by an analysis of alternatives they were positioned against, of fears, needs and desires that promoted their development. Might logic be a cultural answer to a specific configuration of fears, needs, and desires?

Attend to the following passage where Aristotle (trans. 1933) touched on the problem of identity:

> Thus in the first place it is obvious that this at any rate is true: that the term 'to be' or 'not to be' has a definite meaning; so that not everything can be 'so and not so.' Again, if 'man' has one meaning, let this be 'two-footed animal.' [...] If on the other hand it be said that 'man' has an infinite number of meanings, obviously there can be no discourse; for not to have one meaning is to have no meaning, and if words have no meaning there is an end of discourse with others, and even, strictly speaking, with oneself [...]. (1006a–b)

Here, identity appears to be a matter of the fixation of meaning in a social discourse. This perspective suddenly positions logic in the social realm. Identity demands that the meaning of concepts is made independent from individual interpretation, or, in other words, that their meaning becomes dehumanised. But why did Aristotle have to argue for the law of identity in the first place? Were there any alternatives whose legitimacy Aristotle wanted to disprove? What then is the historical background on the basis of which we can explicate the meaning of the logical assumptions listed above? It may seem that we need a genealogy of the very foundations of logic.

Genealogy of logic

Jean-Pierre Vernant's (1982) *Origins of Greek Thought* provides an intriguing account that logic is not inherent but a cultural phenomenon that can, in the Western tradition, be traced back to Ancient Greece. I owe most of the philosophical perspective on this development to Heinrich (1981), whose research took as its objects of study 'the supressed of philosophy, and not the accidentally suppressed but that, which in the systems of thought, in the rationalised systems of occidental thinking, indeed returns' in the form of a compulsive and unconscious formation of thought (p. 173, my translation).

Hesiod's *Theogony* and Homer's epics illustrate the understanding of the world in Ancient Greece in the eighth century BC. It was, as in many other cultures, based on a polytheistic religion. In this worldview, the worldly forces were humanised in the sense of a human-like representation as gods. For example, Ares stood for war, Demeter for agriculture, Dionysus for ecstasy, and Hermes for trade. Crisis in worldly affairs such as droughts, earthquakes, or diseases could be understood through the tempers of and struggles among the gods. Especially, the Greeks believed that all descendants of a god inherited his or her virtues and vices and could never escape this fate – a belief that we will return to.[11]

Historical changes led to doubt about the legitimacy of the polytheistic worldview (Vernant, 1962/1982). The vast trade network of Ancient Greece imported foreign religions, making the polytheistic worldview appear as a mere possibility among others. Wars led to the destruction of kingdoms whose legitimacy was closely connect to the old myth. Democratically organised city states (note that only the male aristocracy belonged to the *dēmos*) such as Athens developed a culture of public discussion where soon not only political but also moral and religious standpoints came to be questioned. Whether the myth was the appropriate way to explain the world became a pressing question. Philosophy developed within this intellectual crisis as the project of finding better explanations. In this context, Heinrich (1981) reported that Plato had Socrates mourn that 'it is the woe of the philosopher to be confused this way, for confusion indeed is the only source of philosophy'.[12]

---

11 Heinrich (1981, p. 99) cited the Curse of the House of Atreus as an illustration, which is documented in the eleventh song of Homer's *Odyssey*: The mythical god-king Tantalus, a son of Zeus, had offered his dismembered son Pelops as a meal to the gods to test their omniscience. They reassembled Pelops, revived him, and cursed the lineage of Tantalus. All descendants of Tantalus, including Pelops, his son Atreus, and his son Agamemnon, were subsequently involved in clan murders, hatred, and conspiracies. No descendant of Tantalus could escape this fate. The curse was inherited, and inheritance was so inescapable that even the gods could not exclude Pelops from the hereditary curse.

12 Heinrich (1981, p. 31) differs from usual German and English translations of the ambiguous Greek original. For example, Harold N. Fowler translated: 'For this

What follows, beginning in the sixth century BC, are philosophical attempts for a reliable theory of the world. Providing an overview of these attempts would carry us too far off, but I will return to the ideas of Anaximander of Miletus, probably a student of Thales, and of Parmenides of Elea, a student of Xenophanes, both living in that time period and laying the intellectual ground for the work of Socrates, Plato, and Aristotle.

Deduction

Looking for the origins for the concept of deduction, one ends up with Anaximander, who composed the philosophical poem 'On Nature' in the first half of the sixth century BC.[13] There, Anaximander (2007) argued that 'everything either is an origin or results from an origin' (p. 35, my translation).[14] Not much is recorded which would further qualify this thought. However, what can be said is that, with Anaximander, the idea was set loose that things do have a *reason*. Anaximander goes on philosophising about the final reason, which we will get back to. For now, it is important to say that Anaximander's worldview was the oldest surviving Greek view not to be built on divine entities. The reason for something to happen was not to be found in the realm of the gods but in nature.

Some scholars say that Anaximander founded physics as he was the first to propose a cosmology that worked without gods and asked for reasons. Aristotle's (trans. 1933) proposition 'that we must obtain

---

feeling of wonder shows that you are a philosopher, since wonder is the only beginning of philosophy' (Plato, trans. 1921b, 155d). Heinrich argues that the Greek word πάθος (*páthos*) does not merely mean 'feeling' but has the connotation of suffering, and that θαυμάζειν (*thaumázein*) is not merely 'wonder' but something negative. I tried to provide an English translation in accordance with Heinrich's interpretation.

13 This poem has only survived through the citations of fragments of it by others. Gemelli Marciano (2007) compiled all the fragments available and offers a good translation into German. For the lack of a compilation with a translation into English, I will refer to the German compilation and offer translations from it.

14 The ambiguous Greek original ἀρχή (*archē*) can be translated to 'beginning', 'origin', 'sovereignty', 'sovereign' or 'principle'. Gemelli Marciano (2007) translates it to '*Prinzip*', which would be 'principle' in English. Following Heinrich (1981), I chose a different translation to emphasise the close connection of Anaximander's thought to deductive thinking.

knowledge of the primary causes, because it is when we think that we understand its primary cause that we claim to know each particular thing' (983a) shows that this idea set a standard even some centuries later. Through the Attic philosophers, the idea of deductive reasoning was set as a standard of Western academia.

It is interesting to note the structural analogy between inheritance and reason: just as the gods passed their traits to their offspring, which then could not escape this fate, deduction presupposes that concepts necessarily hold all the properties of the concepts from which they derive. It is more than a bizarre side note that the move from myth to physics appears to be merely a replacement of gods with natural forces while keeping the overall architecture of argumentation untouched.

Identity

As I have argued earlier, understanding identity as the tautology that 'a is a' does not transport any meaning, needless to say. Instead, the principle of identity should be understood as the plea, the postulation, even the command that there *should* be things that stay the same. We can understand this postulation more socially following Aristotle who argued that people should make sure that they talk about the same things – from person to person and from instance to instance. We can also understand this postulation more religiously as the belief that our world is indeed based on things which do not change, and mathematics might be seen to belong to these things. Indeed, Socrates and Plato followed this essentialist belief, as did Anaximander and Parmenides.

Anaximander (2007) knew that in his cosmos of deductions, the deductive chain would need to start somewhere. He argued that 'there is no origin of the infinite, for otherwise it would be confined' (p. 35, my translation). He continued that this 'infinite' had 'not emerged', was 'imperishable', 'immortal', 'indestructible', 'eternal' and 'not aging', it 'seems to be the origin of all other things' (pp. 34–37, my translation). In the cycle of time, the world emerges from the infinite, only to perish to it again. The inevitability of Anaximander's infinite, on which everything depends, is more relentless than the mythical gods: at least, the latter had, through their humanesque character, a free will and could be fought. In Anaximander's cosmos, fate leaves no hope of being negotiable.

About half a century later, Parmenides (trans. 2009) composed his own poem 'On Nature', in which the author ascends to the gods to hear an epiphany from Dike, the goddess of justice, morals and fair judgement. Therein, Parmenides formulated a first version of the law of identity and introduced the concept of 'truth', mostly referred to as 'being', to the philosophical discussion (pp. 56–57). Parmenides explained that 'that Being is ingenerate and imperishable, entire, unique, unmoved and perfect' (p. 64), 'it never was nor will be, since it is now all together, one, indivisible' (p. 66) and it is 'bound fast by fate to be entire and changeless' (p. 76). The analogies to Anaximander's infinite are obvious. But while Anaximander's infinite is a necessary element of his explanation of the world, Parmenides' truth is an idea that belongs to a discussion of how to reason properly. Consequently, Parmenides has often been considered the founder of logic. It might be added as a side note that his poem also includes the oldest surviving example of a deduction.

What drove the intellectual development that resulted in the invention of truth? When we remember the state of confusion the Ancient Greek aristocracy suffered, the idea of truth offered too good a promise. Heinrich (1981) summarised this promise in the fictional wording: '"Fear not", for there is an existence which remains untouched by fate and death' (pp. 45–46, my translation). We find reassurance for such an interpretation in Parmenides' poem itself. Parmenides (trans. 2009) wrote that Dike did not allow truth 'either to come to be or to be perishing but holds it fast' (p. 68). Dike also asked Parmenides to stay away

> from that on which mortals with no understanding stray two-headed, for perplexity in their own breasts directs their mind astray and they are borne on deaf and blind alike in bewilderment, people without judgement, by whom this has been accepted as both being and not being, the same and not the same [...]. (p. 58)

While all beliefs include the danger of impermanence, truth would, by definition, never disappoint anyone. The price for that security is that truth is also completely independent from humans, that knowledge is dehumanised. In this vein, the vernacular expression of 'dead knowledge' for scientific truths resembles the ideas of Parmenides and his disciples rather well. Indeed, Plato (trans. 1921a) had the death-sentenced Socrates say 'that those who pursue philosophy aright study nothing but dying and being dead' and that 'it would be absurd to be eager for nothing but this all their lives, and then to be troubled when that came for which they had all along been eagerly practicing' (64a). There might be becoming and perishing in life, but nothing but eternal truth in death.

Dichotomies

The last quote from Parmenides (trans. 2009) also gives a hint that he already had an understanding of what earlier I presented as the laws of the excluded middle and the excluded contradiction. In fact, he is often attributed as the first philosopher to ever formulate these laws. In the quote above, the perplexed 'two-headed', who are incapable of judgement, unable to say what is, accept things 'as both being and not being', as both true and false, thus violating the law of the excluded contradiction (p. 58). Elsewhere, Parmenides added that 'mortals' suppose some things 'to be coming to be and perishing, to be and not to be, and to change their place' (p. 78). Here we have the connection in one line: allowing contradiction would invite the forces of becoming and perishing into philosophy, but these are deadly forces that change the face of the earth, that are unstable, and thus no foundation for any stable worldview.

Why start thinking *like this*?

We followed Vernant (1982) in maintaining that a driving force of philosophy in Ancient Greece was to build a more reliable fundament for understanding the world than the polytheistic myth had been able to. Apparently, the idea of truth promised the possibility of a secure understanding in its purest form. Nevertheless, it remains interesting to ask why scholars in Ancient Greece started to think like this, on the grounds of these fundamental assumptions of logic. We should hesitate to explain this development by assuming a logical order of the world or of the human mind, for that would mean that all scholars and cultures who did not follow the assumptions of logic discussed above have not reached the right access to our world or have not developed the right way of thinking. Such a position would make us a complicit of the superiority which logicians such as Parmenides assume for the kind of thinking they present, whereas in research, we should seek to obtain an unbiased distance to what we study. Therefore, it is necessary to ask why logic is organised in this peculiar way, if there would be no other possibilities to explain the world on the basis of a concept of truth.

Now, is there a socio-cultural explanation for why logic assumed the form it did? I know only one such explanation: namely that this form copies an order which was already being lived in the patriarchal society of Ancient Greece. It would have been difficult to come up with an order of thought out of the blue, but it should have been easier to come up with an order that is an abstraction of lived social organisation. The analogy between the fate of gods and the law of reason already gave us a glimpse of such a connection.

When I refer to patriarchy here, I refer to a very specific organisation of society which marks the beginning of Greek history. Pre-patriarchal societies know no fatherhood, no possession, no male superiority, and usually worshipped the holy mother who brought life into the world (Lerner, 1986). Patriarchal societies introduce the ideas of fatherhood, of male rule over women and offspring, of marriage, of property, and of inheriting.

Horkheimer and Adorno (2002) claim that 'the generality of the ideas developed by discursive logic, power [*die Herrschaft*] in the sphere of the concept, is built on the foundation of power in reality' (p. 10). What might they have meant? Throughout their treatise, Horkheimer and Adorno (2002) point to the changes that have come with the introduction of patriarchy but do not illuminate that connection further. Only later, Gerhard Schwarz (2007) demonstrated the analogy between patriarchal and military hierarchy, while Fischer (2001) identified a structural analogy between the patriarchal and the logical order. Imagine a typical visualisation of hierarchies, a root network starting in one point and branching out downwards. At the top, we see the patriarchal father, the military commander, or the most general concept respectively. Branching out, we see the sons of that father and again their sons and grandsons; we see the soldiers second highest in rank, followed by those third highest in rank; we see concepts which are gradually more specific in meaning, for example, the triangle and quadrilateral branch out from the broader category of polygon. The triangle, in turn, divides further into equilateral, isosceles, and scalene triangles. Now note that the logical assumptions described earlier are inscribed already in the historical configuration of the family and the military: the law of identity means that you stay who you are in that configuration, you cannot change your position, cannot become your father's father or your superior's superior.[15] The law of sufficient reason means that everybody has a father, everybody has a direct superior. Admittedly, that might not be true for the founder of a house or for the commander-in-chief, just as Anaximander admitted for his logic that at least one thing cannot have an origin. Finally, the either-or resulting from the laws of the excluded middle and the excluded contradiction means that in regard of any person in the respective orders, this person either is your father or your direct superior respectively, or he is not. It is not possible that somebody is neither your father nor not your father, nor is it possible that somebody is both your father and not your father.

I know of no arguments which would explain why these analogies between the patriarchal family, military organisation, and the assumptions of logic discussed above are necessary. Instead, these analogies are very peculiar. Note that pre-patriarchal societies had no concept of fatherhood at all, and some partisan military groups partly renounce formal ranks. Note also that this logic cannot work if mothers were meant to enter it in a position equal to fathers, or if the paternity of a child is in doubt (which therefore causes a major crisis in the patriarchal order). Ancient Greece had also seen different ontologies which assumed that nothing is fixed and everything is in flux, as expressed by Heraclitus of Ephesus (trans. 1979) who, in the sixth century BC, stated that 'one cannot step twice into the same river' (p. 53), a saying further escalated by Cratylus, whom Aristotle (trans. 1933) reported to have added 'that it cannot be done even once' (1010a). A contemporary example might be the struggles around the erosion of the either-or in the dichotomy of gender (see Chapter 19 in this volume).

The analogous form of these notably particular social systems demands an explanation. An explanation for this apparent coincidence

15 That should be clear for the family. For the military it should be noted that, in Ancient Greece, positions were assigned by birth right and perhaps by economic status, without there being any system of promotion into higher ranks.

would be that the patriarchal family served as a model for the military order and for the relationships between the gods (a system which we could have successfully included in the discussion of analogies above), and that the latter entered philosophical deliberations and was eventually secularised into the system of logic. Then, the assumptions of logic are not only not necessary, they also bear the imprint of a very specific form of social organisation, which legitimises disposing of somebody's life, regarding women and children as property, exchanging individuality against obedience and loyalty, and holding social positions fixed instead of allowing people to become what they want. Besides, that logic also provides the basis for introducing further hierarchies such as those of social classes and of ethnicities. All this proposes that logic developed out of social practices which are at the root of dehumanising people.

Problematising dehumanisation through logic

Despite the inability of logic to provide fallacy-free theories (see Chapter 4 in this volume), the logically organised discipline of mathematics has provided an astounding complexity of insights, which are used in countless applications in our world. From that point of view, the organisation of thought in analogy with the patriarchal social order can be called a success. Most of us would not want to live without the technological achievements of our time, which rely heavily on applications of mathematics and on logically order discourses. However, it has to be acknowledged that this success stands in a dialectical relationship with practices of dehumanisation.

If we look at logic as a social practice, we may ask: What does this practice entail? Following the above analysis, it entails thinking in permanent and universal concepts that are arranged in hierarchies and irreconcilable antagonisms, and assuming the properties of these concepts to necessarily follow from the properties of concepts that stand higher in the hierarchy. It should be acknowledged that this is a very particular form of organising thought and that not all discourses will follow this example. As one choice among many, logical thought will have a specific potential, a specific price to pray, and specific limitations, which altogether deserve critical attention.

Logic is apparently a tool for the dehumanisation of others. Aristotle (trans. 1989) explicitly stated that his discussion of logic followed the purpose of understanding

> what sorts of things one must look to when refuting or establishing, and how one must search for premises concerning whatever is proposed, in the case of any discipline whatever, and finally the route through which we may obtain the principles concerning each subject. (52b–53a)

Aristotle's philosophy of communication did not aim at mutual understanding and amicable compromise; it did not even foreground the discovery of truth. Instead, Aristotle presented logic as a tool of rhetoric dominance. Pointing to inconsistencies in the other's use of concepts, to violations of antagonistic concepts and to contradictions that the other's ideas might result in, are techniques to devalue somebody else's thoughts. Instead, deduction is the attempt to force the other to accept one's own argument. Wittgenstein (1978) reflected on the logical argument as a command directed at the other:

> In what sense is [the] logical argument a compulsion?—'After all you grant *this* and *this*; so you must also grant *this*!' That is the way of compelling someone. That is to say, one can in fact compel people to admit something in this way.—Just as one can e.g. compel someone to go over there by pointing over there with a bidding gesture of the hand. (p. 81)

But why would the other follow the command to organise the discourse logically?[16] Here, the Ancient Greek philosophers fail to provide good reasons and turn to defamation instead. While Aristotle (trans. 1933) merely stated that those questioning logic 'lack education' (1006a), Parmenides (trans. 2009) scolded to keep back from the way

> on which mortals with no understanding stray two-headed, for perplexity in their own breasts directs their mind astray and they are borne on deaf and blind alike in bewilderment, people without judgement, by whom this has been accepted as both being and not being the same and not the same, and for all of whom their journey turns backwards again. (p. 58)

We see that the birth of logic was accompanied by a clear dehumanisation of others who think differently. Whether they still 'lack education' or

16 This is a question that Wittgenstein (1978) was puzzled by.

whether they suffer from mental illness and sensory disabilities, they always lack something to their full humanness which the protagonists of logical thinking do not lack. This is the technique that labels non-logical thought worthless and its wielder voiceless.

Logic is not only dehumanising others by stigmatising non-logical thinking, it itself leaves no possibility for people to express their individuality. Logic is not looking for the always different in the individual, it is looking for that which cannot change. Aristotle (trans. 1934) concluded that scholars following logic

> conceive that a thing which we know scientifically cannot vary; when a thing that can vary is beyond the range of our observation, we do not know whether it exists or not. An object of Scientific Knowledge, therefore, exists of necessity. It is therefore eternal, for everything existing of absolute necessity is eternal; and what is eternal does not come into existence or perish. Again, it is held that all Scientific Knowledge can be communicated by teaching, and that what is scientifically known must be learnt. (1139b)

In his reading of Anaximander, Friedrich Nietzsche (1962) tried to understand the mental state of the scholar. Anaximander's contribution for the appreciation of truth over the dynamics of becoming and perishing can hardly be overestimated. In fact, Anaximander regarded the eternal as the only legitimate existence and, as Nietzsche (1962) formulated, 'all coming-to-be as though it were an illegitimate emancipation from eternal being, a wrong for which destruction is the only penance' (p. 46). The totalising worship of logic bears the danger of devaluating life itself.

Concerning the self, logical thinking, although it might come with the promise of aligning with the eternal (Heinrich, 1981), demands a strict conduct of thought. I know of no psychoanalysis of this conduct of the self,[17] but I find Elizabeth de Freitas' (2008) report of Agnes, a fictional learner indulging in mathematics, to be a good provocation for scholarship:

---

17 Note the following comment by Paul Ernest (2016): 'I do not ask the interesting psychological question as to why persons might feel uncomfortable with uncertainty and have or feel the need for certainty or indeed of the place of uncertainty in the human condition. This would take me in another direction, possibly needing psychoanalytic theory, beyond the scope of my present inquiry' (p. 380).

> The peacefulness of deduction, the lack of dissent or debate, allowed for austere moments of meditation. Agnes indulged in that quiet hard work. She developed a passionate attachment to the symbolic world of mathematics. She saw beauty in mathematics. But the beauty captured in a mathematical proof was a purist's beauty that despised the messiness of the world. Agnes embraced this purist beauty and this method so completely that it crippled her will. She became possessed by reason; her body, emotions, and actions inscripted by logic. What began as tolerance and respect for the truth, devolved into a defensive self-abnegating disposition, a retreat from risk and adventure. An erasure of voice. (pp. 284–285)

De Freitas might have thought about a point made by Horkheimer and Adorno (2002) of how logic related to determinism and thus to the negation of choice:

> The arid wisdom which acknowledges nothing new under the sun, because all the pieces in the meaningless game have been played out, all the great thoughts have been thought, all possible discoveries can be construed in advance, and human beings are defined by self-preservation through adaptation – this barren wisdom merely reproduces the fantastic doctrine it rejects: the sanction of fate which, through retribution, incessantly reinstates what always was. Whatever might be different is made the same. (p. 8)

Consequently, we might argue that logical thinkers deny their roles in changing the world, that they silence their voices, that they confine themselves to discover and proclaim the eternal truths based on logic. This is a way of denying one's own humanity and reducing one's own intellect to what, to an increasing extent, even computers can achieve.[18]

Rehumanising mathematics?

I tried to show that mathematics, through its practices of calculation and logic, aims at a dehumanisation of action and thought. Admittedly, we might ask if other scientific disciplines do not seek a dehumanisation of action and thought themselves, if dehumanisation is not intrinsic to the idea of science producing objective knowledge. If we followed Aristotle's

18 See Chapter 2 for a note on computers proving or refuting mathematical theorems on the basis of logical calculations.

(trans. 1934) idea of science cited above, that would be so. However, science is no thoroughly logical enterprise, as already the co-existence of mutually conflicting theories in physics proves. Disciplines other than mathematics work empirically and have to offer theories that somehow work in practice. In contrast, mathematics, especially but not exclusively in its formalist fashion, reserved the luxury of considering itself a merely intellectual discipline. This ideal explains why mathematics can reject any empiricism, handle its objects in its liking, and mould them in forms that implement the idea of dehumanised action and thought like no other discipline. It is no coincidence that Leibniz (1996), who dreamt of a 'universal characteristic' that could express every scientific question and solve it through computation, was a mathematician.

Post-structuralism has taught us that the meaning of concepts is never fixed but in a state of permanent renegotiation. I guess that this was what Ole Skovsmose (2011) had in mind when he presented 'mathematics education as being undetermined', 'without "essence"', able to 'be acted out in many different ways and come to serve a grand variety of social, political, and economic functions and interests' (p. 2). The same should hold true for mathematics. Mathematics is not imposed on us but what we make of it. Can we alter the *modus operandi* of mathematics so that dehumanisation leaves the equation?

Attempts to present mathematics as a social practice are important, but not sufficient, steps in this direction. Indeed, the mathematical philosophies behind the works of scholars such as George Pólya (1945) and Imre Lakatos (1976) as well as attempts to write a philosophy of mathematics as a social practice as proposed by Davis and Hersh (1980) can be understood as projects to show the human side of producing mathematics. Obviously, this action is not logical in nature, but full of individual ideas, emotions, and conflict. These attempts in mathematics related closely to programs in mathematics education, which lay emphasis on activities such as problem solving and modelling instead of presenting mathematics in its logical structure or as a toolbox of calculative techniques. Although these perspectives help us to understand the doing and learning of mathematics, they do not reject the idea that the final product of all this activity is a logically ordered discourse that provides techniques for calculation. None of these perspectives question, for example, the legitimacy of the ignorance of meaning inscribed in calculative practices or the epistemological consequences of a two-valued logic.

Attempts to alter the inner working of mathematics itself are rare. A few general ideas can be found in feminist perspectives on mathematics and its education (see Chapter 19 in this volume), and a new perspective has secured recent attention in the form of Rochelle Gutiérrez' projects of rehumanising mathematics and mathematx.[19] Gutiérrez (2012) referred to problematisations of academic mathematics as White middle-class masculine knowledge. She argued that for 'most women, the working class, and people of color, a focus on dominant mathematics means that engaging in school mathematics largely require becoming someone else' and demanded a different kind of engagement with mathematics in which 'their participation will somehow change the nature of mathematics as a discipline' (p. 30). In a later publication, Gutiérrez (2018) introduced the idea of 'rehumanising mathematics'[20] in the sense that 'a student should be able to feel whole as a person—to draw upon all of their cultural and linguistic resources—while participating in school mathematics' (p. 1). In a different publication, Gutiérrez (2017a) focused less on education and more on mathematics as a scientific discipline. There, she pleaded for mathematx as 'a radical reimagination of mathematics, a version that embraces the body, emotions, and harmony' (p. 15). Gutiérrez countered Western essentialism with Indigenous epistemologies as the new basis of a practice that is pleasing, aesthetic, action-based, embodied, and diverse. Although she provided some examples of what that might entail, she leaves open the question 'which new forms of mathematics might arise' (p. 20). Elsewhere, Gutiérrez (2017b) commented more

---

19 IPA: [ˌmæθməˈtɛʃ], or *mathe-ma-tesh.*

20 To avoid misunderstanding, it should be noted that what Gutiérrez (2018) meant with 'rehumanising' does not directly respond to what I called 'dehumanisation'. While I presented dehumanisation as a denial of one's full humanity and a prerequisite of mathematics, Gutiérrez assumes 'that people throughout the world already do mathematics in everyday ways that are humane' (p. 2), but that this doing is denied by the hegemonial practices in the mathematics education classroom. I refrain from sharing this position, for Gutiérrez' list of such ways of doing mathematics (p. 4) reveals that we face what Dowling (1998) called 'celebrating non-European cultural practices only by describing them in European mathematical terms' or the recognition of 'a practice as mathematical only by virtue of recognition principles which derive from their own enculturation into European mathematics' (p. 14). Regardless of this point of critique, I find value in Gutiérrez' overall ideas.

carefully that 'we do not have good models for what a feminist, pro-Black/Indigenous/Latinx, socialist mathematics education would look like, or if even such a thing could exist' (p. 12).

I meet Gutiérrez' last comment with a good portion of pessimism. As Fischer (2006) and Bettina Heintz (2000) pointed out, a paramount social function and driving force of mathematics is the production of consensual knowledge and practice. The analyses of calculation and logic presented above concluded that both aspects of mathematics can be understood under the term of reaching consensus. The very nature of these techniques is that the individual is disregarded. Eventually, I would negate my earlier question if we can alter the *modus operandi* of mathematics so that dehumanisation leaves the equation.

In contrast to that, Gutiérrez (2017a) suggested that 'mathematx acknowledges that all persons will seek, acknowledge, and create patterns differently in order to solve problems and experience joy' and that 'multiple knowledges are valued and sought' (pp. 19–20). Apparently, mathematx would not be able to replace mathematics in its function of reaching consensus. A shift from mathematics to mathematx would mean that this practice loses its paramount, maybe even its entire function for society, thus making itself expendable. Could it be that mathematx would turn out as something completely different than mathematics? And if so, why then talk about *mathe*matx and not simply about an alternative epistemology? Or, asked differently, what of mathematics would be conserved in mathematx?

I propose that alternative epistemologies remain important for the study of dehumanisation through mathematics, because they help us to understand that our world can be understood differently. Such insights might not result in new epistemic forms of mathematics, but they might allow us to better capture the epistemological potential, limits, and dangers of mathematics. Admittedly, this perspective does not help us to counter epistemological discrimination in the mathematics classroom as was the initial attempt of Gutiérrez. We might come to find that we cannot wrench mathematics from the quills of White middle-class men that roam the history of the discipline. However, awareness of the particularities and political nature of the epistemology of mathematics, gladly aided by alternative visions of how to approach the world we live it, can help us to understand, support, or confront the ways in which mathematics contributes to dehumanisation in our societies.

References

Anaximander. (2007). Fragmente. In M. L. Gemelli Marciano (Ed.), *Die Vorsokratiker. Band 1* (pp. 32–51). Artemis & Winkler.

Aristotle. (trans. 1933). *Metaphysics*. Harvard University Press.

Aristotle. (trans. 1934). *Nicomachean ethics*. Harvard University Press.

Aristotle. (trans. 1989). *Prior analytics*. Hackett.

Bauman, Z. (1989). *Modernity and the Holocaust*. Polity Press.

Bell, A. (2016). Wem schenken wir das Herz? Gerechtigkeit in der Spenderorganallokation [Whom do we give the heart to? Justice in donor organ allocation]. In P.-C. Chittilappilly (Ed.), *Horizonte gegenwärtiger Ethik* (pp. 553–567). Herder.

Bishop, A. J. (1988). *Mathematical enculturation: A cultural perspective on mathematics education*. Kluwer.

Cervantes, M. de. (1605). *El ingenioso hidalgo don Quijote de la Mancha* [The ingenious gentleman Don Quixote of la Mancha]. Francisco de Robles.

Chace, A. B., Bull, L., & Manning, H. P. (1929). *The Rhind mathematical papyrus*. Oberlin.

Chemla, K. (2012). Historiography and history of mathematical proof: A research programme. In K. Chemla (Ed.), *The history of mathematical proof in ancient traditions* (pp. 1–68). Cambridge University Press.

Davis, P. J. (2012). Applied mathematics old and new. In P. J. Davis & R. Hersh (Eds.), *The mathematical experience* (pp. 490–492). Birkhäuser.

Davis, P. J., & Hersh, R. (1980). *The mathematical experience*. Birkhäuser.

Davis, P. J., & Hersh, R. (1986). *Descartes' dream: The world according to mathematics*. Harvester.

Desrosières, A. (1993). *La politique des grands nombres: Histoire de la raison statistique* [The politics of large numbers: A history of statistical reasoning]. *Anthropologie des sciences et des techniques*. La Découverte.

Dowling, P. (1998). *The sociology of mathematics education: Mathematical myths / pedagogic texts*. Falmer.

Ernest, P. (2016). The problem of certainty in mathematics. *Educational Studies in Mathematics*, *92*(3), 379–393. https://doi.org/10.1007/s10649-015-9651-x

Fischer, R. (2001). Mathematik und Bürokratie [Mathematics and bureaucracy]. In K. Lengnink, S. Prediger, & F. Siebel (Eds.), *Mathematik und Mensch: Sichtweisen der allgemeinen Mathematik* (pp. 53–64). Allgemeine Wissenschaft.

Fischer, R. (2006). Materialization and organization: Towards a cultural anthropology of mathematics. *ZDM Mathematics Education*, *38*(4), 316–322. https://doi.org/10.1007/BF02652791

Foucault, M. (1954). *Maladie mentale et psychologie* [Mental illness and psychology]. PUF.

Foucault, M. (1966). *Les mots et les choses: Une archéologie des sciences humaines* [The order of things: An archaeology of the human sciences]. Gallimard.

Foucault, M. (1984). Nietzsche, genealogy, history. In P. Rabinow (Ed.), *The Foucault reader* (pp. 76–100). Pantheon. (Original work published 1971)

Foucault, M. (2007). *The order of things: An archaeology of the human sciences.* Routledge. (Original work published 1966)

Frank, H. (1939). *Mathematik* für *höhere Schulen* [Mathematics for higher secondary schools]. Coppenrath.

Frege, G. (1960). Frege against the formalists. In P. Geach & M. Black (Eds.), *Translations from the philosophical writings of Gottlob Frege* (pp. 182–233). Blackwell. (Original work published 1903)

Freitas, E. de (2008). Mathematics and its other: (Dis)locating the feminine. *Gender and Education*, *20*(3), 281–290. https://doi.org/10.1080/09540250801964189

Gemelli Marciano, M. L. (Ed.). (2007). *Die Vorsokratiker* [The Presocratics]. Artemis & Winkler.

George, A., & Velleman, D. J. (2002). *Philosophies of mathematics.* Blackwell.

Gutiérrez, R. (2012). Embracing Nepantla: Rethinking 'knowledge' and its use in mathematics teaching. *Journal for Research in Mathematics Education*, *1*(1), 29–56. https://doi.org/10.4471/redimat.2012.02

Gutiérrez, R. (2017a). Living mathematx: Towards a vision for the future. *Philosophy of Mathematics Education*, *32*. http://socialsciences.exeter.ac.uk/education/research/centres/stem/publications/pmej

Gutiérrez, R. (2017b). Why mathematics (education) was late to the backlash party: The need for a revolution. *Journal of Urban Mathematics Education*, *10*(2), 8–24. https://doi.org/10.21423/jume-v10i2a347

Gutiérrez, R. (2018). The need to rehumanize mathematics. In I. Goffney & R. Gutiérrez (Eds.), *Rehumanizing mathematics for Black, Indigenous, and Latinx students* (pp. 1–10). National Council of Teachers of Mathematics.

Haslam, N. (2015). Dehumanization and intergroup relations. In M. Mikulincer, P. R. Shaver, J. F. Dovidio, & J. A. Simpson (Eds.), *APA handbook of personality and social psychology* (Vol. 2, pp. 295–314). American Psychological Association. https://doi.org/10.1037/14342-011

Heinrich, K. (1981). *Tertium datur: Eine religionsphilosophische Einführung in die Logik* [Tertium datur: An introduction to logic from the philosophy of religion]. Stroemfeld.

Heintz, B. (2000). *Die Innenwelt der Mathematik: Zur Kultur und Praxis einer beweisenden Disziplin* [The inner world of mathematics: On the culture and praxis of a proving discipline]. Springer.

Heraclitus. (1979). Fragments. In *The art and thought of Heraclitus*. Cambridge University Press.

Hersh, R. (1997). *What is mathematics, really?* Oxford University Press.

Horkheimer, M. (2004). *Eclipse of reason*. Continuum. (Original work published 1947)

Horkheimer, M., & Adorno, T. W. (2002). *Dialectic of enlightenment: Philosophical fragments*. Stanford University. (Original work published 1944)

Kelman, H. C. (1973). Violence without moral restraint: Reflections on the dehumanization of victims and victimizers. *Journal of Social Issues, 29*(4), 25–61. https://doi.org/10.1111/j.1540-4560.1973.tb00102.x

Kleiner, I. (2001). History of the infinitely small and infinitely large in calculus. *Educational Studies in Mathematics, 48*(2), 137–174. https://doi.org/10.1023/A:1016090528065

Kollosche, D. (2014). Mathematics and power: An alliance in the foundations of mathematics and its teaching. *ZDM Mathematics Education, 46*(7), 1061–1072. https://doi.org/10.1007/s11858-014-0584-0

Krämer, S. (1988a). *Symbolische Maschinen* [Symbolic machines]. Wissenschaftliche Buchgesellschaft.

Krämer, S. (1988b). The 'Universal Thinking Machine' or on the genesis of schematized reasoning in the 17th century. In I. Hronszky, M. Fehér, & B. Dajka (Eds.), *Scientific knowledge socialized* (pp. 179–191). Kluwer.

Krämer, S. (1991). *Berechenbare Vernunft: Kalkül und Rationalismus im 17. Jahrhundert* [Calculable reason: Calculus and rationality in the 17th century]. de Gruyter.

Krämer, S. (1998). Zentralperspektive, Kalkül, Virtuelle Realität: Sieben Thesen über die Weltbildimplikationen symbolischer Formen [Central perspective, calculus, virtual reality: Seven propositions on the implications of symbolic forms for conceptions of the world]. In G. Vattimo & W. Welsch (Eds.), *Medien-Welten Wirklichkeiten* (pp. 27–37). Fink.

Kronfeldner, M. (Ed.). (2021). *The Routledge handbook of dehumanization*. Routledge.

Kütting, H. (2012). Ideologie des Nationalsozialismus im Bildungssystem am Beispiel der Mathematik [Ideology of National Socialism in the educational system exemplified by mathematics]. *Mitteilungen der Gesellschaft für Didaktik der Mathematik,* (93), 6–22.

Lakatos, I. (1976). *Proofs and refutations: The logic of mathematical discovery.* Cambridge University Press. (Original work published 1964)

Leibniz, G. W. (1896). *New essays concerning human understanding* (A. G. Langley, Ed.). Macmillan. (Original work published 1765)

Leibniz, G. W. (1996). Zur allgemeinen Charakteristik [On the universal characteristic]. In E. Cassirer & A. Buchenau (Eds.), *Philosophische Bibliothek: Vol. 496. Philosophische Werke in vier Bänden* (Vol. 1, pp. 16–23). Meiner. (Original work published 1686)

Lerner, G. (1986). *The creation of patriarchy.* Oxford University Press.

Mikkola, M. (2021). Why dehumanization is distinct from objectification. In M. Kronfeldner (Ed.), *The Routledge handbook of dehumanization* (pp. 326–340). Routledge.

Nietzsche, F. (1962). *Philosophy in the tragic age of the Greeks.* Regnery. (Original work published 1923)

Parmenides. (trans. 2009). *Fragments of Parmenides: A critical text with introduction and translation, the Ancient testimonia and a commentary.* Parmenides Publishing.

Plato. (trans. 1921a). Phaedo. In H. N. Fowler (Ed.), *Plato in Twelve Volumes* (Vol. 1). Heinemann.

Plato. (trans. 1921b). Theaetetus. In H. N. Fowler (Ed.), *Plato in Twelve Volumes* (Vol. 12). Heinemann.

Pólya, G. (1945). *How to solve it: A new aspect of mathematical method.* Princeton University Press.

Porter, T. M. (1996). *Trust in numbers: The pursuit of objectivity in science and public life.* Princeton University Press.

Schopenhauer, A. (1903). *On the fourfold root of the principle of sufficient reason.* George Bell & Sons. (Original work published 1813)

Schwarz, G. (2007). *Die „Heilige Ordnung" der Männer: Hierarchie, Gruppendynamik und die neue Rolle der Frauen* [The 'holy order' of men: Hierarchy, group dynamics and the new role of women]. Verlag für Sozialwissenschaften. https://doi.org/10.1007/978-3-531-90824-3 (Original work published 1985)

Skovsmose, O. (1994). *Towards a philosophy of critical mathematics education.* Kluwer. https://doi.org/10.1007/978-94-017-3556-8

Skovsmose, O. (2011). *An invitation to critical mathematics education.* Sense. https://doi.org/10.1007/978-94-6091-442-3

Ullmann, P. (2008). *Mathematik, Moderne, Ideologie: Eine kritische Studie zur Legitimität und Praxis der modernen Mathematik* [Mathematics, modernity, ideology: A critical study on the legitimacy and praxis of modern mathematics]. UVK.

Vernant, J.-P. (1982). *The origins of Greek thought*. Cornell. (Original work published 1962)

Voswinkel, R. (1998). Erzogen und entfremdet: Meine Erfahrungen mit der Mathematik [Educated and alienated: My experiences with mathematics]. *Mathematik Lehren*, 86, 18–19.

Walkerdine, V. (1988). *The mastery of reason: Cognitive development and the production of rationality*. Routledge.

Weber, M. (1978). *Economy and society: An outline of interpretive sociology*. University of California Press. (Original work published 1922)

Wigner, E. (1960). The unreasonable effectiveness of mathematics in the natural sciences. *Communications in Pure and Applied Mathematics*, *13*(1), 1–14. https://doi.org/10.1002/cpa.3160130102

Wittgenstein, L. (1978). *Remarks on the foundations of mathematics*. Blackwell. (Original work published 1956)

6. A short commentary on Kollosche's 'Dehumanisation through mathematics'

Roy Wagner

In this short response to David Kollosche, I briefly point out some complementary historical narratives of mathematics to suggest how mathematics may not only be complemented by more humanized forms of knowledge, but may also be inherently more humanized in itself.

In Chapter 5 of this volume, entitled 'Dehumanisation through mathematics', David Kollosche follows up on a well-known characterisation of mathematics: it is a rule-based, highly technocratic family of practices, which imposes uniform templates on disparate situations, setting aside not only the specific objects that we mathematise, but also the human – as practitioner, object, and addressee.

Kollosche argues coherently and convincingly – perhaps so convincingly, that one might be tempted to consider him as a math-basher. The critical tradition that he builds on served, at the time of Max Horkheimer and Theodor Adorno, to reign in a rampant scientism threatening to subjugate humanity. Today, it may end up playing into the hands of those who disparage science so as to dissuade us from acting on its warnings about the fate of our planet. Indeed, if mathematics is a dehumanising discipline, how can its own models, applied by climate scientists, serve to save humanity from demise? If the diagnostic tool is poisoned, how can we trust the cure that it suggests?

Kollosche's solution, spelled out in his final paragraphs of the paper, is not to reform mathematics – he is explicitly pessimistic about such

©2024 Roy Wagner, CC BY-NC 4.0 https://doi.org/10.11647/OBP.0407.06

a project. Since he acknowledges that we can hardly do without it, he proposes, instead, that we should complement mathematical analyses by less dehumanising forms of knowledge, balancing dehumanising mathematics with other methodologies. This is a fair and realistic prospect. To change mathematics is to move a mountain, and it is not clear if this mountain would serve us better once displaced. Anyway, it will take several generations to move it about.

I, however, am still tempted, if not to move the mountain, at least to try to chip at it, and rearrange some bits and pieces. For that, however, the history of mathematics and logic needs to be retold along lines different from those highlighted by Kollosche. Indeed, not all mathematics is as dehumanising. The mystical-cosmological mathematical speculations that for most of the last three millennia were a hugely popular form of mathematics in Europe, North Africa, and Asia (mathematical astronomical/astrology, number-theoretic numerology – practiced by some of the most celebrated mathematicians, like Johannes Kepler and Isaac Newton), were deeply anchored in how humans experienced numbers and geometric patterns. The same goes for the mathematics of artisans and artists, who were after beautiful patterns, virtuosity, and elegance, and for pre-school children, for whom numbers at least begin (although for an all too short a while) with songs and games. And even contemporary research mathematicians evaluate mathematics in ways that cannot be reduced to dehumanised formal rigour ('there is no permanent place in the world for ugly mathematics', wrote G. H. Hardy).

Deduction, as Kollosche notes, can be seen as analogous to patriarchal structures, and the implied genetic relation definitely deserves attention. But other forms of mathematical justification abound historically.[1] Even today, the styles of mathematical reasoning pursued in various contexts, from engineering to elementary school teaching, are often far removed from strictly deductive ideals.

More specifically, a long tradition in the historiography of mathematics has been taking great pains to show the complexities of the manifold relations between mathematical signs, practices, and forms of knowledge. This tradition problematises the dominant narrative of

1 Just as a tip of the iceberg: Eulerian reasoning as analysed by Ferraro (2004, 2012), Chinese mathematics as analysed by Chemla (2020), and Indian mathematics as analysed by Srinivas (2005, 2015).

'loss of meaning' that appears to connect Greek Antiquity to modern formalism – a narrative that would be difficult to attribute even to David Hilbert himself, who considers only one layer of mathematical signs to be purely syntactic, and assigns the epistemological authority of even that layer to its connection with meaningful signs (Hilbert, 1983).

The above revisionist narratives of mathematics are extremely important to a historian, who, like me, has devoted so much work to them. But I can understand how a grand narrative like 'the loss of meaning' is sufficient for others, as it highlights some of the most salient features of mainstream mathematics today. That these are, crucially, some of mathematics' most objectionable, or at least controversial, features, however, is precisely what would lead the followers of this narrative to give up on the hope of humanising mathematics, which I would like to keep alive.

Toward the very end of the chapter, Kollosche diagnoses the one aspect of mathematics that would block any attempt to humanise it: its rigid and impressive consensus. To maintain it, mathematics has to remain dehumanised, at least in some important senses. But even that is not a universal or necessary feature of mathematics. In fact, I recently argued that as a historical phenomenon, it is quite new and exceptional (Wagner, 2022). The most consensual aspect of mathematics today, namely the agreement on whether a given argument does or does not prove a given theorem in a given mathematical system, was much more open for debate in the past. Since many sciences are highly successful despite (or even because of) their longstanding controversies, a non-consensual or less-consensual mathematics need not be thought of as a dead end.

The mathematics we celebrate today is highly valuable and at the same time often dehumanising. But these are not universal characteristics of all past and present ways of doing mathematics. And while Kollosche is right that the most immediate way to handle dehumanising mathematics is to complement it by other styles of reasoning, I would like to hold on to the possibility of building on past and present practices of mathematics that do not set humans and non-humans aside.

In fact, recent developments in Artificial Intelligence (AI) suggest an interesting possible humanistic future for mathematicians – albeit currently restricted to the realm of science-fiction. If AI could, as is projected by some, learn to write valid mathematical proofs of open problems, and if it would surpass humans in that capacity, then the role of the mathematician may change dramatically from a producer of proofs to that of a commentator. In other words, the mathematician would be charged with the task of making sense of the most important AI-generated mathematical proofs. In the context of this task, intuitive and accessible narration of proofs for the purpose of large- and medium-scale understanding would become more important than fine-grained rigour, since the rigour of proofs would already be guaranteed by the AI that generated them. The mathematician would then become an interpreter-critic and communicator of mathematical ideas, not unlike a literature professor. While some would claim that this is already part of what the best mathematicians implicitly do, in our little science-fiction projection, interpretation, communication and critique would become the very definition of what a mathematician does. Would this open up the way to a (re-)humanized mathematics?

References

Chemla, K. (2020). Different clusters of texts from Ancient China, different mathematical ontologies. In G. E. R. Lloyd & A. Vilaça (Eds.), *Science in the forest, Science in the past* (pp. 121–146). HAU. https://library.oapen.org/handle/20.500.12657/47354

Ferraro, G. (2004). Differentials and differential coefficients in the Eulerian foundations of the calculus. *Historia Mathematica, 31*(1), 34–61. https://doi.org/10.1016/S0315-0860(03)00030-2

Ferraro, G. (2012). Euler, infinitesimals and limits. https://shs.hal.science/halshs-00657694v2

Hilbert, D. (1983). On the infinite. In P. Benacerraf and H. Putnam (Eds.), *Philosophy of mathematics: Selected readings* (2nd edition, pp. 66–76). Cambridge University Press.

Srinivas, M. D. (2005). Proofs in Indian mathematics. In G. G. Emch, R. Sridharan, & M. D. Srinivas (Eds.), *Contributions to the history of Indian mathematics* (pp. 209–248). Hindustan Book Agency.

Srinivas, M. D. (2015). On the nature of mathematics and scientific knowledge in Indian tradition. In J. M. Kanjirakkat, G. McOuat, & S. Sarukkai (Eds.), *Science and narratives of nature: East and West* (pp. 220–238). Routledge. https://doi.org/10.4324/9781315088358-11

Wagner, R. (2022). Mathematical consensus: A research program. *Axiomathes, 32*, 1185–1204. https://doi.org/10.1007/s10516-022-09634-2

7. Intuition revived

Ole Skovsmose

In the preface to Mathematics as an Educational Task, *Hans Freudenthal states that his educational interpretation of mathematics betrays the influence of L. E. J. Brouwer's view on mathematics. In this chapter we explore the nature of this possible influence. According to Brouwer, intuition plays a crucial role in any form of mathematical construction, which he specifies in terms of mental acts. He finds that mathematics does not have any adequate articulation in language, and that mathematical formalisms are nothing but imprecise and mischievous depictions of genuine mathematical processes. Freudenthal characterises mathematics as a* human activity, *thereby subsuming the overall intuitionist outlook that Brouwer had condensed into the notion of* mental activity. *While Brouwer installed intuition in a central position in mathematics, Freudenthal created a vast space for intuition in all kinds of activities in mathematics education. In his writings, Freudenthal does not demonstrate any interest in socio-political issues related to mathematics. Structuralism and the Modern Mathematics Movement are manifestations of the dogma of neutrality, and so is Freudenthal's formulation of mathematics as a human activity. However, although he does not repudiate a dogma of neutrality, he simultaneously provides ideas that help in formulating a critical mathematics education.*

Intuitionism is one prime example of how a conception of mathematics may influence the teaching and learning of mathematics. L. E. J. Brouwer and Hans Freudenthal are two protagonists in this development. Brouwer was a mathematician contributing to a broad range of topics, later focused on formulating an intuitionistic mathematics. For an extended period Freudenthal worked as Brouwer's assistant as a dedicated mathematics researcher, while in the later part of his career he concentrated on mathematics education.

©2024 Ole Skovsmose, CC BY-NC 4.0 https://doi.org/10.11647/OBP.0407.07

In the preface to *Mathematics as an Educational Task*, Freudenthal (1973) makes the following comment: 'My educational interpretation of mathematics betrays the influence of L. E. J. Brouwer's view on mathematics (though not on education)' (p. ix).

Let us first look at the side-remark in the parenthesis. How was Brouwer as a teacher? Bartel van der Waerden, who studied mathematics in Amsterdam, makes the following comment about Brouwer:

> I once interrupted him during a lecture to ask a question. Before the next week's lesson, his assistant came to me to say that Brouwer did not want questions put to him in class. He just did not want them, he was always looking at the blackboard, never towards the students. (O'Connor & Robertson, 2003)

Freudenthal's side-remark might not be at all surprising to those who knew Brouwer as a teacher, and therefore let it remain in the parenthesis. What more does Freudenthal tell us in *Mathematics as an Educational Task* about Brouwer's influence? Surprisingly, nothing.[1] In Freudenthal's other books on mathematics education – *Weeding and Sowing* (1978), *Didactical Phenomenology of Mathematical Structures* (1983), and *Revisiting Mathematics Education* (1991) – one finds almost no mention of Brouwer, except for a couple of references. Thus, in Freudenthal's own texts, one does not find a clarification of the nature of Brouwer's influence.[2] Nevertheless, this influence is the focus of this chapter.

Luitzen Egbertus Jan Brouwer (1881–1966) worked in several mathematical areas, including topology, set theory, and measure theory. Brouwer's (1911) contribution to topology includes a theorem that is referred to as Brouwer's fixed-point theorem. It states that for any continuous function f mapping a compact convex set onto itself there exists a point x_0 such that $f(x_0) = x_0$. The theorem is fascinating. When one stirs a cup of coffee – and we assume that the coffee represents a compact convex set, and that the stirring operates like a continuous function – then at least one of the coffee atoms will end up in the same position as it had before the stirring.

In 1912, Brouwer secured a permanent position at the University of Amsterdam, and in his inaugural lecture 'Intuitionism and Formalism'

1 He refers only once more to Brouwer (p. 40), with respect to a different issue.
2 In la Bastide-van Gemert (2015), I did not find any clarification either.

he started articulating more carefully his conception of intuitionism. Brouwer confronted formalism, in the first instance as represented by David Hilbert. This was a confrontation with many ramifications, for instance with respect to the editorial policy of *Mathematische Annalen*, which was the most important international mathematical research journal of the time. From 1902 to 1939, Hilbert was editor, while Brouwer was a member of the editorial board from 1914 to 1928. Due to Hilbert's initiative, Brouwer was removed from the board; other members of the board protested, among them Albert Einstein. The confrontation between intuitionism and formalism was a clash between research paradigms as well as between personalities.

Hans Freudenthal (1905–1990) was born in Germany. In 1923, he started studying mathematics in Berlin, where, in 1927, he met Brouwer, who was giving a lecture. In 1930, Freudenthal completed his doctoral thesis on topology,[3] and soon after he was invited by Brouwer to come to Amsterdam, where during the 1930s he worked as an assistant for Brouwer. After the German invasion in 1940, Freudenthal was suspended from his position due to his Jewish origins. In 1943, he was sent to a concentration camp, but in 1944 through the support of his Dutch wife he managed to escape, and he went into hiding in Amsterdam until the end of the war. After the war, Brouwer was not interested in offering Freudenthal a position again, and in 1946, he took up a position at the University of Utrecht, where he remained for the rest of his career.

Freudenthal was a dedicated mathematics researcher with a specific focus on algebraic topology.[4] However, he did not show any particular dedication to the detailed mathematical elaborations of intuitionistic mathematics. From the late 1960s, Freudenthal started engaging in mathematics education. In 1968, he founded the journal *Educational Studies in Mathematics*, and in 1971 he became nominated as director of the new research institute IOWO, the Dutch abbreviation for *Institut voor de Ontwikkeling van het Wiskunde Onderwijs* (Institute for the Development of Mathematics Education) in Utrecht. By that time, Freudenthal had published widely in mathematics education, and

3 For an important result of this work, see Freudenthal (1931).

4 He proved what are referred to at Freudenthal's spectral theorem and Freudenthal's suspension theorem. Other mathematical conceptions also carry his name.

many of these publications were brought together and reworked into his monumental work *Mathematics as an Educational Task* that appeared in 1973.[5]

In the following, we explore how Brouwer saw *mathematics as a mental activity*. We move on to explore Freudenthal's conception of *mathematics as a human activity* as it came to be expressed in *Mathematics as an Educational Task*. As an indication of what this conception could mean for mathematics education, we look at the example *Ship Ahoy*. As a conclusion we raise the question: What about socio-political issues?

Mathematics as a mental activity

As a way out of the foundational crises in mathematics, Brouwer launched an approach different from those suggested by logicism and formalism.[6] According to him, both logicism and formalism were wrong in their approaches in trying to eliminate intuition from mathematics. The way out of the crisis had to be found in the opposite direction: intuition had to be installed in its proper position as the core of mathematical thinking.

Brouwer found that the emergence of the paradoxes that brought about the foundational crises indicated that something had gone wrong within mathematics itself, and that this problem was manifest in logicism and formalism. What was needed was a much more radical approach. According to Brouwer, the emergence of paradoxes indicates that mathematics has applied forms of reasoning and proof strategies that are not valid in mathematics. Over time mathematics has incorporated a range of theorems, which should not count as such. It is not surprising, then, that paradoxes do appear. The whole body of mathematics had to be re-examined, and for doing so a revitalisation of intuition was needed. This is what Brouwer suggested by formulating an intuitionist conception of mathematics.

---

5 In 1991, one year after the death of Freudenthal, IOWO was renamed as the Freudenthal Institute. In 2006, due to the integration of more areas, the institute turned into the Freudenthal Institute for Science and Mathematics Education (see van Heuvel-Panhuizen, 2015).

6 For the following presentation of Brouwer's intuitionism, I draw on Ravn and Skovsmose (2019). For a discussion of the foundational crises in mathematics, see Chapter 4 in this volume.

Brouwer saw formalisations as being inaccurate, if not simply misleading. According to him, one can never identify mathematics with any formalism. That would be the same mistake as assuming that a plaster cast of a human being is the actual human being. Mathematics is alive, formalisms are not. Formalisms are only external and imprecise representations of intuitive mental acts, which constitute genuine mathematics.

In 1905, Brouwer (1996) published a short text *Life, Art and Mysticism*, in which he states: 'Always and everywhere truth is in the air, and whenever it breaks through, truth is always the same to those who understand' (p. 404). Brouwer sees truth in absolute terms, and this idea he maintains in his formulation of intuitionistic mathematics. One could think of intuition as being imprecise and open-ended, making space for a variety of interpretations compromising the possible connections between mathematics and certainty. However, Brouwer does not operate with any common-sense interpretation of intuition. He does not relate intuition to uncertainty and ambiguity, but to particular mental acts that bring about mathematical truths with certainty. To him, truth becomes the same to 'those who understand'.

In 1913, Brouwer published his inaugural lecture 'Intuitionism and Formalism'. Here he relates his ideas to those of Immanuel Kant (1973), who in *Critique of Pure Reason*, first published in German in 1781, provided a radical new departure for interpreting mathematics. Kant finds that our experiences become organised according to pre-given categories of understanding, and that mathematics provides the basic structures of the conceptual twins: space and time. That mathematics applies to our experiences of nature is not due to the fact that nature as such operates according to mathematical patterns, but to the fact that mathematics organises our experiences of nature. Brouwer (1913) sees Kant as articulating an intuitionism, but he also highlights that in Kant 'we find an old form of intuitionism, now almost completely abandoned, in which time and space are taken to be forms of conception inherent in human reason' (p. 83).

To Kant, Euclidean geometry reveals details of our category of space. Many interpreted the emergence of non-Euclidean geometries as devastating for Kant's conception of mathematics. Brouwer, however, is not troubled by this critique. He highlights that the position of intuitionism has 'recovered by abandoning Kant's apriority of space but adhering the more resolutely to the apriority of time' (p. 85). For identifying the origin of mathematical intuition, Brouwer put aside any intuition of space, and concentrated on the intuition of time.

From where does an intuition of time emerge? One could think of it in psychological terms. In a *System of Logic*, first published in 1843, John Stuart Mill (1970) argues that all human knowledge, including mathematics, is based on empirical evidence. However, Brouwer does not assume any such psychologism. Like Kant, he sees time as a category for understanding, and not as a psychological notion referring to some particular experiences.

A critical notion to Brouwer is two-oneness. This notion represents the time-specific origin of mathematics. Let us start looking at Brouwer's (1913) own presentation of the notion:

> Neo-intuitionism considers the falling apart of moments of life into qualitatively different parts, to be reunited only while remaining separated by time as the fundamental phenomenon of the human intellect, passing by abstracting from its emotional content into the fundamental phenomenon of mathematical thinking, the intuition of the bare two-oneness. (p. 85)

While Brouwer thinks of Kant's position as an old form of intuitionism, he refers to his own formulation as a neo-intuitionism.[7] He highlights that time is the fundamental phenomenon of the human intellect. Through this formulation he somehow makes space for a psychologism, but immediately distances himself from this position by highlighting that we need to abstract away the emotional content associated with time in order to reach the fundamental phenomenon of mathematical thinking, thus sweeping aside psychological content in order to reach time as a pure category. In this way he gets to the fundamental mathematical phenomenon of mathematical thinking: the bare two-oneness.

7 Brouwer acknowledges that there are several sources of inspiration for this new form of intuitionism, and, with reference to controversies with respect to the interpretation of mathematical laws, he refers to '*intuitionism* (largely French) and *formalism* (largely German)' (Brouwer, 1913, p. 82). Brouwer also makes references to Henri Poincaré and Émile Borel, who together with Henri Lebesgue and several others have been referred to as semi-intuitionists (see Troelstra, 2011).

In 'Intuitionism and Formalism' Brouwer does not give any further explanation of why he uses the expression 'two-oneness', and not, say, 'one-twoness'. It would seem that the latter expression would indicate more directly the start of the counting process. However, there might be linguistic reasons for Brouwer's choice of terminology. He might be alluding to the notion of 'trinity'. In Dutch the word for trinity is *drie-eenheid*, which literally means 'three-oneness'. Later, as for instance in the *Cambridge Lectures*, Brouwer talks about a 'twoity', where the allusion to trinity is even more explicit.

The intuition of movement of time in terms of two-oneness is the basic departure for mathematical thinking. Any mathematical concept becomes created by this intuition:

> This intuition of two-oneness, the basal intuition of mathematics, creates not only the numbers one and two, but also all finite ordinal numbers, inasmuch as one of the elements of the two-oneness may be thought of as a new two-oneness, which process may be repeated indefinitely; this gives rise still further to the smallest infinite ordinal number ω. (pp. 85–86)

Brouwer uses the formulation 'the basic intuition of mathematics, creates…'. The notion 'creates' is crucial, it is a mental process that constructs mathematical entities staring out from the intuition of two-oneness. The two-oneness is not an intuition through which one discovers mathematical truths. It is an intuition through which one constructs mathematical entities and mathematical truths. One can think of the two-oneness as referring to the first step in a process of counting: one, two. This process can be repeated, and one counts: one, two, three. It can be repeated again and again: one, two, three, etc. Brouwer does not accept the concept of actual infinity, but assumes the idea of potential infinity. The sequence of natural numbers can be indefinitely extended. It is in this sense we need to read Brouwer's claim that the counting process gives rise to 'the smallest infinite ordinal number ω'.

What about geometry? Brouwer has put aside intuition of space as being an irrelevant category, as he finds that also geometric notions are also developed from the intuition of time:

> The apriority of time does not only qualify the properties of arithmetic as synthetic a priori judgments, but it does the same for those of geometry,

> and not only for elementary two- and three-dimensional geometry, but for non-Euclidean and n-dimensional geometries as well. (p. 86)

Kant's principal point is that mathematical statements are synthetic *a priori* judgements. Brouwer shares this idea with respect to time. That mathematical statements have a content and simultaneously are independent of empirical observations, is due to the fact that the intuition of time ensures an *a priori* structuring and simultaneously provides mathematical statements with a synthetic content.

After outlining the basic ideas of intuitionism, Brouwer continues in 'Intuitionism and Formalism' to address the paradoxes that provoked the foundational crises in mathematics. He points out that within an intuitionistic approach such paradoxes will evaporate. For instance, the intuitionistic restrictions with respect to the construction of sets will imply that the set-theoretical paradox that was identified by Bertrand Russell and Ernst Zermelo will disappear.[8] Thus Brouwer tries to demonstrate that intuitionism establishes a solid route out of the foundational crises.

After the presentation of 'Intuitionism and Formalism', Brouwer elaborated intensively on all aspects of the intuitionist program, and he gave series of lectures. In 1926, he lectured in Göttingen, which was the most prominent place for mathematical research, directed by Brouwer's principal opponent, Hilbert. In 1927, he lectured in Berlin, where Freudenthal was in the audience. In 1928, he lectured in Vienna, where Ludwig Wittgenstein was attending and got inspired to return to philosophy. In 1934, Brouwer lectured in Geneva, and during the years 1947–1951, he gave a series of lectures in Cambridge. His intention was to organise these lectures in a book, and he completed five of the planned six chapters. They became published posthumously as *Brouwer's Cambridge Lectures on Intuitionism* (Brouwer, 1981).

In 'Intuitionism and Formalism' (1913), he gave an opening outline of intuitionism, while the *Cambridge Lectures* can be read as his more reflected formulations. Here Brouwer (1981) uses the terminology that mathematics develops through particular acts. In this way, he highlighted explicitly the constructivist nature of intuitionism. He presents what he refers to as the *first act of intuitionism* in the following way:

8 See Chapter 4 in this volume, for a presentation of this paradox.

> Intuitionistic mathematics is an essentially languageless activity of the mind having its origin in the perception of a move of time. This perception of a move of time may be described as the falling apart of a life moment into two distinct things, one of which gives way to the other, but is retained by memory. If the twoity thus born is divested of all quality, it passes into the empty form of the common substratum of all twoities. And it is this common substratum, this empty form, which is the basic intuition of mathematics. (pp. 4–5)

As in 'Intuitionism and Formalism', Brouwer refers to a 'falling apart of a life moment' as constituting the origin of mathematics. He talks about a twoity that when stripped of particular emotional qualities, turns into an 'empty form of the common substratum of all twoities'. We are dealing with a pure twoity, which represents the basic intuition of mathematics. It signifies the first mental act of intuitionism. It is the same intuition that Brouwer previously had referred to as a two-oneness.

Brouwer claims that intuitionistic mathematics is essentially a languageless activity. However, intuitionistic mathematics also becomes expressed through symbols, and I assume that Brouwer did write something at the blackboard when giving his Cambridge lectures. But still, according to intuitionism, this is just chalky shadows of what mathematics really is: *a languageless activity of the mind*.

In the *Cambridge Lectures*, Brouwer presents a *second act of intuitionism*, which is also a way of creating new mathematical entities:

> In the shape of mathematical species, i.e. properties supposable for mathematical entities previously acquired, satisfying the condition that if they hold for a certain mathematical entity, they also hold for all mathematical entities which have been defined to be 'equal' to it, definitions of equality having to satisfy the conditions of symmetry, reflexivity and transitivity. (p. 8)

This second act refers to ways of creating species of already created mathematical entities. Brouwer does not use the notion of set, but one can think of species as a collection of entities being 'equal' to each other.

Brouwer claims that all mathematics can be constructed through the two acts of intuitionism; no other pattern of construction is necessary. This is the clue to Brouwer's constructive interpretation of mathematics.

Many traditional forms of mathematical inferences are not valid from an intuitionist point of view. Mathematics has been all too tolerant by applying inferences which are not guided by the acts of intuitionism. Let us consider a classic proof of the theorem T: *There exist infinitely many prime numbers*. The negation $\neg T$ states: *There exists a maximum prime number* that we can refer to as P. Let us assume $\neg T$. Let the sequence of prime numbers smaller than P be $p_1, p_2, \ldots, p_n$. We define a new number N as $N = p_1 \times p_2 \times \ldots \times p_n \times P + 1$. As for any number, N can be uniquely factorised as the product of prime numbers. Consider one of these prime numbers, which we can call Q. Q cannot be any of the numbers $p_1, p_2, \ldots, p_n, P$, as a prime number cannot be a factor in two consecutive numbers. It follows that Q must be bigger than P. By assuming $\neg T$, we reach a contradiction. As a consequence, we conclude T: There exist infinitely many prime numbers.

Brouwer does not accept indirect proving, as this does not represent a constructive way of binging about a mathematical entity or a mathematical truth. Assuming a Platonist position, the set of natural numbers is a pre-existing entity, and so is the set of prime numbers. Either the set of prime numbers is finite, or it is infinite. Only these two alternatives are possible. If one assumes that there exists a maximum prime, and this leads to a contradiction, the alternative must be true. But this is not a constructive proof, according to Brouwer. If one wants to prove T, then one has to provide a construction that leads to T. One could easily be in a situation where one cannot prove T or $\neg T$, and according to Brouwer, neither T nor $\neg T$ is true until one of them has been proved constructively.[9]

According to intuitionism, then, the whole body of existing mathematical theories needs a careful revision, which includes three elements. First, one needs to consider what classic mathematical results can be considered valid within an intuitionistic mathematics.

9 Brouwer (1981) makes the following observation: 'The belief in the universal validity of the principle of the excluded third in mathematics is considered by the intuitionists as a phenomenon of the history of civilisation of the same kind as the former belief in the rationality of π, or in the rotation of the firmament about the Earth' (p. 7). The validity of is nothing but a cultural phenomenon that can be explained along the same lines as many other superstitions. There is nothing in this logical formula except long-lasting preconceptions. However, Brouwer does acknowledge that in some particular domains the principle of the excluded middle does work: it could be in everyday situations; it could also be in some more particular mathematical cases. But as a general principle to be used in mathematics, it is illegitimate.

Second, one needs to consider which classic proofs can be reformulated and given new constructive formats. Third, one has to consider what parts of classic mathematics cannot be transferred into intuitionistic mathematics. Georg Cantor's (1874) theory of sets, which leads to the idea of an infinity of infinities, is an obvious candidate. Through such a re-examination, mathematics will be cleansed of invalid results, and possible paradoxes will be eliminated.

Let us consider again the classic proof of the existence of infinitely many prime numbers. It applies the principle of excluded middle, and is therefore not constructive. But the theorem can be reformulated and the proof reorganised to meet constructivist standards. The theorem can be stated as: *For any prime number P, it is always possible to construct a prime number Q that is bigger than P*. Define *N* as in the non-constructivist proof above and let *Q* be a prime factor of *N*. It follows that *Q* must be bigger than *P*. This formation is in accordance with intuitionism, not assuming any actual infinity. Through the very proving, we have constructed the prime number *Q* bigger than *P*, and we can conclude: For any prime number *P*, it is always possible to construct a prime number *Q* bigger than *P*.[10]

In 1975 and 1976, Brouwer's collected works appeared in two volumes. The first, *Collected Works, Vol 1: Philosophy and Foundations of Mathematics* is edited by Arend Heyting.[11] The second, *Collected Works, Vol. 2: Geometry, Analysis, Topology and Mechanics* is edited by Freudenthal. The two editors, Heyting and Freudenthal, are real insiders of intuitionistic mathematics.[12]

10 The reformulation of the classic proof for the infinity of prime numbers was not a big deal, as the classic proof already contained the constructive features; it just had to be reformulated. However, there are mathematical theorems that are much trickier. For instance, what about Brouwer's own fixed-point theorem? He made the proof according to classic standards; however, Kellogg, Li and Yorke (1976) 'saved' the theorem by giving a constructive proof. Brouwer's fixed-point theorem makes part of intuitionistic mathematics.

11 Heyting (1971) provides a captivating introduction to intuitionism.

12 Intuitionistic mathematics has had a tremendous development. Mathematical analyses have been developed according to an intuitionistic outlook (Bishop, 1967; Lorenzen, 1971; Martin-Löf ,1968). It has turned out that this approach has a particular significance for computing (Martin-Löf, 1982). Intuitionism has paved the way for a new richness of philosophic discussions (Dummett, 1977; Lorenzen 1969). Intuitionistic logic, as formalised by Heyting (1930) to the great consternation of Brouwer, got related to other logical structures by Gödel (1933), and came to play a crucial role as a logic relevant for computer science (see Reeves

Summary of Brouwer's conception of mathematics

No mathematical entity or mathematical truth exists before it has been constructed. This claim opposes the ontology of any form of Platonism, which assumes that mathematical entities have a real existence, independent of human intervention. To Brouwer, processes of obtaining mathematical knowledge are processes of construction, not processes of discovery.

According to Brouwer, intuition plays a crucial role in any form of mathematical construction. This intuition he specifies in terms of two mental acts. Brouwer does not think of such acts as taking place in a specific mind. He does not present mental acts in psychological terms, and does not suggest any form of what could be referred to as psychological constructivism.[13] Nor does Brouwer's constructivism includes any trace of social constructivism. The mental acts Brouwer has in mind do not presuppose any interaction; they are idealised individual acts; and they bring about the same entities and the same truths for 'those who understand'.

An intuition of time is a fundamental phenomenon in human life, and after abstracting away all emotional features of the movement of time, we reach the fundamental phenomenon of mathematical acting: the intuition of the naked two-oneness, also referred to as a twoity. While the first act of intuitionism takes the form of counting, the second act takes the form of groupings of already constructed mathematical entities. According to Brouwer, all mathematics can be constructed through these two acts.

Mathematics is languageless. Mathematics does not have any adequate articulation in language, and mathematical formalisms are nothing but imprecise and at times mischievous depictions of genuine mathematical processes. Mathematical processes are alive, while mathematical formalisms are dead and distorted copies. Mathematics is a languageless activity of the mind.

---

and Clarke, 2003). For a general overview of the development of intuitionism, see Troelstra and Dalen (1988).

13 Compared to Brouwer's constructivism, Jean Piaget's constructivism is psychological by highlighting the importance of the mental processes of assimilation and accommodation for the construction of knowledge.

Brouwer's conception of mathematics means a revitalisation of intuition as a crucial feature of mathematics. By doing so, Brouwer confronts formalism, which tried to eliminate intuition from mathematics. Formalism saw intuition as the cause of the foundational crisis in mathematics, Brouwer sees intuition in terms of well-defined mental acts, as saving mathematics from contradictions.

Mathematics as a human activity

Brouwer's ideas did not directly bring changes to mathematics education. However, some of his ideas became re-elaborated by Freudenthal, who opened a new terrain for activities in mathematics education by making plenty of space in which for intuition to operate. Before Freudenthal, other Dutch mathematics educators sought inspiration in intuitionistic ideas, and such visions for mathematics education had been presented in the Dutch mathematics teacher education journal *Euclides*.[14] However, these visions faded away, while Freudenthal's elaboration turned out to have a profound impact.

We are going to consider Freudenthal's conception of mathematics as expressed in *Mathematics as an Educational Task*.[15] Freudenthal sees mathematics as a human activity, while Brouwer sees it as mental acts.[16] We will point out similarities and differences between these two conceptions.[17] We will try to clarify what Freudenthal referred to when, in the Preface, he mentioned that his educational interpretation of mathematics betrays the influence of Brouwer's view of mathematics.

---

14 Let me refer to two publications: Rootselaar (1957) and Heyting (1957). Heyting observes that intuitionism might have an educational relevance, as several intuitionistic concepts come close to students' natural perceptions. Both papers focus on intuitionism as a source of inspiration for mathematics teachers, not as a proper goal in mathematics education. I do not read Dutch, but Danny Beckers has provided me with these references and a short summary of them.

15 Other important contributions by Freudenthal that we also could address are Freudenthal (1978, 1983, 1991).

16 See Gravemeijer and Terwel (2000) for a careful presentation of what Freudenthal means by mathematics being a human activity.

17 We have to be aware of a principal difference in the presentation of the two conceptions. While Brouwer presents his conception explicitly, as in 'Intuitionism and Formalism' and in the *Cambridge Lectures*, Freudenthal's main focus in *Mathematics as an Educational Task* is to formulate a view on mathematics education, rather than present an explicit conception of mathematics.

When Brouwer launched his view, formalism was in powerful development, establishing itself not only as a philosophy of mathematics, but also as an emerging mathematical research paradigm. Brouwer's intuitionism was up against this powerful opponent identifying formal structures with mathematics itself. Freudenthal was also up against formalism, specifically in the form of structuralism as advocated by Bourbaki and acted out through the Modern Mathematics Movement. Freudenthal did not see formal structures as providing a proper departure for mathematics education; instead, students should be involved in mathematical activities.

Freudenthal refers to the Socratic method, which highlights the importance of developing understanding through the students' own activities. He formulates this idea in the following way:

> I will suppose as Socrates did that the teaching matter is re-invention or re-discovery in the course of teaching. Rather than being dogmatically presented, the subject matter originated before the students' eyes. (p. 101)

Freudenthal's critique of a delivery-education can be compared to Paulo Freire's (1972) critique of banking education. Freire criticises profoundly the idea that education means bringing parcels of assumed knowledge to the students, and Freudenthal expresses a similar critique.

The Socratic method is presented in Plato's dialogue *Menon*, in which Socrates talks with Menon's slave.[18] The point of the dialogue is that Socrates does not teach the slave anything. Socrates only puts questions, so no 'transfer' of knowledge is taking place. Starting from these questions, the slave reaches a mathematical insight. This dialogue illustrates Plato's idea that learning means remembering. We can interpret the example as embedded in a Platonic outlook, according to which any kind of obtaining mathematical knowledge takes the form of discovering some truths about an already exiting mathematical reality. This means that any form of mathematical learning becomes a re-discovery, or a dis-covery.[19]

18 See *The Internet Classics Archive*, http://classics.mit.edu/Plato/meno.html

19 Kollosche (2017) provides a detailed analysis of the notion of discovery and dis-covery addressing the Platonic features that might be included in these notions.

While Brouwer would certainly oppose any such interpretation of learning mathematics, Freudenthal is not explicit in formulating an anti-Platonic position. However, I am tempted to interpret Freudenthal's reference to the Socratic method not as an assumption of any Platonism, but more as a general interpretation of learning as being resourced by interaction, communication, and dialogue. In making such an interpretation, Freudenthal certainly distances himself from Brouwer. One can think of Freudenthal as assuming a social interpretation of constructivism, contrary to Brouwer's individual constructivism. I see Freudenthal's reference to the Socratic method in this light. However, we also have to be aware that Freudenthal does not refer to his own interpretation of learning mathematics in terms of constructivism. This is a label, however, that I feel tempted to apply.

Being constructivist does not imply being relativist. To Brouwer, mental constructions of mathematics will lead to the same mathematics. There is only one form of intuitionistic mathematics. Brouwer has inserted an absolutism into his anti-Platonic constructivism. It might be possible to find shades of the same absolutism in Freudenthal's conception of mathematics. This absolutism appears when Freudenthal presents learning as a guided activity, which leads to an insight in already established mathematical knowledge. Freudenthal not only uses the notion of re-discovery, but also the notion of re-invention. By talking about re-invention and not just about invention, Freudenthal makes clear that he does not think of learning mathematics as a process that brings about new mathematical insight in any objective interpretation, but in a subjective. This process brings about new mathematical insight for the students.

Freudenthal does not use the notion of construction, but other related notions – such as activity, creative inventions, direct invention, and re-invention – that bring the message:

> Today, I believe, most people would agree that no teaching matter should be imposed upon the students as a ready-made product. Most present-day educators look on teaching as initiation into certain activities. Science at its summit has always been creative inventions, and today it is even so at levels lower than that of masters. The learning process has to include phases of direct invention, that is, of invention not in the objective but in the subjective sense, seen from the perspective of the students. (p. 118)

Freudenthal talks about a ready-made product being imposed on students, and with such a remark he points his finger at the Modern Mathematics Movement. Through this movement, the whole curriculum became predefined through the structural architecture of mathematics. Again and again, Freudenthal criticises this approach. I suspect he is being ironic when he states that today it is broadly agreed that 'no teaching matter should be imposed upon the students as a ready-made project'. When *Mathematics as an Educational Task* was published in 1973, the Modern Mathematics Movement was still in full swing, although difficulties in its implementation had become recognised.

When Freudenthal describes processes of learning mathematics, he uses several expressions referring, not to the final and polished mathematical structures, but to the processes that can lead to mathematical understanding. Freudenthal changes the focus from 'what to teach' to 'how to learn'. He highlights that the 'learning process has to include phases of direct invention'. Invention, however, is not to be understood in absolute terms, but always with reference to the students' horizons. Freudenthal finds it crucial that students experience that mathematical insight becomes developed from within, and not imposed on them.[20]

Brouwer also concentrates on mathematical processes and refers to mental acts. However, Freudenthal has a much broader conception of mathematical activity in mind. I have no doubt that he was fully aware of the very specific interpretation of mathematical construction provided by Brouwer, and that he did not want to assume Brouwer's metaphysics with respect to the nature of mental acts. To Brouwer the mathematics-creating mental acts are individual; no trace of social interaction can be located in these acts. Freudenthal's conception of mathematics as human activity is different. Formulating arguments, addressing possibilities, evaluating results are all features of mathematical activities, seen as

20 Gravemeijer and Terwell (2000) make this point clearly in the following way: 'As a research mathematician, doing mathematics was more important to Freudenthal than mathematics as a ready-made product. In his view, the same should hold true for mathematics education: mathematics education was a process of doing mathematics that led to a result, mathematics-as-a-product. In traditional mathematics education, the result of the mathematical activities of *others* was taken as a starting point for instruction, and Freudenthal (1973) characterised this as an *anti-didactical inversion*. Things were upside down if one started by teaching the result of an activity rather than by teaching the activity itself' (p. 780).

social processes among students and teachers. This is pointed out by Freudenthal through his reference to the Socratic method. Freudenthal sees the role of the mathematics teacher, not as being a lecturer, rather as being a supervisor helping the students to come to participate in mathematical activities.

By highlighting that we are dealing with a human activity, Freudenthal also stresses that mathematics is not an activity presupposing some particular abilities. It is a common activity. Everybody can participate in a mathematical activity. Freudenthal provides the conception of activity with a broad inclusivity, while Brouwer's mental acts appear exclusive, reserved for 'those who understand'.

Freudenthal talks about 'connected mathematics', and 'lived-through realities', which is very different form talking about mathematical structures:

> To teach connected mathematics it is not wise to start out looking for direct connections; they should rather be found between the contact points where mathematics is attached to the lived-through reality of the learner. Reality is the framework to which mathematics attaches itself, and though these are initially seemingly unrelated elements of mathematics, in due process of maturation connections will develop. Let the mathematicians enjoy the freewheeling system of mathematics – for the non-mathematicians the relations with the lived-through reality are incomparably more momentous. (p. 77)

Freudenthal's clue is that it is not wise to start out looking for direct connections. The point of departure is not any mathematical structures already elaborated by others, but the students' lived-through realities that include mathematics fraught with relations.

> When speaking about mathematics fraught with relations, I stressed the relations with a lived-through reality rather that with a dead mock reality that has been invented with the only purpose of serving as an example of application. This is what often happens even in arithmetic teaching. I do not repudiate play realities. At a low level games may be useful means of motivation. But it is dangerous to rely too much on games. Ephemeral games are no substitutes for lived-through reality. The rules of games that are not daily exercised are easily forgotten as mathematics or even faster. The lived-through reality should be the backbone which joins mathematical experiences together. (pp. 78–79)

By referring to a 'dead mock reality', Freudenthal not only criticises the Modern Mathematics Movement, but also the school mathematics tradition.[21] In this tradition exercises invented by textbook authors play a particular role: Peter has to buy 4.5 kilos of apples ... A family is driving on holiday with the average speed of 70 km per hour ... The shadow of the flag post is 4.6 meter long ... All such exercises are pure inventions; they do not represent any lived-through realities, rather stereotypical didactical inventions.

Brouwer is a radical anti-Platonist. The existence of any mathematical entity or mathematical truth has to be constructed. Before being constructed, nothing exists. This claim brought him to abandon classic logic. Apparently, Freudenthal shares Brouwer's disregard for formal logic. But while Brouwer is very specific in his critique of formal logic, Freudenthal simply makes space for all kinds of reasoning as forming part of mathematical activities.

Summary of Freudenthal's conception of mathematics

If mathematics is an activity, it is not defined by any Platonic reality, nor by any logical or formal structures. Instead of activity, one can also try to use the notion of construction and think of mathematics as a human construction. I find that Freudenthal operates with a constructivist perspective on mathematics, although he does not use this label.

Brouwer did not include any relativism in his version of constructivism, nor does Freudenthal seem to. While the construction of mathematics through research might represent objective inventions, the construction established through education represents subjective inventions.

Whereas Brouwer confronted formalism as represented by David Hilbert, Freudenthal confronted structuralism as represented by Bourbaki and the Modern Mathematics Movement. Confronting formalism and structuralism means giving value to intuition, and both Brouwer and Freudenthal do so. While Brouwer installed intuition in

21 For a characterisation of the school mathematics tradition, see Skovsmose and Penteado (2016).

a central position in mathematics, Freudenthal made a vast space for intuition in all kinds of educational activities.

Brouwer did not see formal logic as capturing the nature of mathematical reasoning. Freudenthal shared this idea, however in *Mathematics as an Educational Task* I do not see traces of Brouwer's way of arguing for this position. Freudenthal acknowledges the different patterns of mathematical reasoning, but he never shows interest in trying to capture a universal pattern of this reasoning. Freudenthal is rather interested in exploring a broad spectrum of intuitive mathematical reasoning in educational contexts.

Many times, Freudenthal characterises mathematics as a human activity. By talking about human *activity*, he assumes the overall intuitionist outlook that Brouwer had condensed in the notion for mental *activity*. By talking about *human* activity and not about *mental* activity, Freudenthal also distances himself from Brouwer. While mental activity refers to highly idealised constructive processes, human activity refers to real-life interactive processes of creating mathematical understanding.

Freudenthal's conception of mathematics means a revitalisation of intuition in mathematics education. It might be this revitalisation that Freudenthal had in mind when in the preface to *Mathematics as an Educational Task*, he mentioned that his educational interpretation of mathematics betrays the influence of Brouwer's view on mathematics.

Ship Ahoy

The Modern Mathematics Movement was guided by a well-defined conception of mathematics: mathematics is formed by its structures, and three basis structures, also referred to as mother structures, had been identified by the Bourbaki group. According to Jean Piaget, three similar structures characterise children's operations with objects, which brought him to assume that he had identified the genetic routs of mathematics. This assumption provided the whole Modern Mathematics Movement with an outstanding legitimisation: the structural organisation of mathematics shows also the natural way of learning mathematics. Freudenthal considered this justification to be nonsense.[22]

Seeing mathematics as human activity opposes directly the conception of mathematics that guided the Modern Mathematics Movement. As an illustration of what this could mean, I refer to an example published in *Five Years IOWO*, published as a special issue of *Educational Studies* in 1976 when Freudenthal retired (Freudenthal et al., 1976).

Ship Ahoy is for children around ten to eleven years old. The whole project is planned to last for about ten lessons. *Ship Ahoy* starts with the children listening to a communication between two ships, Bermuda (B), a yacht, and Constance (C), a tug. The storm makes it sometimes difficult to hear what is said:

> C: Do not read you. Repeat. Over.
> B: This is Bermuda. This is Bermuda. We are in danger, in danger. The motor has failed ... (noise) ... Cast the anchor, but the chain can break any moment. Over.
> C: I read you. What is your position? Over.
> B: Do not know, do not know. Wemelringe area. Probably Wemelringe area. No vision. Over.
> C: Do you see the coast? Over?
> B: Yes, we ... (noise) ...
> C: I do not read you. I do not read you. Over.
> B: We see a lighthouse in the distance, lighthouse in the distance. Over.
> C: We read you. Do you see a church tower? A church tower? Over.
> B: Only water. Only water. Over.
> C: Keep looking and call in. Over.
> B: Yes. A church tower to the left of the lighthouse! Over.
> C: Good, we have your approximate position. We are on our way. On our way. Over.
> B: Thank you. Please hurry. Over.
> C: We are on our way. Keep looking. There is a small house to the right of the lighthouse. Keep looking. Over and out.

In 1977, when I first time read the presentation of *Ship Ahoy*, I was surprised: Could this be mathematics? I am sure that I was not the only one being surprised. At that time, the perception of mathematics was dominated by the Modern Mathematics Moment, which operated with

22 See Chapter 4 in this volume, for a short presentation of Piaget's position and of Freudenthal's critique of Piaget.

a clear idea of what counted as mathematics. This idea was shaken by this and other examples presented by IOWO. Freudenthal's conception of mathematics, as formulated in *Mathematics as an Educational Task*, become both concrete and provocative.

The work in the classroom begins: What is the situation? What could happen? Why is Bermuda in difficulties? What can they see from Bermuda? The children are presented with some pictures showing the lighthouse, the church, and the small house in different positions. Could any of these pictures show the situation as observed from Bermuda? A map of the area is handed out. It shows the position of the lighthouse, the church, and the small house. The map has to be read and properly understood, and then comes the question: Where might Bermuda be located?

Could readings of maps and spatial reasoning be considered mathematical tasks? In 1976, this was hardly considered mathematics. In Denmark, a short textbook for students around fifteen years old had been published, giving a strict axiomatic presentation of incidence geometry. Here lines were defined as sets of points and illustrated as sets conventionally are, within egg-shaped circles. Two non-overlapping eggs illustrated two parallel lines, and so on. The deduction from the presented axioms observed strict formalities. No intuition with respect to points and lines were necessary; such intuitions were in fact considered disturbing for the deduction. An initial part of incidence geometry was carefully elaborated, and the majority of students were completely lost. Compared to such an approach to geometry, looking at maps and speculating about possible perspectives expand the scope of mathematical activities enormously. From being marginalised, intuition moves to the centre of mathematical reasoning.

The intuition cultivated in *Ship Ahoy* concerns three-dimensional space and three-dimensional geometry. The general assumption, associated with traditional mathematics education as well as with the Modern Mathematics Moment, was that one needed to start with two-dimensional geometry and only later get to three-dimensional geometry. When paying particular attention to intuition and not to any axiomatic organisation of geometry, this order turns artificial. We live in a three-dimensional space. All our daily-life experiences are located in such a space. So why not start out with issues related to our three-dimensional space of life? That is precisely what *Ship Ahoy* does.

The rescue work continues. Bermuda is found, and Constance takes her on tow. However, it has become night before they reach the harbour. How to keep the right course? From Constance, one can see the two lights in the Harbour. How the two lights are placed in the harbour can be seen on a map of the harbour also handed out to the children. One light is positioned higher up than the other. How should the captain on Constance see the positions of the two lights in order to keep the right course? The children become engaged in such discussions, and the rescue work continues. Freudenthal talked about starting from situations fraught with relations, and *Ship Ahoy* is an illustration of what this could mean.

The inspiration from Freudenthal and IOWO spread world-wide. By the late 1970s, the inspiration had reached Denmark, where the Modern Mathematics Movement had been broadly implemented. The Freudenthal and IOWO approach showed alternatives, and intuition got revitalised in mathematics education.[23]

What about socio-political issues?

In 1967, I graduated from a teacher education college in Denmark, where I had been carefully introduced to the Modern Mathematics Movement. In 1968, I started studying mathematics at university, and here I encountered a structuralist approach where, for instance, the introductory course in mathematical analysis began with abstract topology.

In 1977, I was accepted as a PhD student at the Royal Danish School of Educational Studies, which concentrated on in-service training of teachers. Since the beginning of the 1960s, the Modern Mathematics Movement had been broadly introduced in Denmark, not least due to the dedicated work of Bent Christiansen from that institution. However, Christiansen became much inspired by Freudenthal's work, and he directed a major change in mathematics education in Denmark.

23 The notion of *realistic mathematics* has been coined and elaborated in detail at the Freudenthal Institute. See, for instance Gravemeijer (1994), De Lange (1987), and Streefland (1991).

Christiansen and Tage Werner were my supervisors, and Christiansen told me about Freudenthal and about IOWO, and he showed me a copy of *Five Years IOWO*. During his whole career, Werner had been a consistent anti-formalist, providing a range of suggestions for engaging students in mathematical activities. He was in line with IOWO even before *Five Years IOWO* was published.

The aim of my PhD project was to formulate a critical mathematics education, and soon after I got started my supervisors made it possible for me to visit IOWO in Utrecht and to meet with Freudenthal. I was anxious. At that time my English was not very good, Freudenthal was so famous, and I was overawed.

Freudenthal met me with a welcoming smile, and I felt relaxed. His enthusiasm was evident when he shared various possible mathematical activities. When I tried to explain about my project and wanted to ask how he viewed the connection between socio-political issues and mathematics education, he seemed, however, uninterested. I did not insist, so our conversation remained focused on possible mathematical activities. Through this interaction, I experienced the richness of educational ideas that emerge from viewing mathematics as a human activity.

I would have liked to insist on my question. Freudenthal uses the notion of *lived-through reality*, which I find to be powerful. It can be given a range of interpretations. The reality for whom? One could think of a *lived-through socio-political reality*. Such a conception can be related to Paulo Freire's notion of *generative themes*, which opens towards a huge variety of mathematical activities with political significance. While Freudenthal talks about mathematics as a *human activity*, one could consider what it could mean to talk about mathematics as a *political activity*.

The notion of lived-through reality can be related to critical mathematics education, but in my meeting with Freudenthal, he was not interested in addressing any such possibility. Nor do I locate any interest in his writings. Structuralism and the Modern Mathematics Movement are manifestations of the dogma of neutrality. They operate *as if* mathematics is neutral and mathematics education can be kept separate from socio-political issues. Freudenthal operates with the same *as if*. He formulates mathematics as an educational task within an apolitical outlook. However, although he embraces a dogma of neutrality, he simultaneously provides notions and ideas that help in formulating a critical mathematics education.

Acknowledgements

I want to thank Danny Beckers for detailed comments and suggestions. He has also drawn my attention to some publications in Dutch, a language that I do not read.

References

la Bastide-van Gemert, S. (2015). *All positive action stars with criticism: Hans Freudenthal and the didactics of Mathematics.* Springer.

Bishop, E. (1967). *Foundations of constructive analysis.* McGraw-Hill.

Brouwer, L. E. J. (1911). Über Abbildungen von Mannigfaltigkeiten [About mappings of manifolds]. *Mathematische Annalen, 71,* 97–115. https://doi.org/10.1007/BF01456931

Brouwer, L. E. J. (1913). Intuitionism and formalism. *Bulletin of the American Mathematical Society, 20*(2), 81–96. https://doi.org/10.1090/S0002-9904-1913-02440-6

Brouwer, L. E. J. (1981). *Brouwer's Cambridge lectures on intuitionism* (D. van Dalen, Ed.). Cambridge University Press.

Brouwer, L. E. J. (1996). Life, art, and mysticism. *Notre Dame Journal of Formal Logic, 37*(3), 389–429. https://doi.org/10.1305/ndjfl/1039886518

Cantor, G. (1874). Über eine Eigenschaft des Inbegriffes aller reelen algebraischen Zahlen [About a property of the epitome of all real algebraic numbers]. *Journal für die reine und angewandte Mathematik, 77,* 258–262. https://doi.org/10.1515/crll.1874.77.258

Dummett, M. (1977). *Elements of intuitionism.* Clarendon Press.

Freire, P. (1972). *Pedagogy of the oppressed.* Penguin.

Freudenthal, H. (1931). Über die Enden topologischer Räume und Gruppen. *Mathematische Zeitschrift, 33,* 692–713. https://doi.org/10.1007/BF01174375

Freudenthal, H. (1973). *Mathematics as an educational task.* Reidel.

Freudenthal, H. (1978). *Weeding and sowing.* Reidel.

Freudenthal, H. (1983). *Didactical phenomenology of mathematical structures.* Reidel.

Freudenthal, H. (1991). *Revisiting mathematics education: China lectures.* Kluwer.

Freudenthal, H., Janssen, G. M., Sweers, W. J., van Barneveld, G. B., Bosman, J. N., van den Brink, F. J., van Bruggen, J. C., Gilissen, L. J. M., Goffree, F., de Gooijer-Quint, J. F., ter Heege, J., Jansen, H. M. M., de Jong, R. A., Kremers, W. H. M., Leenders, C. P., Meijer, G. H., de Moor, E. W. A., Schoemaker, G., Streefland, L., ... Wijdeveld, E. J. (1976). Five Years IOWO [Special issue]. *Educational Studies in Mathematics, 7*(3).

Gödel, K. (1933). Eine Interpretation des intuitionistischen Aussagenkalküls [An interpretation of the intuitionistic propositional calculus]. *Ergebnisse eines mathematischen Kolloquiums, 4*(39–40), 300–301.

Gravemeijer, K. (1994). *Developing realistic mathematics education.* CD-ß Press, Freudenthal Institute.

Gravemeijer, K. J., & Terwel, J. (2000). Hans Freudenthal: A mathematician on didactics and curriculum theory. *Journal of Curriculum Studies, 32*(6), 777–796. https://doi.org/10.1080/00220270050167170

Heuvel-Panhuizen, M. van den. (2015). Freudenthal's work continues. In C. Cho (Ed.), *Selected Regular Lectures from the 12th International Congress on Mathematical Education* (pp. 309–331). Springer. https://doi.org/10.1007/978-3-319-17187-6_18

Heyting, A. (1930). Die formalen Regeln der intuitionistischen Logik [The formal rules of intuitionist logic]. *Sitzungsberichte der Preussischen Akademie der Wissenschaften*, 42–56, 57–71, 158–169.

Heyting, A. (1957). Intuïtionisme en schoolwiskunde [Intuitionism and school mathematics]. *Euclides, 33*(1), 1–12.

Heyting, A. (1971). *Intuitionism: An introduction.* North Holland.

Kant, I. (1973). *Critique of pure reason.* MacMillan.

Kellogg, R. B., Li, T. Y., & Yorke, J. A. (1976). A constructive proof of the Brouwer fixed-point theorem and computational results. *SIAM Journal on Numerical Analysis, 13*(4), 473–483.

Kollosche, D. (2017). Entdeckendes Lernen: Eine Problematisierung [Discovery learning: A problematisation]. *Journal für Mathematik-Didaktik, 38*(2), 209–237. https://doi.org/10.1007/s13138-017-0116-x

Lange, J. de. (1987). *Mathematics, insight and meaning.* Utrecht University.

Lorenzen, P. (1969). *Normative logic and ethics.* Bibliographisches Institute.

Lorenzen, P. (1971). *Differential and integral: A constructive introduction to classical analysis.* University of Texas Press.

Martin-Löf, P. (1968). *Notes on constructive analysis.* Almquist & Wiksell.

Martin-Löf, P. (1982). Constructive mathematics and computer programming. In L. J. Cohen, J. Los, H. Pfeiffer, & K.-P. Podewski (Eds.), *Logic, methodology and philosophy of sciences VI*. North-Holland.

Mill, J. S. (1970). A system of logic. Longman.

O'Connor, J. J., & Robertson, E. F. (2003) *Luitzen Egbertus Jan Brouwer*. https://mathshistory.st-andrews.ac.uk/Biographies/Brouwer

Ravn, O., & Skovsmose, O. (2019). *Connecting humans to equations: A reinterpretation of the philosophy of mathematics*. Springer. https://doi.org/10.1007/978-3-030-01337-0

Reeves, S., & Clarke, M. (2003). *Logic for computer science*. Department of Computer Science, Queen May and Westfield College, University of London, and Department of Computer Science, University of Waikato, New Zealand. http://www.cs.ru.nl/~herman/onderwijs/soflp2013/reeves-clarke-lcs.pdf

Rootselaar, B. van. (1957). Intuïtionisme en rekenkunde [Intuitionism and arithmetic]. *Euclides, 32*(6), 193–198.

Skovsmose, O., & Penteado, M. G. (2016). Mathematics education and democracy: An open landscape of tensions, uncertainties, and challenges. In L. D. English & D. Kirshner (Eds.), *Handbook of international research in mathematics education* (pp. 359–373). Routledge. https://doi.org/10.4324/9780203448946-22

Streefland, L. (1991). *Fractions in realistic mathematics education: A paradigm of developmental research*. Kluwer. https://doi.org/10.1007/978-94-011-3168-1

Troelstra, A. S. (2011). History of constructivism in the 20th century. In J. Kennedy & R. Kossak (Eds.), *Set theory, arithmetic, and foundations of mathematics: Theorems, philosophies* (pp. 150–179). Cambridge University Press. https://doi.org/10.1017/CBO9780511910616.009

Troelstra, A. S., & Dalen, D. van. (1988). *Constructivism in mathematics: An introduction*. North-Holland.

8. Human mathematics

Ole Ravn

This chapter discusses how we can think about mathematics as a human enterprise. It takes as its starting point the portrait of a European tradition that has considered mathematics as essentially a non-human realm. As a challenge to this tradition, a Wittgensteinian interpretation of mathematics as a special type of language among all the human languages is outlined and used to develop a platform for understanding mathematics as 'human mathematics'. This conception is finally given shape through two discussions, first through a challenge to the positioning of mathematics in our contemporary universities in close proximity to the natural and technological sciences. Instead, a narrowing of the gap between the sciences and the humanities with a consequent repositioning of mathematics in the epistemological landscape of our knowledge institutions is advocated. Secondly, a human mathematics conception is discussed in relation to learning and teaching. Connections are made to socio-cultural learning theory, and it is argued that the concepts of 'fog of mathematics' and 'centreless mathematics' can help in reconfiguring how to think about the learning of mathematics.

Introduction

Dominant stories told about mathematics are often linked to science and the certainty of scientific knowledge. Other more socially oriented stories about mathematics are related to our everyday practices in schools, homes, or the workplace. From a Wittgensteinian perspective – a perspective I shall discuss in the following – these socially oriented stories and practices, in conjunction, hold the truth about what mathematics is. This difference in perspectives and understandings of mathematics is the axis around which this chapter revolves.

©2024 Ole Ravn, CC BY-NC 4.0 https://doi.org/10.11647/OBP.0407.08

I will attempt to portray these two different perspectives in the following and discuss how there is an argument that the meaning of mathematics is nothing more than a social agreement in the use of signs we have developed over centuries – a purely human mathematics. This perspective highlights that non-human ontological ideas about mathematics could be utterly misleading and opens the discussion about socially oriented reinterpretations in the epistemology and ontology of mathematics. In this sense, the chapter deals with the question: What implications could a human and socially centred interpretation of mathematics have for our current practices?

My approach to getting closer to answering this question involves the following steps:

1. Give a short historical account of some of the central roles mathematics has played in our thinking about science and universities. This is a story dominated by the view that mathematics is non-human and represents the eternal structures of the world.

2. Present a language-centred philosophical position that argues mathematics can be understood as a multitude of human language constructions with many different types of uses and functions in our lives.

3. Discuss the opening of perspectives that presents mathematics as a completely social construction. I highlight two cases to illustrate this. The first case concerns university and mathematics – that is, how should mathematics be positioned in our epistemological and ontological landscape of sciences? In this case, the attempt is to give mathematics new interpretations in relation to the humanities as an expression of human creativity along the lines of poetry and literature. This is potentially a story of mathematics as exploring the limitations of (constructed) reasoning in the process of developing ever new and complex mathematical measures. The second case discusses how thinking about learning mathematics from the human mathematics perspective in general will differ somewhat from many traditional approaches. Thinking about the learning and teaching of mathematics under the

assumption that mathematics is a 100% social construction means that some principles can be highlighted to give direction for educational development.

With respect to point 2., I will draw upon the Wittgensteinian argument that mathematics consists of language games, which play many different roles in our lives, particularly in how we use, develop, and reach agreements on mathematical concepts. I will draw on Stuart Shanker's interpretation of Wittgenstein's social turn in the philosophy of mathematics, which carves out a specific position in the interpretation of Wittgenstein's writings on mathematics (Shanker, 1987). This is not a chapter that aims to persuade all critics of a thoroughly social interpretation of mathematics, but I will point to the main ideas and reasoning behind the position I call 'a human mathematics' in what follows.

Order of the galaxy

Historical configurations of knowledge and mathematics have a huge impact on our understanding of the role played by mathematics today. Consequently, it seems reasonable to start thinking about the positioning of mathematics in our scientific worldview with an outline of some of the historical constructions that have surrounded mathematics. In order to trace some of the routes mathematics has traveled until today I will discuss aspects of its institutional connections to science and knowledge in a European context. The aim is to highlight dominant patterns of thinking about mathematics in the European history of ideas, being fully aware that there are many non-European historical stories that go even further back in time. The author of this chapter is of European origin and this unfortunately puts some limits on his insights into other historical trajectories. Accordingly, the following should be thought of as a local perspective about past ideas related to mathematics and how we could conceivably think differently about them in the future.

The local story inevitably connects to the highly influential interpretation by the Pythagoreans in Ancient Greece and the Academy built by Plato later on. The influence of this early interpretation of mathematics within a larger ontological framework can be traced in the medieval university structure, as presently discussed. And, in today's university, the idea of mathematics as an especially important element in exploring the world is, for example, reflected in the acronym STEM (Science, Technology, Engineering, and Mathematics) which at the same time disconnects mathematics from the sciences of language and social sciences.

By echoing this classical interpretation, one runs the risk of oversimplifying the actual historical complexities and entanglements and the connections between both modern and medieval scholarship environments and their Ancient Greek counterparts (see, for example, Høyrup, 1996, for an interesting account of these complexities). However, as Jens Høyrup suggests, there is no doubt about the dominant narrative concerning Ancient Greek mathematics and its impact on the European interpretation of the development of mathematics:

> This tale, more or less biased or false as an historical account, has none the less become material truth in the sense that it has contributed to the self-understanding and thereby to the cultural identity of the European mathematical community/communities for centuries. (Høyrup, 1996, p. 103)

However flawed this narrative is in representing the vast complexities of European and non-European origins of ideas about mathematics, its domination as a narrative is what matters for the argument of this chapter.

The Pythagorean interpretation of mathematics is often ridiculed as coming from a very speculative and religious environment. Nevertheless, the Pythagoreans are very importantly famous for connecting numbers and the relation between numbers to the heavenly spheres; in this way, they started a long tradition of relating mathematics to the structures of the universe and to the field of astronomy.

Archytas (428–347 BCE), at the time of Plato, explained how the Pythagoreans were the inspiration to connect the study of numbers to the study of the universe, as they 'handed down clear knowledge of the speed of stars and their rising and setting, and of geometry, arithmetic, and spherics and not least music, for these studies turned out to be sisters' (Archytas, cited in Pedersen, 1979, p. 20). Spherics was closely associated with what we would refer to as astronomy today and it was thought of as the materialisation of numbers in nature in its continuous form. Music was thought of as numbers in nature in their discrete form, while geometry and arithmetic were thought of as numbers in themselves—in both discrete and continuous forms (see Pedersen, 1979, p. 20).

The impact of the Pythagorean ideas about mathematics was established most forcefully by Plato. When he constructed his Academy, which later became an influential inspiration for the early medieval university, he found a central position for mathematics as a field that was of the utmost importance in the formation of thinking among his students. In many of Plato's writings, and especially in *The Republic* (Plato, 2022a), he outlines how the road towards a deeper insight into the many aspects of life can be furthered by prolonged studies of mathematics. In this way, Plato set the course towards putting mathematics on a pedestal among the sciences as the discipline that will train and strengthen reasoning and logical deduction. And it was notably a form of mathematics that was also considered as metaphysically connected to the order of things in the physical world.

When the first European universities were established in the eleventh century they were inspired by the Ancient Greek constellation and understanding of knowledge and their structuring of the different fields of study. In these universities, the faculties were normally the philosophical, the judicial, the medical, and the theological. To access one of the higher faculties one had to pass the bachelor exams in the philosophical faculty. Based on the tradition from the Pythagorean division of knowledge classification, these were ordered into 'seven liberal arts', divided between the study of Number (Quadrivium), with four subdisciplines, and the study of Letter (Trivium), which focused on grammar, logic, and rhetoric (Grane, 1991, p. 23).

Mathematics was connected by the Pythagoreans to the study of the universe, implying that the building blocks of the universe are of a mathematical nature. This conception underlines the idea that Letters are about human matters whereas Numbers are about the matters of the universe. Studies related to the Letter were, on the other hand, directly associated with handling human life and the social sphere. In this way, the deep gap in the scientific community today between STEM and not-STEM areas can be thought of as having been nurtured from this specific and speculative ontology in relation to the power of Numbers.

Plato himself was an active constructor in the reification of this constellation of knowledge. The key mathematical work to be handed down through history from Ancient Greece is Euclid's *Elements* and much of its content was inspired by scholars educated or situated in Plato's Academy. Archytas has already been mentioned and other famous examples are Theaetetus (417–369 BCE) and Eudoxus of Cnidus (395–342 BCE) who both researched at the Academy (O'Connor & Robertson, 1999).

To highlight the philosophical significance of the *Elements* for the times to come one can make several observations. First of all, within the specific Euclidean framework of mathematics – a celestial mathematics, you might say – the only permitted construction methods for constructing objects were the use of a ruler and a pair of compasses. In the Euclidean-Platonic epistemology of mathematics, these constructions represent eternal objects. Many students for centuries afterwards have been trained in these basic skills and the ruler and a pair of compasses were certainly to be found in the mathematics pupil's toolkit.

The first book of Euclid's *Elements* starts out by proving that one can construct an equilateral triangle from some basic actions of construction using the ruler and a pair of compasses. Many volumes later, the final proof in the final book of the *Elements* establishes the construction of the so-called Platonic Solids (Euclid, 1998, Book XIII). It is proven that there are exactly five of these solids, interpreted as representing the 'elements' (fire, earth, air, and water), with the fifth representing heaven and its twelve constellations. In other places, Plato relates the Platonic elements to the building blocks of all things and in this way makes a transparent connection between Euclid's *Elements* and his own Pythagorean and mathematically inspired ontology (Plato, 2022b).

In the last decades it has been more and more acknowledged that the early modern scientists like Isaac Newton and Johannes Kepler were much inspired by similar thoughts on mathematics. It is telling how Kepler describes how the regular polyhedra can be understood as the structure of the universe and here very much brings the ontology of the Ancient Greeks into the core construction of modern science. In his early work *Mysterium Cosmographicum*, he describes how the regular polyhedra in Euclid's *Elements* are to be conceptualised in an astronomical sense as the spherical structures surrounding earth (Kepler, 1596). In this way, he establishes a direct line to Plato's ideas about mathematics. He is echoed by the insights and ideas of Galileo Galilei on the study of nature as a realm where only mathematics can reach the deepest insights (Galilei, 1957, pp. 237–238).

After having stripped away the metaphysical connotations, universities today are very much aligned with the conclusions of this story. In the modern Humboldtian inspired universities, mathematics is often located next to physics and the natural sciences. This means that mathematics is still connected to the idea that it is the main tool for describing and understanding the physical world around us. The essence of this conception of knowledge amounts to something like the following: to understand human actions you must study letters and natural language and to understand the physical world, including the human body and its behaviours, you must use the numbers to get to the truth. And this is a relatively moderate interpretation of matters. The stronger interpretation goes along the lines that if you *really* want to understand any field of study you need the *hard* sciences defined by their use of quantification of the world through the Number.

A social enterprise

In the previous section, I have tried to portray how mathematics has been interpreted as connected to the building blocks of the universe. It is a deep cultural heritage in Western inspired universities that mathematics is the language that can tell you the most about the world.

However, a contrasting perspective does exist, though it is much less dominant. In fact, numerous challenges have been raised against the idea of mathematics as a mirror of real-world structures, inherently tied to fields like physics, chemistry, engineering, and technology more than to other fields of knowledge. Among these challenges I will try to highlight a Wittgensteinian perspective that suggests that mathematics is a human construction through and through. From this perspective mathematics is not about mirroring the logical structure of the world but instead about creating a diverse mathematical language to use in a multitude of different types of social practices. From the many interpretations of mathematics as a social structure, I have consistently found Wittgenstein's interpretation, in his later works, to be both the most radical and the most credible. It is developed in the posthumously published *Remarks on the Foundations of Mathematics* but is also closely connected to his later principal work *Philosophical Investigations*. Wittgenstein's interpretation of anything is always subject to heavy debate and, as mentioned above, I will follow the interpretation presented by Shanker (1987) in *Wittgenstein and the Turning-Point in the Philosophy of Mathematics*. Explaining Wittgenstein's position in detail is beyond the scope of this chapter; the discussion is developed further in Ravn and Skovsmose (2019, 2020).

A social interpretation of language revolves around the idea that our words and sentences can only have meaning from a group of language users. In fact, Wittgenstein is famous for his argument that no single or isolated human would ever be able to pinpoint meaning within a word or symbol because there would be no group of users to discuss and reflect to what degree the use of the symbol or word would be correct. This is known as the private language argument in Wittgensteinian research, and it has been heavily discussed through the years (see, e.g., Candlish, 1998, and his discussion of Saul Kripke's notorious interpretation).

The argument is that only the use of symbols or words in a social group can establish the meaning of the symbol. In the Wittgenstein literature, this is known as the 'meaning is use' principle; to illustrate this, consider one of my favourite examples: the sign we make when we point our finger in a certain direction. This is a simple concept for adults to understand, and often across cultures. However, a young child might not grasp the symbolic meaning of the gesture in their early years, and may simply look just at your finger, regardless of the direction you are indicating. They do not know how to use this part of language and only gradually will they learn how to use this symbolic gesture.

The situation is similar in mathematics, according to Wittgenstein. When we are told to repeatedly add 2, we feel forced to write '2, 4, 6, 8, ...' But what is it that compels us to do so? In Wittgenstein's interpretation, the only force at stake is the social training and large-scale practice in a community of mathematics users that, in the end, creates the sensation of the forced conclusion as being the most natural endeavour imaginable (Wittgenstein, 1967, p. 3e–6e). If we imagine the sequence '2, 4, 6, 8, ...' written in chalk on a blackboard, Wittgenstein's point is that there is nothing hidden behind the chalk. The symbols themselves do not have any concealed meaning that could force us to act as we all do. Often this feeling of force has been attributed to logic, but according to Wittgenstein there is nothing supernatural occurring in logic or mathematics. When he rhetorically asks, 'In what sense is logic something sublime?' the answer is clearly 'in no way', and instead he presents the idea that mathematics is essentially collective agreements about rule-following in connection to specific symbols (Wittgenstein, 1997, p. 42e):

> Let us remember that in mathematics we are convinced of grammatical propositions; so the expression, the result, of our being convinced is that we accept a rule. I am trying to say something like this: even if the proved mathematical proposition seems to point to a reality outside itself, still it is only the expression of acceptance of a new measure (of reality). (Wittgenstein 1978, pp. 162–163)

But is this not a flawed position, as it might suggest that mathematics could then be arbitrarily agreed to mean anything? This arbitrariness is actually a cornerstone in Wittgenstein's interpretation. It highlights that his view of mathematics is as a language that, in its development, is not constrained or dictated by a sublime logic, nature, the universe, or anything else:

> But then doesn't it (mathematics) need a sanction for this? Can it extend the network arbitrarily? Well, I could say: mathematicians are always inventing new forms of description. Some stimulated by practical needs, others from aesthetic needs—and yet others in a variety of ways. And here imagine a landscape gardener designing paths for the layout of a garden; it may well be that he draws them on a drawing-board merely as ornamental strips without the slightest thought of someone's sometime walking on them. (Wittgenstein 1978, p. 99)

The mathematician is an inventor – a poet of the numbers one might say – one that slowly, in co-operation with a collective of other mathematicians, pushes the boundaries of what can be thought of as being rational in terms of measuring the world. Wittgenstein therefore agrees that mathematics is special, but not in the supernatural sense of revealing deeper or hidden dimensions of reality, unlike natural language.

> We feel that mathematics stands on a pedestal – this pedestal it has because of a particular role that its propositions play in our language games. What is proved by a mathematical proof is set up as an internal relation and withdrawn from doubt. (Wittgenstein 1978, p. 363)

What is special about mathematics is that it represents knowledge or grammar that has been established and removed from doubt. The task of the professional mathematical community is to push the limits for the ways in which we can measure reality and, in this process, to resolve any doubts about the rationality of these approaches. In this way the professional community of mathematicians has gradually constructed a grammar that supports science and calculations in everyday aspects of our lives.

This interpretation of mathematics is somewhat different from some of the other social interpretations of symbols of mathematics. In those views, mathematics is seen as resembling the empirical sciences by being based on extremely large amounts of empirical data and is, in principle, fallible, much like theories in physics or biology (as is well-known from the tradition following Lakatos – see, e.g., Hersh, 1998). Wittgenstein disagrees with this interpretation of mathematics. Mathematics does have a deep history based on human experiences, but this does not mean it has been forced upon us in any way by our surroundings. It would be more in line with Wittgenstein's ideas to say that mathematics is a tool in a large language toolbox that, for example, enables us to express empirical statements in the sciences. Mathematics is itself a measure or grammar, rather than the thing being measured.

This toolbox of mathematics has no clear boundaries – it can be about all sorts of measures relating to surfaces or statistics, strange types of numbers, and the many things that we cannot even imagine today that will come about in decades to come. The image of mathematics in this interpretation is one without a centre – mathematical concepts are continuously being developed and enriched by the structures and concepts surrounding them, and concepts of mathematics are constantly being renegotiated in minute details in everyday practices of mathematics users, but sometimes also on a major scale when new types of numbers, or the like, are introduced.

Wittgenstein was opposed to the interpretation provided by the influential logicians of the early twentieth century, including himself in the form of the so-called 'early' Wittgenstein. The outline presented above of mathematical development through gradual playing with the concepts is in stark contrast to his earlier thought patterns about logic and mathematics (Wittgenstein, 1983, first published 1922).

> It is interesting to compare the multiplicity of the tools in language and of the ways they are used, the multiplicity of kinds of word and sentence, with what logicians have said about the structure of language. (Including the author of the *Tractatus Logico-Philosophicus.*) (Wittgenstein, 1997, p. 12e)

The aim of these early twentieth-century logicians was to show that (scientific) rationality could have only one form and that this form was definitely not a human form but something humans had access to (in contrast to animals) through labour or through talent etc. By denouncing his own earlier work in the *Tractatus*, the later Wittgenstein presents a much more vivid and organically developing image of mathematics that is open for new paths and absolutely freed from an axiomatic limitation on how mathematics can develop as we know it from Euclid's *Elements*.

If it's a language?

The above arguments and discussions are all of a historical and philosophical nature and one meets many of them in research communities again and again. Taking the position that mathematics is social through and through, will it really make a difference? That is the question that will be pursued now.

As described in the introduction I will try to imagine what difference could be associated with our mathematical practices. I am deeply inspired by the Wittgensteinian interpretation of mathematics; however, in the following I will go far beyond what Wittgenstein (or Shanker) could be held accountable for. I will delve into two aspects of what I shall call a 'human mathematics'. The first is the positioning of mathematics in the landscape of sciences in universities. The other aspect relates to the many learning situations that could be directly influenced by the social interpretation of mathematics.

The position of mathematics in the university

Let us imagine that university faculties were up for reconstruction. Where would mathematics fit within the new landscape? We have seen how mathematics was historically tied to the description of the universe and nature and therefore placed close to physics and other natural and technological sciences.

Considering a human mathematics reconstruction of the university the positioning of mathematics could be quite different. The myriads of possible uses of mathematics today are related not only to the description of nature and the universe but, perhaps even more so, to human affairs and the structuring of the social sphere. This goes for the economy, infrastructure, working hours, tax systems, online presence, traffic, and so on. In many ways, the shift suggests that mathematics in practice also has a tremendous impact on almost all branches of the humanities and social sciences and that human life practices are flooded with numbers and measurements.

According to the human interpretation of mathematics, we should consider mathematics as being equally connected to the humanities, social sciences, and the natural sciences. Some might find this a disturbing or even threateningly invasive approach to repositioning mathematics, but it could also point towards something more fruitful. In the following we can consider different sub-elements of the discussion to qualify the issue.

First of all, there is a very famous cousin of mathematics often positioned in the humanities, namely logic (in accordance with the Trivium disciplines). Logic in many universities has been positioned within philosophy, where it is also closely related to the area referred to as theories of argumentation in philosophy. These studies within the humanities are not initially alienated from what we might call formal systems, including logic, mathematics, programming languages etc. From a Wittgensteinian perspective they could rightfully be called 'grammars' and in this way share family resemblances with the studies of the natural languages.

Second, some parts of the humanities and social sciences are very far from using formulae and mathematical expressions in their practices of research. This might be the case for literature studies, and some language and cultural studies, even though statistics or other applications of mathematics might be used in some approaches of these fields. However, while the scientific approach used in natural sciences focuses on numbers and quantifiable experiments, research in the humanities focuses on qualitative approaches. Humanistic research is never satisfied with counting or measuring but is in essence focused on establishing detailed narratives and rich interpretations about human culture under specific circumstances. This is known from approaches in phenomenology and hermeneutics as well as, for example, in organisational studies from post-structural perspectives. This means that Numbers can never be the focus or goal of all research. Mathematics in itself is a strong and diverse toolbox but it also has immense boundaries to what kind of knowledge and insights it can produce. Using only quantifiable measures in the world in research is only an extreme case of doing science that can reveal some things, but simultaneously it hides a lot of other things. In a possible narrowing of the gap between humanities, social sciences, and mathematics, mathematics must be given a clearer role in the scientific toolbox that holds a myriad of qualitative as well as quantitative approaches and attempts to merge or overlap approaches from these two main categories. To do better research overall, a landscape of scientific approaches much broader than mathematical tools is highly needed for both traditional studies in the STEM area as well as in the humanities and social sciences.

Third, it is interesting to discuss mathematics as an outdated science. I am hinting at the perspective that with the invention of computers – and the use of them in, so to speak, all practices in modern societies – new sciences have been constructed that are closely related to formal languages in new ways. Computer science is the broad term for the many logical studies that bring formal languages closer to practical use, whether it is used in a hardware or software product, or even in the theoretical underpinnings beneath the World Wide Web and other platforms of social interaction. It is from this perspective quite easy to get the idea that mathematics is more than anything else a cultural and historical phenomenon.

Fourth, there are ways in which mathematics portrays family resemblances less with formal approaches but more with the creation of language as we know it from literature and poetry. In the human-oriented interpretation of mathematics, the research component has the task first and foremost of developing new measures to be used in different aspects of life and of demonstrating their practical sensibility. In a way this resembles treading new paths in literature and poetry. Literature and poetry are developed both within well-known schematics or new forms of media that are used to help us think about reality by challenging everyday conceptions or playing with new meanings of concepts. In many ways, this could be thought of in parallel to the developments in mathematics. In this way, mathematics could be understood as a language in constant development, in parallel to literature where we essentially search for new ways of interpreting the world around us.

Finally, the division between the humanities and social sciences, on the one hand, and the natural sciences together with the technological disciplines, on the other, is a well-known issue of two scientific cultures. Often reference is made to the work of C. P. Snow (1993) who discussed how these cultures are in opposition to each other. The argument I will make is that the interpretation and positioning of mathematics is at the centre of this cultural dispute. As long as mathematics is seen as a secure foundation beneath all 'real' (read STEM) knowledge, then the humanities and social sciences will remain in the periphery of what is recognised as truly scientific science. The argument that mathematics is a purely human language formed in connection to the world around us, implies that the humanities are, in fact, a natural destination for thinking about mathematics. Languages have histories and are produced under the pressure of political and social circumstances that need to be understood in order to understand languages and their use – even in the case of formal languages.

This imaginative discussion of the positioning of mathematics leads me to the conclusion that narrowing the gap between the two scientific cultures necessitates a deeper reflection about the nature of science itself. This reflection should incorporate both mathematics as we know it today and qualitatively oriented research approaches in a collective narrative about the diversity of science and scientific research approaches. Wittgenstein describes how mathematics is placed on a pedestal for a specific reason. However, we need a much broader area of expertise than formal languages to be put on that pedestal in order to establish the most insightful knowledge creation and dissemination in universities and beyond.

Learning mathematics from a social perspective

In the landscape of learning theories, a dominant position is the so-called socio-cultural learning theory championed perhaps most forcefully by Etienne Wenger (1999) with the concept of 'communities of practice'. In many ways this theory of learning can be understood as an extension of the social aspects of learning that a human mathematics could propose. According to this theory, key dimensions in a learning process relate to identity building from participation in a community of learners. Each individual needs to travel the distance from the periphery of the social practice to its centre in order to become more and more proficient in the specific practices of the community. Learning in this framework has an extremely high focus on the community of learners as opposed to the individual. To learn something is to become a member of a certain social group and know their ways and behaviours. This is in stark contrast to an idea of teaching and learning that is focused on delivering clear logical packages of knowledge to a sole and rational learner. Instead, the idea is that all knowledge is so incorporated into social practices that learning content itself cannot be disconnected from being an active member of the practice. Learning mathematics is identity construction and is about becoming enculturated into the practices of mathematics.

In everyday school practices the social understanding of learning and mathematics will inevitably entail a strong emphasis on participating actively in mathematical practices. It is crucial to speak the language oneself in co-operation with fellow students and guided by teachers who are individuals that carry the social practices of mathematics as part of their identity. This way of thinking about mathematics learning and teaching is therefore alienated from an approach that tells students the 'result', so to speak, on a blackboard, based on the hope that a logical ability located in the skull of the individual will give them an 'aha' experience about the right way to prove or calculate something. The direct route to learning a particular part of mathematics is involvement in the actual practice of transforming chalk while discussing and evaluating with peers and strong community members.

One principle that should be highlighted is what I will term the *fog of mathematics* (drawing on the idea of the 'fog of war' in many computer games, where only parts of the map are visible at any given moment – the rest of the map is hidden in fog, and you can only guess who is where). Under the social interpretation of human mathematics outlined above, students have no access to a logical faculty of the brain or something of the sort. We have touched upon the example of pointing a finger in a certain direction in parallel to the situation where you need to 'add 2'. For the student unfamiliar with adding 2 ad infinitum this could mean many things. At first, the teacher may experience that things are going as planned – 2, 4, 6, ... – but then when the student reaches 20, she starts to add 2 twice – 18, 20, 24, 28, ... etc. There is absolutely nothing except the community of practitioners that can tell how the fog of war should be cleared. When one first enters a new practice, only imagination and familiarity with similar practices can advance understanding beyond solitary efforts. The only and final test of truly grasping a concept or practice lies in how it stands up to scrutiny and feedback from the experienced language users in the mathematics community.

Wittgenstein's argument on this topic is that no sign itself holds information about its own meaning. Even the simplest of signs in mathematics like '1', '2', '3', ... are completely open for interpretation in so far as a community has not clearly stated how to proceed and, even then, there will be millions of possibilities for misinterpretation of the proposed decided meaning of a concept or symbol.

In this way the human interpretation of mathematics deletes any notion of the contemplative approach to learning mathematics or studying mathematics. Mathematics is not located in the individual. Instead, being capable of doing mathematics means to be able to participate in communities of uses of different kinds.

This participation in a mathematical community points towards what can be called the principle of a *centreless mathematics* (in the sense of there being many equally important and complex mathematical practices). One community of mathematics is located in first grade, and another is found in a discussion on vector spaces at an international conference. These communities have family resemblances in the way they possibly share some symbol transformation, argue by writing on the blackboard, or present mathematical themes to their peers. But they are all also obviously incredibly different. Wittgenstein tries to highlight different practices that we relate under the same area or concept as for example 'mathematics' as having family resemblances (Wittgenstein, 1997, p. 32e). In this way doing mathematics is something that is deeply dependent on the context in which it is conducted.

Even in the much more closely related communities of mathematical practice, such as first grade mathematics and seventh grade mathematics, the concepts and the use of symbols do not have the same meaning. In first grade, '2' is the number focused on. In the seventh grade, the meaning of '2' has been developed into '+2' because it has been incorporated into a context including negative numbers. The meaning of mathematical concepts in this sense differs across the many practices where they are used even in the fairly similar contexts of school classes. And remember that in the interpretation of human mathematics there is no 'real' version or story about the number '2'. The meaning of '2' in the first-grade classroom is just as valid and just as valuable as the meaning of '2' in the international conference room. There is epistemologically and ontologically no true '2' to gravitate towards. You might see the conference '2' as more complex or further developed or more precise but there is no non-human reality to measure against, and this puts mathematical practices on an even footing, ontologically speaking.

Summary

In this chapter I have discussed several aspects about mathematics under a human mathematics interpretation. I have explored reconfiguring mathematics in relation to its position in the broader landscape of sciences and its influence on how we think about learning mathematics. The discussions are obviously only initial steps towards reshaping our notion of mathematics. They are also connected to many other discussions, for example to the problems in academia and beyond of putting quantitative research on a pedestal.

Another connected discussion relates to what, in a Danish context (the author's main frame of reference), is referred to as the distinction between the hard and the soft sciences. This normative description of sciences is used without any hesitation far too often. This chapter is also an attempt to reflect on what constitutes 'soft' versus 'hard'. According to my interpretation of human mathematics, disciplines like mathematics, physics, and technology may be considered a lot 'softer' than typically assumed, especially in the context of funding allocations between STEM fields and the humanities and social sciences. In fact, 'soft' might actually encompass some of the most challenging aspects of both life and research, following a human mathematics interpretation. We need a more balanced scientific landscape that will make dichotomies like these irrelevant and here our understanding of mathematics plays a key role.

References

Candlish, S. (1998). Private language argument. *Routledge Encyclopedia of Philosophy*. Routledge. https://doi.org/10.4324/9780415249126-V027-1

Euclid. (1998). *Elements* (D. E. Joyce, Ed.). http://aleph0.clarku.edu/~djoyce/java/elements/elements.html

Galilei, G. (1957). The assayer. In S. Drake (Ed.), *Discoveries and opinions of Galileo* (pp. 229–281). Anchor.

Grane, L. (1991). *Københavns Universitet 1479–1979: Vol. 1. Almindelig historie 1479–1788*. Gads.

Hersh, R. (1998). *What is mathematics, really?* Vintage.

Høyrup, J. (1996). The formation of a myth: Greek mathematics – our mathematics. In J. Ritter, C. Goldstein, & J. Gray (Eds.), *L'Europe mathématique: Mathematical Europe* (pp. 103–119). Maison des Sciences de l'Homme.

Kepler (1596). *Mysterium cosmographicum* [The mystery of the cosmos]. Georg Gruppenbach. https://doi.org/10.3931/e-rara-445

O'Connor, J. J., & Robertson, E. F. (1999). *Euclid of Alexandria*. https://mathshistory.st-andrews.ac.uk/Biographies/Euclid

Pedersen, O. (1979). *Studium generale: De europæiske universiteters tilblivelse* [Studium generale: The emergence of the European universities]. Gyldendal.

Plato. (2022a). *The Republic* (B. Jowett, Trans.). http://classics.mit.edu/Plato/republic.7.vi.html

Plato. (2022b). *Timaeus* (B. Jowett, Trans.). http://classics.mit.edu/Plato/timaeus.html

Ravn, O., & Skovsmose, O. (2019). *Connecting humans and equations: A reinterpretation of the philosophy of mathematics*. Springer. https://doi.org/10.1007/978-3-030-01337-0

Ravn, O., & Skovsmose, O. (2020). Mathematics as measure. *Revista Brasileira de História da Matemática*. https://www.rbhm.org.br/index.php/RBHM/article/view/293

Shanker, S. (1987). *Wittgenstein and the turning-point in the philosophy of mathematics*. Croom Helm.

Snow, C. P. (1993). *The two cultures*. Cambridge University Press.

Wenger, E. (1999). *Communities of practice: Learning, meaning, and identity*. Cambridge University Press.

Wittgenstein, L. (1983). *Tractatus logico-philosophicus*. Routledge & Kegan Paul.

Wittgenstein, L. (1967). *Remarks on the foundations of mathematics*. Blackwell.

Wittgenstein, L. (1978). *Remarks on the foundations of mathematics*. Blackwell.

Wittgenstein, L. (1997). *Philosophical investigations*. Blackwell.

9. The case of Ramanujan: Investigating social and sociomathematical norms outside the mathematics classroom

Felix Lensing

Ever since mathematics education research has 'divorced' from the discipline of mathematics and set out to become a discipline in its own right, there has been a constant debate about what can and should be understood by mathematics education research. In this chapter, I start from the assumption that mathematics education research necessarily takes a 'reflexive stance' towards its objects of study: mathematics education research is not simply engaged with mathematics, but rather with the engagement with mathematics. It investigates the complex interplay of bodily, cognitive, and social processes that are involved in the genesis of mathematical knowledge – especially (but by no means only) when this genesis occurs in educational contexts. Against this background, I will examine the particular role that the distinction between social and sociomathematical norms may play in the empirical study of the social aspects of this genesis. To do so, I will proceed in two steps: I will first detach the distinction between social and sociomathematical norms from its 'conceptual tie' to mathematics classroom practice. Then, I will use the famous correspondence between mathematicians Srinivasa Ramanujan and G. H. Hardy as an example to show how the distinction may offer a fresh perspective on mathematical practices outside the mathematics classroom.

©2024 Felix Lensing, CC BY-NC 4.0 https://doi.org/10.11647/OBP.0407.09

Introduction

In the second half of the last century, mathematics education research emancipated itself from the discipline of mathematics and set out to become a research discipline in its own right. An important insight that paved the way for this emancipation was the recognition that it is not mathematics itself, but rather the doing of mathematics, encompassing mathematical activity in all its different forms and contexts, that constitutes the field of study of mathematics education research. The mathematics education researcher does not simply see what those who participate in mathematical practices see, he or she does not focus on mathematical objects and their manifold relations, but rather examines the underlying 'processes of objectification' (Radford, 2013), that is, the processes in which these very objects and relations are constituted in the first place. As a consequence, mathematics education research does not produce mathematical knowledge, but knowledge about the production of mathematical knowledge. It does not, for instance, formulate and substantiate knowledge claims about mathematical objects, but it seeks to better understand the bodily, cognitive, and social conditions of these formulations and substantiations. It could perhaps be said that mathematics education research facilitates a reflection of mathematical practice upon itself. And it is, of course, particularly interested in mathematical activities as they take place in educational contexts. Once one adopts this 'reflexive stance' and no longer focuses only on the mathematics but rather on the bodily, cognitive, and social processes that underlie it, a whole new field of inquiry opens up. Now all sorts of extra-mathematical factors come into view that regulate these processes and thus also influence what 'comes out' as mathematics in the end.

In this chapter, I want to show how the distinction between social and sociomathematical norms (Voigt, 1995; Yackel & Cobb, 1996), a conceptual tool originally designed for analysing mathematical classroom practice, can be used to examine some of these extra-mathematical factors. Taking the social practice of mathematical research as an example, I will attempt to show that said distinction is also appropriate for the analysis of mathematical practices outside the educational context. Such an analysis, however, requires a generalisation of the distinction between social and sociomathematical norms. So, before it can be applied to all kinds of mathematical practices, the distinction must first be detached from its 'conceptual tie' to the mathematics classroom. In order to achieve this, I will begin with some theoretical considerations concerning the question of what can be understood by norms in general (Section 2). Then, I will introduce the distinction between social and sociomathematical norms as a further subdivision in the realm of norms, thus removing the restriction of the distinction to mathematical classroom practice (Section 3). Finally, I will take the famous correspondence between mathematicians Srinivasa Ramanujan and G. H. Hardy from the beginning of the last century (Berndt & Rankin, 1997) as an example to show how the distinction between social and sociomathematical norms may offer a fresh perspective on mathematical research practice (Section 4).[1]

On the concept of norm

In the attempt at pinpointing the concept of norm, one will inevitably be faced with the problem that norms appear in the most diverse forms. There are cultural, legal, political, educational, linguistic, industrial, and moral norms, to name just a few. But what is the pattern that connects? What, for instance, do linguistic norms have in common with industrial norms? And what do these two share with moral norms? A common answer to these questions is: Whether linguistic, industrial, or moral, all these norms determine the way in which certain other things should exist. Norms do not say how things *are*, but how they *ought to be*. Linguistic norms dictate how signs of a language *ought to be* used, industrial norms prescribe how products *ought to be* manufactured, and moral norms determine how we *ought to be* acting. It is quite tempting to simply define norms by the factual presence of this peculiar 'ought character': Whenever one comes across something that determines how something else *ought to be*, let's call it a norm. Such definition is of course

1 Note that the aim of this chapter is not to reconstruct the story of Srinivasa Ramanujan's life, but to learn something general about the practice of mathematical research from the individual case of Ramanujan. What aspects of the *person* Ramanujan are relevant to my analysis, and thus what constitutes the *case* of Ramanujan (in the sense intended here) will become clear over the course of this chapter.

possible and also frequently being used.[2] But it leaves the social genesis as well as the social function of norms unexplored. It leaves unexplored how and under which circumstances (e.g., in which social relations of power and control) norms acquire their peculiar 'ought character' and what is gained thereby.

Niklas Luhmann, hence, has proposed to define the concept of norm in a different way (Luhmann, 1995, pp. 319–325). He begins with a more general concept – that of mutual expectation – and then asks: What is the essential quality that is added when a mutual expectation becomes a norm? In what way is it altered by its normalisation? His surprising answer is: not at all. Whether or not a mutual expectation is a norm, cannot be decided by any analysis – however detailed – of its qualities. Rather, it depends exclusively on how the mutual expectation is treated in the case of its disappointment: while mutual expectations of *cognitive character* are abandoned or, at least, altered in case of their disappointment, *normative* ones are being retained even when disappointed (see Luhmann, 1995, pp. 320–321). The normativity of a norm lies in its counterfactual stabilisation: whether or not the world events correspond to it, the norm is left unchanged. Normative expectations have a sort of 'built-in safeguard' that prevents them from being modified. It can thus be anticipated what to do in case of their disappointment, namely: hold on to them. From this analysis it follows that the peculiar 'ought character' of norms is merely a consequence of a more fundamental property, that of counterfactual stabilisation. Norms specify how something ought to be because factual violations have no consequences on them, i.e., do not lead to their alteration.

What this analysis has not yet addressed is the question of what norms are for: What is their social function? Which social problem is solved by protecting mutual expectations against their alteration? Luhmann's answer to those questions is: through the technique of normalisation, even highly uncertain expectations are able to obtain social validity. If one appeals to norms, then one can assure in the here and now 'that one will not be left helpless by disappointment or reveal oneself as someone who simply does not know the world and harbored false expectations'

2 For example, Hans Kelsen (1959) writes: 'Now, what is a norm? A norm is a specific meaning, the meaning that something ought be, or ought to be done, although actually it may not' (p. 107).

(Luhmann, 1995, p. 320). Instead, the trajectories for how things might continue are already clearly mapped out.

An example may illustrate that: a common norm in mathematics classrooms is that students ought to pay close attention to class. But teaching experience shows over and over again that this mutual expectation is being disappointed. Despite this obvious uncertainty of expectation, though, normalisation allows teachers and students to be prepared for these events of disappointment: Teachers, for instance, may think of disciplinary measures and, in addition, can be sure in advance to be able to justify having taken those measures. Likewise, students who are being held responsible for a classroom interruption can assume that it will be sufficient to indicate their readiness to reinstate the very norm they have just disappointed. All that is required is an apology after the fact and accepting imposed measures to rehabilitate the violated norm. Even denying a norm violation – or at any rate, its personal attribution ('Gee, but it wasn't me, Mrs. Baker') – ultimately only confirms the violated norm and thus serves to reinforce it.

On the distinction between social and sociomathematical norms

Now that I have discussed some of the aspects that characterise norms in general, I want to introduce a further subdivision into the realm of norms, namely the distinction between *social* and *sociomathematical norms*. Whether in family life, educational contexts, or mathematical science, whenever a *mathematical practice* arises two types of norms can be delineated within the norms that govern the behaviour occurring in that practice: 1) those norms that regulate the behaviour with reference to its mathematical content, and 2) those norms that regulate it without such reference. While I will refer to the first type of norms as *sociomathematical norms*, I will call the second type *social norms*. To take up the above example: the norm that students ought to pay close attention to class is a *social norm* because it regulates the classroom practice without any reference to its mathematical content. It defines a general boundary between legitimate and illegitimate behaviour in the classroom, albeit a kind of 'generalised' one which is valid not only within the mathematics classrooms but across all school subjects. In contrast, the question of what counts as a mathematical argument in a particular classroom, for instance, refers to a *sociomathematical norm*. This question can only be answered by recourse to the mathematical content as it is thematised in classroom communication.

The distinction between social and sociomathematical norms was originally introduced as a conceptual tool to investigate norms in school mathematics classrooms (Voigt, 1995; Yackel & Cobb, 1996).[3] Later, it was also used to seek for normative orders in mathematics education contexts at university level (Yackel et al., 2000). With my determination of the distinction above, however, I am aiming at giving up its 'conceptual tie' to the educational context altogether. Naturally, mathematics instruction at all different educational levels remains a potential field of application for the distinction; but I am convinced that, in principle, any mathematical practice can be examined for its social and sociomathematical norms. In the remainder of this article, I will support this conviction with an exemplary analysis of some social and sociomathematical norms of mathematical research.

Before I turn to this exemplary analysis it is, however, necessary to highlight an important methodological implication from the preceding theoretical considerations: if norms can be characterised by their counterfactual stabilisation, then situations in which they are violated are of particular interest in reconstructing norms. This point was also highlighted by Anna Sfard:

> A norm becomes explicit and most visible when violated. Violation evokes interlocutors' spontaneous attempts at correction, often accompanied by a condemnation of the transgressor's illegitimate behavior. (Sfard, 2010, p. 204)

In short, anyone who wants to investigate the social and sociomathematical norms of a particular mathematical practice should be looking for situations in which a norm violation occurs.

---

3 To be more precise: For Cobb, Yackel, and colleagues, the distinction played more than a purely analytical role; from the outset, it was linked to questions of how to develop new forms of mathematics instruction. As a consequence, they were, for example, also concerned with the question of how to give social validity to certain norms to which they wanted to orient the instruction in their project classrooms. In this chapter, however, I am primarily concerned with the distinction as an analytical tool for the empirical study of mathematical practice.

The case of Ramanujan

An extreme case of norm violation occurred at the beginning of the last century. In January 1913, Godfrey Harold Hardy, at the time a professor at the University of Cambridge and one of the leading mathematicians in the fields of calculus and number theory, received a letter from a young Indian mathematician named Srinivasa Ramanujan. The letter began with the following words:

> Dear Sir,
>
> I beg to introduce myself to you as a clerk in the Accounts Department of the Port Trust Office at Madras on a salary of only £20 per annum. I am now about 23 years of age. I have had no University education but I have undergone the ordinary school course. After leaving school I have been employing the spare time at my disposal to work at Mathematics. [...] I have made a special investigation of divergent series in general and the results I get are termed by the local mathematicians as ›startling‹. [...] I would request you to go through the enclosed papers. Being poor, if you are convinced that there is anything of value I would like to have my theorems published. I have not given the actual investigations nor the expressions that I get but I have indicated the lines on which I proceed. [...]
>
> I remain, Dear Sir, Yours truly,
>
> S. Ramanujan (Berndt & Rankin, 1997, pp. 21–22)

Ramanujan was a mathematical genius without any direct exposure to the specialised culture of European mathematics ('I have had no University education'). The 'enclosed papers' consisted of nine densely written pages on which Ramanujan presented a selection of his mathematical findings. He had arrived at his – as it later turned out, groundbreaking – findings mainly through self-study.[4] Only a few mathematics books served him as a base (Berndt & Rankin, 2000). Ramanujan's explicitly formulated goal was to publish his mathematical findings, probably also to earn some money ('Being poor, if you are convinced that there is anything of value I would like to have my theorems published'). Ramanujan's position as 'mathematical outsider', that is to say, his role as someone who had barely experienced guided forms of mathematical enculturation, makes this case an ideal object

4 Accounts of his life can be found in Rao (1998) and Kanigel (1992).

of study.[5] Because who could commit greater norm violations than someone who has encountered prevailing norms only implicitly, namely through the study of a few selected books?

Hardy replied to Ramanujan's letter the following month:

> I was exceedingly interested by your letter and by the theorems which you state. You will however understand that, before I can judge properly of the value of what you have done, it is essential that I should see proofs of some of your assertions.
>
> Your results roughly seem to fall into three classes:
>
> 1. there are a number of results which are already known, or are easily deducible from known theorems;
> 2. there are results which, so far as I know, are new and interesting, but interesting rather from their curiosity and apparent difficulty than their importance;
> 3. there are results which appear to be new and important, but in which almost everything depends on the precise rigour of the methods of proof which you have used. (Berndt & Rankin, 1997, p. 46)

The short excerpt of Hardy's letter shows that in the presentation of his findings Ramanujan had violated several norms at once: some of his findings were not new or, at least, easily derivable from known theorems. His mathematical results were missing proofs. And Ramanujan himself did not seem to know which of his findings were merely interesting, and which were of great mathematical importance.

I am now going to consider these three aspects with regard to the social and sociomathematical norms that regulate the acceptance of mathematical findings for publication: I will deal with the norm of mathematical novelty first (a), then turn to the question of what counts as a valid result in mathematics (b), and finally deal with the question of how a particular finding can obtain mathematical importance or significance (c).

On the novelty of mathematical findings

5 Note that it is this one 'abstract' aspect of the *person* Ramanujan (his being a 'mathematical outsider' in the sense described above) that is relevant to the analysis conducted here, namely, to the reconstruction of some of the social and sociomathematical norms being valid in mathematical research at the turn of the century.

A major requirement for mathematical findings in order to be accepted for publication is *mathematical novelty*. Findings that are published in mathematical journals should neither be already known nor should they be direct consequences from what is already known ('there are a number of results which are already known, or are easily deducible from known theorems'). Novelty practically means to disappoint expectations. For something new to emerge, one must deviate from the paths already walked in the epistemic processes. Without such deviation from the expected, no new mathematical knowledge can evolve. By elevating mathematical novelty into a necessary condition for publication, mathematics embraces the unexpected. In a sense, it forces itself to learn. Mathematics cannot reject mathematical findings because it does not know anything about them yet. If new mathematical findings arise, then mathematics is compelled to expand its knowledge. The boundary between the known and the unknown is redrawn with every mathematical publication. In this successive advancement of knowledge, not only mathematical knowledge increases, but also what is yet mathematically unknown. Each newly developed mathematical theory leads to further mathematical problems. Every solved mathematical problem generates a multitude of resultant problems. As David Hilbert (1902) once put it: 'It is by the solution of problems that the investigator tests the temper of his steel; he finds new methods and new outlooks, and gains a wider and freer horizon' (p. 438).

Since any decision concerning the novelty of a mathematical finding can only be made on the basis of the current state of mathematical knowledge, mathematical novelty is a sociomathematical norm. However, this sociomathematical norm is based on a social norm: it holds true for science in general that it forces itself to shift the boundary between the known and the unknown in consideration of new findings. That mathematics aims at surprising itself is thus a norm that it shares with other sciences.

On the validity of mathematical findings

However, their mathematical novelty is not enough for mathematical findings to be published. Another important question is how to decide upon the validity of mathematical results.[6] To answer this question, first hints can be drawn from the excerpt of Hardy's letter. Hardy points out that he can only make a final judgment about the scientific value of Ramanujan's mathematical work with access to the proofs for his findings ('You will however understand that before I can judge properly of the value of what you have done, it is essential that I should see proofs of some of your assertions').[7] It becomes evident from his choice of words that Hardy is not referring to a mere personal expectation here, but to a norm generally valid in mathematics. By prefacing his argument with the phrase 'You will however understand that', it becomes clear that Ramanujan is expected to be able to rehabilitate the norm he violated. The said norm can perhaps be formulated as follows:

> (1) Whether a mathematical finding is valid or not ought to be decided on the basis of mathematical proof.

First, it must be emphasised that this norm is an evolutionary achievement of mathematics: by no means has it always been the case that the validation of a mathematical finding had to be carried out on the basis of proof.[8] Moreover, a comparison with other scientific disciplines

---

6 This question, of course, is at the heart of the traditional understanding of philosophy of mathematics, or more precisely: epistemology of mathematics, and one could fill entire libraries with books written on the question of the justification of mathematical knowledge. Hence, in the following I will limit myself to only those few aspects that appear in the correspondence, and I ask the reader's indulgence for falling far short of the level of discussion reached in the philosophy of mathematics. Indeed, my goal in this chapter is not so much to contribute to this discussion, but simply to show that the distinction between social and sociomathematical norms can be used as a conceptual tool in the empirical study of mathematical research practice.

7 Comparing Hardy's response with Ramanujan's letter, it is noticeable that the term 'value' has undergone a subtle semantic transformation: while in Ramanujan's letter 'value' also seems to be linked to an economic aspect ('Being poor, if you are convinced that there is anything of value I would like to have my theorems published'), this aspect no longer appears in Hardy's answer (see above). Here, the term 'value' seems to be used solely in the sense of 'scientific value'.

8 For example, one reads in Kleiner (1991) about Babylonian mathematics: 'Babylonian mathematics is the most advanced and sophisticated of pre-Greek mathematics, but it lacks the concept of proof. There are no general statements in Babylonian mathematics and there is no attempt at deduction, or even at reasonable explanation, of the validity of the results. This mathematics deals with specific problems, and the solutions are prescriptive – do this and that and you will get the answer' (p. 292). For a more comprehensive account of Babylonian mathematics, see also the work of Jens Høyrup, particularly Høyrup (2002).

shows that this norm is a sociomathematical norm. It is true for science in general that findings always require a scientific justification. Just like in the case of mathematical novelty, the sociomathematical norm refers to a more general social norm. But there are very few disciplines besides mathematics in which proof plays such a prominent role in the context of justification. If one asks for the validity of a mathematical finding, one is thus referred to a mathematical proof. But if it is the proof upon which the validity of the mathematical finding rests, then the question arises as to when a mathematical proof can be considered as valid. The sociomathematical norm as expressed in (1) may thus be specified as follows:

> (2) Whether a mathematical finding is valid or not ought to be decided by examining the validity of the associated mathematical proof.[9]

But what are the criteria for a mathematical proof to be valid? The answer can only be: it depends. Even a brief look at the history of mathematics leads to the conclusion that the criteria for validity of a proof have changed again and again in the socio-cultural evolution of mathematical research (Calude et al., 2003; Chemla, 2015; Kleiner, 1991; MacKenzie, 1999). History of mathematics is rife with 'incomplete' proofs that were 'completed' by mathematicians of a following generation, only to be exposed as incomplete again and so on.[10] But if leading mathematicians of any generation repeatedly come up with incomplete proofs, this can only be a sign that the underlying validity criteria change over time.

---

9 This could be a starting point for historical studies that reconstruct the transformation of the validity criteria for mathematical proofs over the course of time. For the purposes of this analysis, however, such a purely formal description of the sociomathematical norm shall suffice.

10 As an example may serve the proof history of the Fundamental Theorem of Algebra. The Fundamental Theorem of Algebra states that every non-constant single-variable polynomial with complex coefficients has at least one complex root. With regard to its proof history, Kline (1990) says: 'Proofs offered by d'Alembert and Euler were incomplete. In 1772, Lagrange, in a long and detailed argument, 'completed' Euler's proof. But Lagrange, like Euler and his contemporaries, applied freely the ordinary properties of numbers to what were supposedly the roots without establishing that the roots must at worst be complex numbers. Since the nature of the roots was unknown, the proof was actually incomplete. The first substantial proof of the fundamental theorem, though not rigorous by modern standards, was given by Gauss in his doctoral thesis of 1799 at Helmstädt' (p. 598).

> The notion of proof is not absolute. Mathematicians' views of what constitutes an acceptable proof have evolved. [...] The validity of a proof is a reflection of the overall mathematical climate at any given time. (Kleiner, 1991, p. 291)

Based on the previous considerations, this metaphorical description of validity criteria as 'a reflection of the overall mathematical climate at a given time' can be further clarified: validity criteria are consolidated as sociomathematical norms on the ground of the ongoing acceptance and rejection of mathematical findings in mathematical publication practice. The normative requirements that must be met in the presentation of mathematical findings are not imposed on mathematics from the outside. Rather, they arise from within. It is mathematics itself that writes the norms to the sky that guide the publication process. These norms do not have an absolute character, but their normative character means precisely that they can only be changed in the longer term. And only because these expectations are always already found as valid norms by every mathematician who wants to participate in mathematical communication, mathematicians can also use them as a sort of self-control device. Hardy, for instance, can expect himself to have certain expectations about the presentation of mathematical findings by other mathematicians only because the relevant norms have already acquired validity in the social practice of mathematics. Regular participation in this social practice (e.g., the reading, writing, and reviewing of mathematical papers) leads to socialisation effects in the minds of participants. In this way, mathematicians learn what is expected from them when they present their mathematical results to other mathematicians and vice versa.

On the importance of mathematical findings

The acceptance of a mathematical finding for publication is thus conditioned by at least two normative aspects: first, the finding must be a mathematical novelty, and second, it must be accompanied by a valid mathematical proof. There is, however, a third aspect mentioned in the passage of Hardy's letter that influences the publication process: the *mathematical importance* of a finding. In his classification of Ramanujan's new mathematical results, Hardy distinguishes between merely interesting and important results ('there are results which, so far as I know, are new and interesting, but interesting rather from their curiosity and apparent difficulty than their importance'). There are countless new and interesting truths in mathematics, but only a few of them are also mathematically important. But what exactly does this mean? What is the mathematical importance of a finding? 'We may say, roughly, that a mathematical idea is ›significant‹ if it can be connected, in a natural and illuminating way, with a large complex of other mathematical ideas' (Hardy & Snow, 2004, §11). Hardy thus argues that the yardstick for the mathematical importance of a mathematical finding is therefore its 'mathematical connectivity'. The greater the number of mathematically important ideas to which a certain finding can be connected, the greater its mathematical importance. Whether a particular finding is mathematically important or not, thus, can often only be decided in retrospect. It depends on how and to what extent further mathematically important results can be connected to it. While the novelty or validity of a mathematical finding can already be judged with a certain degree of certainty in the here and now, many times its importance becomes apparent only in the future. Mathematical results will have been important. They often acquire their importance only from a certain point in the future, from which it becomes clear that they were the basis for a multitude of further mathematically important results. This inscribed reference to the future prevents the value of mathematical importance from becoming a necessary decision criterion in the publication process. Concrete examples are the works of Évariste Galois on the theory of polynomial equations, Hermann Grassmann's grounding of what was later called linear algebra, or Gottlob Frege's founding of modern logic, all of which had in common that they were hardly noticed, let alone appreciated, by their contemporaries.

With such an analysis, however, the relation between mathematics and time is still insufficiently grasped. Mathematics constantly projects findings from the present into the future. Mathematical relationships are permanently explored on the basis of hypotheses. If a given hypothesis were true, then this set of propositions could be derived from it. This way, in many cases one can already know in the present that the proof of a certain mathematical hypothesis in the future would be of greatest mathematical importance. Whoever solves one of the so-called millennium problems, for example, is guaranteed mathematical fame. This 'anticipated' importance of a mathematical finding has an influence on the publication process. Although mathematical importance cannot be elevated to a necessary condition for publication due to its immanent reference to the future, it can, at least, influence where (i.e., in which mathematical journals) a particular result is published. In contemporary mathematics it makes a considerable difference whether a finding is published in the *Annals of Mathematics* or in the *Mathematische Annalen*. Mathematics introduces a rank order within its field of acceptance: it establishes a hierarchy of mathematical findings through the distribution of publications among the various journals.

Conclusion

I have set out to show that the distinction between social and sociomathematical norms can shed light on some of the extra-mathematical factors involved in the production of mathematical knowledge. For this purpose, I took a series of steps: I first criticised the common practice of defining norms by their peculiar 'ought character' and argued that norms are better understood as a specific kind of mutual expectation. Norms differ from all other kinds of mutual expectations in that they are retained in cases of disappointment. This characterisation does not simply replace the 'ought character' of norms but explains it. If a mutual expectation is counterfactually stabilised, i.e., if factual violations do not lead to a norm's modification, then one is able to know in advance (and independently of what is actually done) what is ought to be done. Based on these theoretical considerations about norms in general, I then introduced the distinction between social and sociomathematical norms: while sociomathematical norms are those norms of a mathematical practice that regulate the participants' behaviour with reference to some mathematical content, social norms do so without such reference. Compared to the original conception of this distinction, which limited its scope to educational contexts, this characterisation is an attempt at extending the distinction to all kinds of mathematical practices. This extension was based on the following assumption: while there may be significant differences between different mathematical practices in terms of *which* social and sociomathematical norms are established, *that* such norms emerge is a universal feature that is common to all mathematical practices. To support this assumption, I then analysed the correspondence between Ramanujan and Hardy and showed that social and sociomathematical norms are 'active' not only in mathematics classrooms but also in mathematical research practice. By analysing the role that sociomathematical norms of mathematical novelty and validity play in the evaluation of mathematical findings, I showed that social and sociomathematical norms are often intertwined with each other without, however, coinciding. It is, for example, a social norm common to all scientific disciplines *that* findings are in need of justification, but *how* they are to be justified is governed by sociomathematical norms specific to mathematical research practice. This peculiar relationship was also noted in the context of mathematics education. Erna Yackel and colleagues (2000), for instance, provide the following examples:

> The expectation *that* one is to give an explanation falls within an analysis of social norms, but *what* is taken as constituting an acceptable mathematical explanation falls within an analysis of sociomathematical norms. Likewise, the expectation *that* one is to offer a solution only if it is different from those already offered falls within the realm of social norms, but *what* is taken as constituting mathematical difference falls within the realm of sociomathematical norms. (p. 282)

So we always have, on the one hand, a social norm that says *that* something should be done and, on the other hand, a correlated sociomathematical norm that tells us *how* it should be done. In all of these cases, sociomathematical norms specify social norms for the particular context, and, conversely, sociomathematical norms are 'backed up', so to speak, by more general social norms. But since there are also social norms, such as the norm that students should follow class attentively, that can stand for themselves, that do not require any further specification by a sociomathematical norm, the question arises: under what conditions does this special relationship between social and sociomathematical norms occur? The empirical analyses have led us to the conceptual limits of the distinction; they revealed that further distinctions are needed to account for all facets of the normative orders of mathematical practices.

Moreover, the analysis of the correspondence between Ramanujan and Hardy has also led to the conclusion that by no means all social structures of mathematical research practice are norms. The example of the value of mathematical importance made it quite clear that the distinction between social and sociomathematical norms captures only a small 'section' of the extra-mathematical factors involved in the production of mathematical knowledge. It is thus an important question for further research addressing the distinction between social and sociomathematical norms to focus on the relation of these two types of norms to other kinds of social structures (e.g., to what Sfard, 2010, pp. 200–208, calls 'metadiscursive rules' or what Voigt, 1985, 1995, calls 'patterns of interaction').

References

Berndt, B. C., & Rankin, R. A. (1997). *Ramanujan: Letters and commentary*. American Mathematical Society.

Berndt, B. C., & Rankin, R. A. (2000). The books studied by Ramanujan in India. *The American Mathematical Monthly, 107*(7), 595–601. https://doi.org/10.1080/00029890.2000.12005244

Calude, C. S., Calude, E., & Marcus, S. (2003). *Passages of proof.* ArXiv. https://doi.org/10.48550/arXiv.math/0305213

Chemla, K. (Ed.). (2015). *The history of mathematical proof in ancient traditions*. Cambridge University Press.

Hardy, G. H., & Snow, C. P. (2004). *A mathematician's apology*. Cambridge University Press.

Hilbert, D. (1902). Mathematical problems. *Bulletin of the American Mathematical Society, 8*(10), 437–479. https://doi.org/10.1090/S0002-9904-1902-00923-3

Høyrup, J. (2002). *Lengths, widths, surfaces: A portrait of Old Babylonian algebra and its kin*. Springer. https://doi.org/10.1007/978-1-4757-3685-4

Kanigel, R. (1992). *The man who knew infinity: A life of the genius Ramanujan*. Washington Square.

Kelsen, H. (1959). *On the basic norm. California Law Review, 47*(1), 107–110.

Kleiner, I. (1991). Rigor and proof in mathematics: A historical perspective. *Mathematics Magazine, 64*(5), 291–314.

Kline, M. (1990). *Mathematical thought from ancient to modern times*. Oxford University Press.

Luhmann, N. (1995). *Social systems*. Stanford University Press.

MacKenzie, D. (1999). Slaying the Kraken: The sociohistory of a mathematical proof. *Social Studies of Science*, *29*(1), 7–60. https://doi.org/10.1177/030631299029001002

Radford, L. (2013). Three key concepts of the theory of objectification: Knowledge, knowing, and learning. *Journal of Research in Mathematics Education*, 2(1), 7–44. https://doi.org/10.4471/redimat.2013.19

Rao, S. K. (1998). *Srinivasa Ramanujan: A mathematical genius*. East West.

Sfard, A. (2010). *Thinking as communicating: Human development, the growth of discourses, and mathematizing*. Cambridge University Press. https://doi.org/10.1017/CBO9780511499944

Voigt, J. (1985). Patterns and routines in classroom interaction. *Recherches en Didactique des Mathématiques*, *6*(1), 69–118.

Voigt, J. (1995). Thematic patterns of interaction and sociomathematical norms. In P. Cobb & H. Bauersfeld (Eds.), *The emergence of mathematical meaning: Interaction in classroom cultures* (pp. 163–201). Erlbaum. https://doi.org/10.4324/9780203053140-9

Yackel, E., & Cobb, P. (1996). Sociomathematical norms, argumentation, and autonomy in mathematics. *Journal for Research in Mathematics Education*, *27*(4), 458–477. https://doi.org/10.5951/jresematheduc.27.4.0458

Yackel, E., Rasmussen, C., & King, K. (2000). Social and sociomathematical norms in an advanced undergraduate mathematics course. *The Journal of Mathematical Behavior*, *19*(3), 275–287. https://doi.org/10.1016/S0732-3123(00)00051-1

10. A performative and relational Ethnomathematics

Aldo Parra

This chapter presents a reconceptualisation of the Ethnomathematics research field, as composed of a series of contingent and purposefully constructed relations between mathematics and culture. This reconceptualisation is useful in formulating a non-essentialist positioning on the nature of mathematics without adhering to cultural relativism. An overview of research experiences on Ethnomathematics is made to illustrate how the reconceptualisation emerged and what are its implications and potentialities, particularly on controversial issues for Ethnomathematics, like mathematics ontology, research validation, and the roles of researchers.

Since its inception as a movement, those working on Ethnomathematics have expressed two of their aims: to question the modernist narrative of triumphalism and uniqueness that surrounds academic mathematics, and to address the geopolitical repercussions that such epistemic prestige has had. Ubiratan D'Ambrosio stated: 'We should not forget that colonialism grew together in a symbiotic relationship with modern science, in particular with mathematics and technology' (D'Ambrosio, 1985, p. 47).

The intense history of definitions and redefinitions of Ethnomathematics is also a history of how the field has tried to achieve those two aims, sometimes providing accounts of mathematical knowledge and its nature. This has created a kind of expectancy towards Ethnomathematics to address and solve an ontological question on mathematics. Such an intended definition would provide a clear-cut characterisation of what is (or is not) mathematical, and would allow us to recognise several cultural knowledge(s) and practices

©2024 Aldo Parra, CC BY-NC 4.0 https://doi.org/10.11647/OBP.0407.10

as having significant mathematical elements embedded in them. As Ethnomathematics has not reached such a definition yet, it can be discussed to what extent the field has been successful in contesting the narrative of supremacy of academic mathematics through the methods and frameworks that the field employs.

Criticisms of a philosophical nature have been raised against Ethnomathematics. In particular, it has been accused of not having a clear account of mathematical knowledge, and falling into a cultural relativism (Horsthemke & Schafer, 2007; Rowlands & Carson, 2002). It is natural to ask, therefore, if it is possible to reject an essentialist account of mathematical knowledge without adhering to cultural relativism. To what extent does the recognition of multiple and incommensurable forms of mathematical knowledge entail a 'particularism that precludes the possibility of construction of translocal relations' (Savransky, 2012, p. 358)?

Criticisms of a political nature have been raised as well, particularly about the ways of empowering and dignifying populations being discriminated against and minoritised through mathematics (Pais, 2013; Vithal & Skovsmose, 1997). The debate emerges within the tension between the rights of minorities not to be marginalised and their rights to be treated differentially.

Researchers such as Bill Barton (1996b, 2008), and, more recently, Roger Miarka and Maria Bicudo (2011) have characterised the ways in which representative ethnomathematicians conceptualise mathematics and its relation to the field. Such characterisations not only highlight the diversity of backgrounds and positionings of influential practitioners, but also help to prefigure an ethnomathematical approach towards mathematics that can address expectations and respond to pertinent questions.

In this chapter, I attempt to summarise such an approach for Ethnomathematics,[1] asserting that, in order to succeed in its goal of overcoming modernist accounts of mathematics, the field does not need to produce a stable, finished, non-self-referential, and free-of-doubt definition of what mathematics is. I contend that rather than look for a modernist answer, Ethnomathematics can forge a decolonial answer.

1 A fully detailed version of the theoretical position was presented elsewhere (Parra, 2018).

Decolonial studies meet Ethnomathematics in their concern for the assemblages of knowledge and power. Within this chapter, I formulate a particular understanding of the Ethnomathematics program as a decolonising program, that vindicates performativity, interaction, and non-essentialism as values to be promoted in the quest for new insights on mathematics as a manifold of culture-based practices that change over time. An overview of recent experiences of Ethnomathematics research is provided to exemplify how such values operate.

Mirar y no tocar se llama respetar[2]

During the Q&A session at a Colombian seminar on Ethnomathematics in 2020, a concerned student asked: 'As Ethnomathematics aims to extract the mathematics present in a certain community, is it okay to transform it? Should I change it?' His sincere question reveals an underlying assumption that I want to highlight: a certain apprehension about intervening in ancestral or vernacular knowledge that is seen as mathematical. This fear emerges because the ethnomathematician's regular work is supposed to be a recognition of knowledge, wisdom, and practices of mathematical nature, that occur in non-conventional or non-academic environments, and are collected to be reported within academic agoras. Such regular work is visible in two types of studies intensely practiced in Ethnomathematics.

A first type is composed of studies revealing knowledge and practices of the past, through descriptions of the characteristics and circumstances of a past event or issue (e.g., how land was measured in a certain place or how an object or food was manufactured). Such methodological procedures generate products that resemble necropsies or archaeological reports and therefore the work of the ethnomathematician emulates that of the forensic scientist or archaeologist.

The second type consists of the non-participant observation of practices that are currently occurring in some group or community (e.g., the preparation of a typical meal, the mathematical knowledge of a gardener or a locksmith, or the locating practices of blind people

2 This is a popular saying in Colombia that can be translated as 'respect is about watching without touching'.

or Indigenous children). Here the documents are similar to a chronicle issued by a special correspondent who reports 'from the scene'. Thus, the profession of the ethnomathematician is related to that of the journalist, or the classical ethnographer, who brings news from exotic places.

The two types of studies converge in that the knowledge and practices investigated pre-exist the specific research. That is to say, when the ethnomathematical researcher arrives in a community, the emergence of non-classical mathematical ideas has already occurred within that community and he comes *a posteriori* to identify and record that exceptional practice. The event, which is neither scientific nor academic, is mathematical to some extent, in some particular sense. The researcher comes to admire and *contemplate* epistemological diversity, but not to create or extend it.

This bucolic image of contemplative admiration has been taken for granted by many scholars when describing Ethnomathematics. My task in this chapter is to break that image, by noticing some of its limitations and also by presenting: (i) an alternative interpretation of what kind of action underlies an Ethnomathematics research; (ii) an exploration of what things research could make possible by making such actions deliberate.

In order to contest the perspective that conceptualises non-Western rationalities as obsolete and in need of upgrading, by imagining a modernist *I* and an inferior *Other*, ethnomathematicians look for the recognition and appraisal of types of knowledge coming from diverse worldviews. A distinctive feature of contemplative admiration is the set of ways in which assumed knowledge diversity becomes recognised. According to this admiration stance, Ethnomathematics is performed by a 'civilised *I*' that needs to expand their scientific/positivist notions of mathematics and knowledge. Meanwhile, the *Other* (Indigenous, peasant, etc.) is discouraged to engage in the reciprocal move, because they will be harmed in their way of life when coming into contact with disciplinary practices (formal mathematics). They must be confined in a source of epistemic purity, an otherness that is not polluted with Cartesian categories.

Such muted Others and their sublimated otherness are necessary to claim that ethnomathematicians are 'giving voice', 'recognising wisdom', and 'valuing knowledge'. They are also needed to establish the debates around whether the 'literate *I*' can/should use their mathematical gaze to describe cultural practices and whether the academic audience can trust in their descriptions and representations. What underlies the reflexivity debates unleashed by contemplative admiration is the conception of Ethnomathematics as an academic endeavour, in which local communities gets a social impact due to the in/exclusion led by the scholarly trained ethnomathematician. The latter has the main role, as the one who contemplates, defining the times, methods, goals, and written results of the research, who will represent the targeted otherness and finally grasp the 'emic' perspective, by the ethnographical procedure of 'being there and writing here'.

Besides the reductions of Ethnomathematics to the academy, and of ethnomathematicians to classic ethnographers, contemplative admiration brings us another limitation, namely the framing of the ontology of mathematics. This aspect is also related to the sublimation of otherness and becomes explicit when certain techniques or models are designed to extract and elucidate the mathematical component of a cultural practice (Albertí Palmer, 2007; Uribe Suarez, 2021), as if certain mathematical attributes were hidden in an artefact, ritual, or practice. Therefore, the discussion is ontological, in order to establish whether the studied cultural practice is mathematical or not. No matter if by 'mathematical' we understand a disciplinary object that belongs to a 'near-universal, conventional mathematics' or a notion coming from a local 'system for dealing with quantitative, relational, or spatial aspects of human experience' (Barton, 2008), the central task is to contend that the practice possesses the attribute.[3]

If ethnomathematicians assume that their work is to contemplate an event until they can determine whether it represents 'the mathematics of cultural practices', then they must be very clear which essence they are seeking. Thus, they need to delimitate what qualifies as mathematical knowledge, or, more precisely, what constitutes 'doing mathematics'.

In order to not fall into a quasi-Platonic universalism, some ethnomathematicians have opted to look for an epistemic relativism in which there are several ways of knowing, according to each culture. Then, such ways would be the '(Ethno)mathematics' of the place and

3 This means that the two theoretical views of the Ethnomathematics that Albanese et al. (2017) proposed are just variations of the same contemplative approach.

the field of Ethnomathematics is the study of those many ways of knowing. When trying to explain how different cultures have found similar results, this trend takes the Wittgensteinian notion of 'family resemblances' to avoid a ground zero for mathematical knowledge, and formulates that several forms of life share and communicate some characteristics, and therefore, their ways of knowing can converge in some matters. Although this theoretical displacement is interesting, as it avoids certain forms of essentialisation, it cannot entirely escape a sense of contemplative admiration, since it leaves open the question of how (and why) a family resemblance is noticed. I wonder if Wittgenstein would agree that a resemblance can be discovered.

The last thing to say in this section about contemplative admiration is that even if its two types of studies (archeological and non-participant) seem to represent an important amount of Ethnomathematics research, they hardly cover the vast diversity of approaches and methodologies explored. Many astonishing works (e.g., Alangui, 2010; Borden, 2013; Cauty, 1999, 2001; Knijnik, 1998; Meaney, Trinick, & Fairhall, 2012; Mesquita, 2010; Oliveira, 2013) go far beyond the idea of reporting exotic practices. All of these long-term and well-established research programs share a commitment to working with local communities around mathematics and local knowledge in a creative way that puts communities' interests and agendas face-to-face with academic goals. Interestingly, it is remarkable how critiques of Ethnomathematics do not comment on work of this nature, although authors like Gelsa Knijnik and Mônica Mesquita have been presenting them since the beginnings of Ethnomathematics, and scholars such as Wilfredo Alangui and Tamsin Meaney are increasingly being referenced within the field nowadays. I wonder if these works are left aside because they would be counterexamples to the idea of cultural statism and the sublimation of otherness, which is so much projected onto Ethnomathematics by its critics.

Outsiders at the centre

Considering the handful of experiences that disrupt contemplative admiration, in my search for a different explanation of Ethnomathematics and its affordances, I realise that a salient characteristic has been the human relationship between researchers and the communities over time. Authors and communities become engaged and collaborate in diverse affairs that are often not bound by the limited scope of an isolated research project; they build and sustain a bond not mediated by reports, deadlines, or institutional funding.

This depth of engagement shapes the researcher's perspective in such a manner that the published research constitutes just a brief part of a broader life experience rooted in partnership. In these collaborations, communities seek to develop and produce a variety of outcomes – such as loan applications to banks for some Movimento Sem-Terra (MST, Landless People Movement) peasants, demonstrations advocating for unpolluted water in Costa de Caparica, or school commemorations of ancestors' arrival by boat in Aotearoa – that demand the use of some academic disciplinary mathematics, prompting the scholar to offer explanations. Scholars do not have a god´s eye view, nor can they alone decide the times, spaces, participants, goals, and outputs of the experience. They need to understand each community´s aims, desires, and ways of thinking about the situation of interest and also the official mathematics it embodies; then, they must translate, mediate, and articulate the two types of perspectives in order to collaborate within a collective learning environment that enhances the community's abilities to participate in social and public debates according to their own organisational style and perspective.

An important point to note here is that scholars do not 'represent' or act 'on behalf of' communities. Instead, they must find out a way to collaborate in the creation of a new, organic communitarian knowledge that can effectively embrace the external (institutional) knowledge required by the situation. They are urged to provide neither a mathematical model of a vernacular practice, to be understood by public officers, nor an insider model of the same practice. Accordingly, concepts such as emic/etic or situated mathematical interpretation seem insufficient, as the tasks are not about representing local ideas for a global audience, or the reverse, and not about one individual acting an enlightened medium. Instead, the focus is on about fostering collective learning, igniting local processes of interpretation, discussion, adaptation, appropriation, and mainly *creation*.

It can be many things, but not anything

If contemplative admiration falls short in describing these styles of research and interaction, and the works are definitely fulfilling the ethical, social, and political call of Ubiratan D'Ambrosio, what is it, then, that they are doing with the knowledge and mathematics involved? Well, by trying to answer this question I have found an idea that can be useful to theorise the activity of the ethnomathematical field, in a way that can overcome some of the limitations and address new challenges.

One guiding notion cuts across these experiences. When the Kwibi Urraga Laboratory in Colombia was trying to translate an algebra book from Spanish to an Indigenous language, and when the Maori Language Commission of New Zealand was engaged in creating a dictionary of mathematical terms in Te Reo Maori, they were not creating meanings for words that would express in a complete way all the formal syntax and semantics of the mathematical formal objects they were dealing with. Rather than hoping to solve the epistemic mismatch between a formal discipline and an Indigenous knowledge system, they attempted to build a mathematical register, a trustable framework of communication within which to negotiate meanings. They did not achieve the submission of one worldview to another 'more complete' worldview. Better than that, they succeeded in establishing *relations* among elements coming from diverse knowledge systems.

In a similar manner, when peasants of the Brazilian MST compare and contrast their own techniques for measuring the volume of wood planks with official techniques (Knijnik, 2007), they put their ways of knowing into *relation* with those coming from 'book mathematics'. Not to substitute one technique with the other, but to 'broaden not only their mathematical world, but also their ways of seeing the complex social relations involved in different *forms of life* that produce such different *language games*' (Knijnik, 2007, p. 16).

To summarise, Ethnomathematics builds relations. Relations among institutional disciplinary mathematics and local ways of knowing. Relations among their languages. Relations among knowledgeable people of diverse traditions and places.

Thus, one ethnomathematical research project is an effort to raise, explain, and share relations among certain objects or practices of two different knowledge systems. Relations can be traced from the cultural knowledge system of a group to the realm of mathematics, for instance when Alangui connects Philippine rice terracing practices with a system of ordinary differential equations (Alangui, 2010). Relations can also be proposed from the realm of mathematics to the cultural knowledge system of a group, as in the description of geometrical relations made by Miguel Andrés Gutiérrez (2019) through concepts of Colombian folk dancers.

When researchers formulate a relation, they need to engage in a debate on the plausibility, pertinence, and utility of the relation. Debates of this nature demand the interaction of several stakeholders and scenarios that have the legitimacy to sanction the relation as acceptable or useful. During the examination of the potential relation, new kinds of learning emerge, and re-elaborations and translations are needed (an epistemic re-arrangement). New personal relations among people of diverse backgrounds, values, interests, skills, certainties, and foregrounds are also established (a political re-arrangement). These re-arrangements are the most substantial part of the research process, more important than the original relation proposed. Because what is at stake in the process is the power to change cultural and mathematical practices. When the examination ends, whether or not the relation was accepted or rejected, an educational experience has occurred, expanding the boundaries of what is assumed to be meant by mathematics and culture, at personal and collective levels. In short, we are less interested in the prey (an ethnomodel suspected of being pareidolic) than in the hunt (a deterritorialization that certainly happened).

Contemplative admiration has misled us into thinking that Ethnomathematics was about uncovering hidden attributes, an act of discovery, a quest to answer ontological questions like: Is this cultural practice a mathematical object? Is this mathematical practice legitimately cultural? By contrast, a relational approach knows that relations are neither given nor automatic because they live in the realm of potentiality. Their formulation implies a creative act, a performative challenge of what could happen if we operate with this? What understandings are unleashed by accepting this relation?

After this explanation of the main features of a relational theorisation, I will describe some values associated with it that are useful for addressing the limitations of the contemplative admiration approach.

Re-visiting ontology

As stated earlier when describing the question of mathematical ontology, ethnomathematicians have rejected the metaphysics of Platonist accounts by appealing to a conventionalist approach that leads them to assume cultural/epistemic relativism. The growing interest in Wittgenstein (Albanese, Adamuz-Povedano, & Bracho-López, 2017; Barton, 1999; Knijnik, 2012; Vilela, 2010) proves that appeal. The notions of *mathematical language games* and *family resemblances* have emerged as useful to explain the convergence of different cultures to some mathematical results. Barton gives us a key insight here into the role of relations in his resemblances:

> What happens when different mathematical systems meet? Wittgenstein's answer is that there are no 'gaps' in mathematics. Each system is complete at any moment. It is not waiting to be added to with new mathematics. Thus (Shanker, 1987, p. 329), any connection between two worlds is not in the same space as either of the worlds. The interconnections are not waiting to be discovered. We choose whether or not to make connections between systems, and if we do then the connections create a new system. (Barton, 2008, p. 130)

With this excerpt, Barton helps us to understand how family resemblances among mathematical language games work and what they produce. We identify resemblances because we want to do so, because we have the will to find them, the need to use them, and an interest in making comparisons among seemingly unrelated things. The resultant thing is a new system, an expanded version of mathematics, and/or an enhanced cultural practice. This insight is crucial because it stresses how culture and mathematics are historical, an idea that Luis Radford has also stressed:

> There is no regulatory, universal reason. The reason is historical and cultural. Their specific forms, what Foucault calls *epistemes*, are conditioned in a way that is not causal or mechanical, by its nesting in

> the social and political practices of the individuals. (Radford, 2016, p. 36, italics given)

> The concept of culture that Marx elaborates indirectly in his writings is, in fact, profoundly historical and transformative. Individuals create culture and, in a reverse or dialectical movement, culture offers the conditions for individuals to create systems of thought whether scientific, aesthetic, legal, etc., and to create themselves. That is why, from a materialist dialectical perspective, human cultures are much more than reified and static entities. (Radford, 2014, p. 56, my translation)

So, by noticing the historicity of mathematical knowledge we can accept that relations have the potential to become reconfigured mathematics. However, those relations have constraints, as they do not operate freely in a void. Some of these constraints are noted by Barton and Denise Vilela:

> This does not mean that mathematics is arbitrary, and thereby open the way for mathematical anarchy. We are free to construct the grammatical rules of mathematics, but not 'blindly or capriciously' (Shanker, 1987, p. 319). The arbitrariness Wittgenstein refers to is its autonomy. [...] Cultural mathematics' are not arbitrary in the sense that they could be anyhow. They are arbitrary in that any culture is free to make its own sense of the world. Mathematics is the way it chooses to express that sense. (Barton, 1996a, p. 182)

> In particular, mathematics or Euclidean geometry, as a set of grammatical rules, are applied because these rules must have an empirical origin and became rules, or forms of intelligibility. (Vilela, 2010, p. 352)

Is it possible that the 'empirical origin' and the limits of a 'mathematical anarchy' reside in a non-human realm? In addition, how do we explain the universal human capacity for playing language games? There are issues with the post-metaphysical turn that deserve more elaboration and Ethnomathematics needs to address these questions without a return to essentialism. In fact, André Cauty and Barton already warned about the ontological dead-end:

> We defend a thesis based on the observation of the historical construction of mathematics, as well as on the observation of the epigenetic time of the formation of a mathematician. This thesis prevents us from fully adhering to the most extreme doctrines: idealism and positivism. Therefore, neither do we believe only in the reality of ideas (Conceptus),

> like the too much idealistic doctrines do, nor believe only in the reality of things (Res), as the too much materialistic doctrines do. A classic solution consists in considering a third order of reality, the one of signs (Vox) and representations. That is, to address entities that are neither things nor ideas, but substitutes for references, both imaginary and real. (Cauty, 2001, p. 77, my translation)

> There would be no question about whether they [the mathematical objects] exist independently or about how we come to know them. We mathematise, and therefore we create the objects by our thought, and attempt to communicate them to one another. The ontology and epistemology of mathematics simply is not a problem anymore. (Barton, 2012, p. 228)

At this point, it becomes clear how important it is to find a way to blur divisions between ontology and epistemology and escape the dichotomy of essentialism/conventionalism. Scholars from the trend of Science and Technology Studies have developed some insightful ideas on this matter. Inspired by the Latourian interactions among the human and the non-human, Andrew Pickering introduced the idea of disciplinary agency, assumed as the 'agency of a discipline that leads us through a series of manipulations within an established conceptual system' (Pickering, 2010, p. 115). Such agency interacts with human agency, producing a dynamic of accommodation and resistance among agencies. Then, mathematical knowledge would be the result of that dynamic, explaining why mathematics is neither arbitrary nor predetermined.

This is very useful to Ethnomathematics because it explains how some mathematical results have been known by different groups throughout history. It is not due to the existence of some essence or structure, but rather a result of groups responding similarly to a non-human agency that presented constraints. In the same manner that sculptors working with the same raw material produce different statues, cultural groups approach the metaphysical and produce different mathematics. Just as we do not equate sculptures with rocks, because we can recognise and value the human agency in the resultant sculpture, we should not equate mathematics with the metaphysics. That is the crucial point here.

Aligned with that, an understanding of reality as continuously transforming and becoming allows us to see controversies about diversities among cultural/mathematical practices as examples of *ontopolitics* in which entities and worlds are 'shaped, sustained and transformed by the social, technical and material practices that take place – and make place – in them' (Savransky, 2012, p. 360). If these cultural/mathematical practices are assumed to be inventive practices, then power and knowledge become entangled and human agency cannot abdicate the political responsibility in an uncertain and unstable reality of multiple worldviews colliding and interacting through relations.

Multiplicity and interaction lead Ethnomathematics to the terrain of decoloniality, namely the ecology of knowledges proposed by Boaventura de Sousa Santos (2012):

> Granting credibility to non-scientific knowledge does not imply discrediting scientific knowledge. What it does imply is using it in a counterhegemonic way. This consists, on the one hand, in exploring alternative scientific practices made visible through plural epistemological scientific practices and, on the other, in promoting interdependence between scientific and non-scientific knowledge. (p. 57)

Re-visiting validation

Contemplative admiration is, at the deepest level, an instance of validation: a mechanism by which non-academic and informal practices become certified as mathematical by academic institutions. This is why the Ethnomathematics produced under this contemplative spirit is basically concerned with how the academy can generate conceptions about mathematics that include the cultural and linguistic diversity that societies possess; therefore, its target audience is the academy itself.

The role of scholarly trained ethnomathematicians as validators is highly problematic, since, on the one hand, some of them get confused by the presence of their own disciplinary gaze and struggle to not see 'with their own eyes' (the paradox of Millroy, 1992), while, on the other hand, critics wonder to what extent this certification procedure enthrones even more the modern rationality that the field promised to problematise (Pais, 2013). Such problems are just another enactment of the reflexivity debates deriving from anthropology (Salzman, 2002; Woolgar, 1988), and arise due to the extended use of classic ethnography as the 'natural' method for Ethnomathematics.

Reflexivity issues lose their importance when classic ethnographical methods are problematised and when knowledgeable people, not necessarily working for (or trained in) academic institutions, are considered Ethnomathematics researchers in their own right, intervening and collaborating in each part of the research experience, being accountable for the results of the research. Within a decolonial perspective, insiders' insights can no longer be made invisible or subjected to the realm of 'Not-being', or impersonated through dubious emic representations. As de Sousa Santos states: 'Non-existence is produced whenever a certain entity is discredited, and considered invisible, non-intelligible, or discardable' (de Sousa Santos, 2012, p. 52).

In decolonial studies, the concept of sociology of absences is used to refer to the type of research that unveils the ways in which denial and non-existence is actively produced. A sociology of absences 'amplifies the present by adding to the existing reality what was subtracted from it' (de Sousa Santos, 2012, p. 56), and for our discussion, such amplification is expected to be a regular procedure for a relational Ethnomathematics.

The reconceptualisation of Ethnomathematics unfolded in this chapter is interested in promoting a broader vision of mathematical knowledge within other social contexts. But what is knowledge other than an interconnected system of people, beliefs, values, institutions, and instances that constitute it in a certain time and space? In that sense, to push the boundaries of what is sanctioned as mathematical is an attempt at intervening in such a system, calling into question the exclusivity of some instances (e.g., the academic ones) as being legitimate.

For that matter, it implies that Ethnomathematics needs to conceive itself as accountable to scenarios other than the academy, otherwise it will not be able to make a strong academic reading of the epistemological/political dimensions of mathematics. A broad idea of validation needs to consider new agents, scenarios, and procedures in such a way that Ethnomathematics become a sociology of absences (de Sousa Santos, 2012).

Knijnik and Alangui envisaged first the agency of local insiders as validators proposed by a relational Ethnomathematics:

> Ethnomathematics offers an arena where indigenous peoples can assert their alternative views and knowledge about the world. (Alangui, 2010, p. 25)

> I use the expression *ethnomathematical approach* to designate *research* into the conceptions, traditions, and mathematical practices of a specific

> subordinated social group and *pedagogical work* involved in making the group members realize that:
> 1. They do have knowledge;
> 2. They can codify and interpret their knowledge;
> 3. They are capable of acquiring academic knowledge;
> 4. They are capable of establishing comparisons between these two different types of knowledge in order to choose the more suitable one when they have real problems to solve. (Knijnik, 1993, p. 24)

More recently, authors like Natalia Caicedo et al., (2012), Cristiane Coppe and Mesquita (2015), and Mesquita (2013) have presented experiences of communitarian research within Ethnomathematics. Those are examples of collaborative research as a co-theorisation process (Rappaport, 2008) that pursues a decolonisation of research methods (Smith, 2013).

Performativity

In order to look for alternative ways to embrace the call of Ethnomathematics to appreciate cultural and epistemic diversity, a useful question emerges: What is the antonym of 'difference'? A quick response would be: 'similarity', but I want to point to 'indifference'. That is the major threat to cultural diversity, as it comes with uncommunication, apathy, passiveness, and inactivity. Conversely, in this line of thought, communication, empathy, engagement, and initiative are values that surround and enhance difference and diversity in an active manner.

To illustrate how this manner breaks the contemplative image, I can mention the Wittgensteinian understanding of mathematics as a social practice and the assumption that the meaning of a word/concept is given by the use of such word/concept within the social practice (Knijnik, 2012). Rather than merely using such Wittgensteinian insight as an analytical tool to describe or interpret mathematical knowledge, we can assume it also as a performative tool, emphasising that people can intervene and operate within social practices; and people can therefore impact what is assumed to be mathematical. The cultural historicity of mathematical knowledge became a place in which we can operate, we can perform.

In the previous two sections we embraced the culture-dependence and historicity of (mathematical) knowledge and appreciated the agency of groups and communities in the constitution of new forms of knowledge. A coherent consequence of that appraisal is to make a reorientation of the regular practices and the expected outcomes of ethnomathematical activity, around a relational perspective that emphasises the importance of interpretation and interaction when proposing connections among different domains. Interpretation and interaction are necessarily performative.

An ethnomathematical work under this relational perspective necessarily comprises a performative condition, in which relations cannot be stated once and for all. They detonate collective processes of meaning-making and, because of that, relations constantly demand rephrasing, reframing, and reassessment. They are to be lived, re-enacted again and again. As the research results are ephemeral and vanish, Ethnomathematics research becomes a type of performance that is different in each instantiation.

Concerns with the agency of communities within the research also entail a performative demand. Agreements, responsibilities, and commitments need to be established differently with each community, every time, and evolve throughout the research process.

This performative character of Ethnomathematics is an enactment of the decolonial notion of sociology of emergence, because:

> The sociology of emergences consists in undertaking a symbolic enlargement of knowledges, practices, and agents in order to identify therein the tendencies of the future (the Not Yet) upon which it is possible to intervene so as to maximize the probability of hope vis-à-vis the probability of frustration. (de Sousa Santos, 2012, p. 56)

This quotation allows me to ask how Ethnomathematics can deserve to be called a proper research program if it does not assume performativity. By revolving around contemplative schemes, it will never be a sociology of emergences.

Finding new places

Although this relational approach was originally built to explain a reduced set of contemporary works that involves communitarian participation, it turned out to be an entire reconceptualisation of the ethnomathematical field. As I said elsewhere:

> According to this view, ethnomathematical research basically traces connections between cultural practices and mathematical objects, to show how culturally embedded is knowledge production. Any modelling or mathematical description of cultural practices is a connection. Cultural contextualizations of mathematical practices are also connections. (...) No matter if they are defrosting mathematics (Gerdes, 2003), finding a family resemblance among practices (Knijnik, 2012), or describing the QRS-systems of a group (Barton, 2008). (Parra, 2018, p. 215)

This means that Ethnomathematics has always been relational. The crucial point is to what extent we have been aware of that condition and how purposefully we have developed concepts and methodological procedures aligned with relationality. In the same way, there is no doubt that every piece of ethnomathematical research attempts to expand the frontiers of what is accepted as mathematical knowledge and culture. The question is which agents and scenarios have been privileged to establish the success of each attempt.

An open and conscious embrace of a relational and performative approach for Ethnomathematics can change many things for the field. Some political and epistemological dilemmas and critiques get dissolved. Also, many new theoretical and methodological concerns can appear, through notions like translation, symmetry, barter, minga, propio, locus of enunciation (Parra, 2018), deconstructionist therapy, and deterritorialisation (Tamayo-Osorio, 2017). Pedagogical consequences of this perspective need to be developed as well: I am currently exploring them through notions like repertoire and jurisdiction (Parra, 2024).

To close the chapter, I contend that the current image of contemplative admiration needs to be refined by a relational one that is more politically driven and can help Ethnomathematics celebrate diversity by multiplying it, and not merely by registering it.

References

Alangui, W. v. (2010). *Stone walls and water flows: Interrogating cultural practice and mathematics* [Doctoral dissertation, University of Auckland].

Albanese, V., Adamuz-Povedano, N., & Bracho-López, R. (2017). The evolution of ethnomathematics: Two theoretical views and two approaches to education. In M. Rosa, L. Shirley, M. E. Gavarrete, & W. v Alangui (Eds.), *Ethnomathematics and its diverse approaches for mathematics education* (pp. 307–328). Springer. https://doi.org/10.1007/978-3-319-59220-6_13

Albertí Palmer, M. (2007). *Interpretación matemática situada de una práctica artesanal* [Situated mathematical interpretation of a craft practice, Doctoral dissertation, Universitat Autónoma de Barcelona].

Barton, B. (1996a). *Ethnomathematics: Exploring cultural diversity in mathematics* [Doctoral dissertation, The University of Auckland].

Barton, B. (1996b). Making sense of ethnomathematics: Ethnomathematics is making sense. *Educational Studies in Mathematics, 31*(1–2), 201–233. https://doi.org/10.1007/BF00143932

Barton, B. (1999). Ethnomathematics and philosophy. *ZDM Mathematics Education, 31*(2), 54–58. https://doi.org/10.1007/s11858-999-0009-7

Barton, B. (2008). *The language of mathematics: Telling mathematical tales.* Springer. https://doi.org/10.1007/978-0-387-72859-9

Barton, B. (2012). Preface to 'Ethnomathematics and philosophy'. In H. Forgasz & F. Rivera (Eds.), *Towards equity in mathematics education: Gender, culture, and diversity* (pp. 227–229). Springer. https://doi.org/10.1007/978-3-642-27702-3_19

Borden, L. L. (2013). What's the word for...? Is there a word for...? How understanding Mi'kmaw language can help support Mi'kmaw learners in mathematics. *Mathematics Education Research Journal, 25*(1), 5–22. https://doi.org/10.1007/s13394-012-0042-7

Caicedo, N., Guegia, G., Parra, A., Guegia, A., Guegia, C., Calambas, L., Castro, H., Pacho, C., Diaz, E., Caicedo, N., & Parra, A. (2012). *Matemáticas en el mundo Nasa* [Mathematics in the Nasa world]. CIIIT.

Cauty, A. (1999). Etnomatemáticas: El laboratorio Kwibi Urraga de la Universidad de la Guajira [Ethnomathematics: The Kwibi Urraga laboratory of the University of Guajira]. In *Congreso de Antropología: Simposio de Etnoeducación* (Vol. 7, pp. 267–365). Fondo de Publicaciones de la Universidad del Atlántico

Cauty, A. (2001). Matemática y lenguajes: Como seguir siendo amerindio y aprender la matemáticas que necesitara? [Mathematics and language arts: How to stay Amerindian and learn the mathematics you need?] In A. L. G. Zapata (Ed.), *Pluriculturalidad y aprendizaje de la matemática en américa latina: Experiencias y desafíos* (pp. 49–87). Ediciones Morata.

Coppe, C., & Mesquita, M. (2015). Fronteiras urbanas: Perspectivas para as investigações em etnomatemática [Urban frontiers: Perspectives for ethnomathematics research]. *Boletim de Educação Matemática, 29*(53), 828–844. https://doi.org/10.1590/1980-4415v29n53a03

D'Ambrosio, U. (1985). Ethnomathematics and its place in the history and pedagogy of mathematics. *For the Learning of Mathematics*, *5*(1), 44–48.

de Sousa Santos, B. (2012). Public sphere and epistemologies of the South. *Africa Development*, *37*(1), 43–67.

Gerdes, P. (2003). *Awakening of geometrical thought in early culture.* MEP.

Gutiérrez, M. (2019). Etnomatemática al aula: La danza como medio en la relación cutura y escuela [Ethnomathematics in the classroom: Dance as a medium in the relationship between culture and school, Master's thesis, Universidad Distrital Francisco José de Caldas]. http://hdl.handle.net/11349/15636

Horsthemke, K., & Schafer, M. (2007). Does 'African mathematics' facilitate access to mathematics? Towards an ongoing critical analysis of ethnomathematics in a South African context. *Pythagoras*, *2007*(65), 2–9. https://hdl.handle.net/10520/EJC20885

Knijnik, G. (1993). An ethnomathematical approach in mathematical education: A matter of political power. *For the Learning of Mathematics*, *13*(2), 23–25.

Knijnik, G. (1998). Ethnomathematics and political struggles. *ZDM Mathematics Education*, *30*(6), 188–194. https://doi.org/10.1007/s11858-998-0009-z

Knijnik, G. (2007). Mathematics education and the Brazilian landless movement: Three different mathematics in the context of the struggle for social justice. *Philosophy of Mathematics Education Journal*, *21*, 1–18. http://people.exeter.ac.uk/PErnest/pome21/index.htm

Knijnik, G. (2012). Differentially positioned language games: Ethnomathematics from a philosophical perspective. *Educational Studies in Mathematics*, *80*(1–2), 87–100. https://doi.org/10.1007/s10649-012-9396-8

Meaney, T., Trinick, T., & Fairhall, U. (2012). *Collaborating to meet language challenges in indigenous mathematics classrooms.* Springer. https://doi.org/10.1007/978-94-007-1994-1

Mesquita, M. (2010). *Urban ethnomathematics and ethnogenesis: Community projects in Caparica* [Conference presentation]. Fourth International Congress of Ethnomathematics, Towson, MD, United States.

Mesquita, M. (2013). *Urban boundaries and love: The emergency of the communitarian education in a transcultural and transdisciplinar movement* [Conference presentation]. The Day of the Social Education Course.

Miarka, R. (2011). *Etnomatemática : Do ôntico ao ontológico* [Ethnomathematics: From the ontic to the ontological, Doctoral dissertation, Universidade Estadual Paulista]. http://hdl.handle.net/11449/102101

Millroy, W. L. (1992). *An ethnographic study of the mathematical ideas of a group of carpenters.* National Council of Teachers of Mathematics. https://doi.org/10.2307/749904

Oliveira, S. De. (2013). *O saber/fazer/ser e conviver dos educadores indígenas Apinayé* [The Apinayé indigenous educators' knowledge/doing/being and living together, Doctoral dissertation, Universidade Estadual Paulista]. http://hdl.handle.net/11449/102113

Pais, A. (2013). Ethnomathematics and the limits of culture. *For the Learning of Mathematics, 33*(3), 2–6.

Parra, A. I. (2024). Algunos lugares comunes en las investigaciones etnomatemáticas [Some commonplaces in ethnomathematical research]. In I.-A. Londoño-Agudelo & H. Blanco-Alvarez (Eds.), *Reflexiones sobre educación matemática desde la Etnomatemática Editorial Universidad de los Llanos* (pp. 131–143). Editorial Universidad de los Llanos

Parra, A. I. (2018). *Curupira's walk: Prowling ethnomathematics theory through decoloniality.* Aalborg Universitetsforlag.

Pickering, A. (2010). *The mangle of practice: Time, agency, and science.* University of Chicago Press.

Radford, L. (2014). Cultura e historia: Dos conceptos difíciles y controversiales en aproximaciones contemporáneas en la educación matemática [Culture and history: Two difficult and controversial concepts in contemporary approaches to mathematics education]. In I. Mendes & C. Farias (Eds.), *Práticas socioculturais e educação matemática* (pp. 49–68). Livraria da Física.

Radford, L. (2016). Epistemology as a research category in mathematics teaching and learning. In B. Hodgson, A. Kuzniak, & J. B. Lagrange (Eds.), *The didactics of mathematics: Approaches and issues* (pp. 31–36). Springer. https://doi.org/10.1007/978-3-319-26047-1_2

Rappaport, J. (2008). Beyond participant observation: Collaborative ethnography as theoretical innovation. *Collaborative Anthropologies, 1*(1), 1–31. https://doi.org/10.1353/cla.0.0014

Rowlands, S., & Carson, R. (2002). Where would formal, academic mathematics stand in a curriculum informed by ethnomathematics? A critical review of ethnomathematics. *Educational Studies in Mathematics, 50*(1), 79–102. https://doi.org/10.1023/A:1020532926983

Salzman, P. C. (2002). On reflexivity. *American Anthropologist, 104*(3), 805–811.

Savransky, M. (2012). Worlds in the making: Social sciences and the ontopolitics of knowledge. *Postcolonial Studies, 15*(3), 351–368. https://doi.org/10.1080/13688790.2012.753572

Shanker, S. (1987). *Wittgenstein and the turning-point in the philosophy of mathematics.* Croom Helm.

Smith, L. T. (2013). *Decolonizing methodologies: Research and indigenous peoples.* Zed.

Tamayo-Osorio, C. (2017). *Vení, vamos hamacar el mundo, hasta que te asustes: uma terapia do desejo de escolarização moderna* [Come on, let's rock the world until you get scared: A therapy of modern schooling desire, Doctoral dissertation, Universidade Estadual de Campinas].

Uribe Suarez, D. E. (2021). *Modelo metodológico comparativo para estudios etnomatemáticos* [Comparative methodological model for ethnomathematical studies, Doctoral dissertation, Universidad Antonio Nariño]. http://repositorio.uan.edu.co/handle/123456789/1947

Vilela, D. (2010). Discussing a philosophical background for the ethnomathematical program. *Educational Studies in Mathematics, 75*(3), 345–358. https://doi.org/10.1007/s10649-010-9261-6

Vithal, R., & Skovsmose, O. (1997). The end of innocence: A critique of 'Ethnomathematics'. *Educational Studies in Mathematics, 34*(2), 131–157. https://doi.org/10.1023/A:1002971922833

Woolgar, S. (1988). *Knowledge and reflexivity: New frontiers in the sociology of knowledge.* Sage.

11. A critical conception of mathematics

Ole Skovsmose

A critical conception of mathematics emerged through several routes: this chapter takes a closer look at three of them. First, we follow how the students' movement, beginning in the late 1960s, inspired a critique of university studies in mathematics. This analysis turned into a critique of mathematical modelling, emphasising that it is an illusion that mathematics ensures objectivity and neutrality. It became recognised that, when brought into action, mathematics may have all kinds of technological, economic, and political impacts, including many of the most questionable kind. Second, we see how mathematics becomes recognised as a plurality of constructions. I show that mathematics is shaped through social, historical, cultural, and political – in short, human – processes, and that any uniform conception of mathematics is a deception, if not a falsification. Third, I illustrate how mathematics can be developed as a critical resource and become a means for identifying forms of economic and political oppression. Mathematics can play a part in the struggle for social justice.

Since Antiquity, mathematics has been admired and celebrated, while a critical conception of mathematics has become clearly formulated only within the last century.

Plato put mathematics on a pedestal, as to him it revealed what it could mean to enter the world of ideas. Euclid's *Elements* brought together an axiomatisation of geometry that, right up to the late nineteenth century, was celebrated as showing a perfect exposition of mathematics as well as a pattern for presenting knowledge in general. In Europe the admiration for mathematics gained additional fuel through the so-called scientific revolution. The people contributing to this – Nicolas Copernicus, Galileo Galilei, Johannes Kepler, and Isaac Newton, to name a few – were all

©2024 Ole Skovsmose, CC BY-NC 4.0 https://doi.org/10.11647/OBP.0407.11

deep believers in God. They saw the world as created by God, implying that an insight into and understanding of nature are an insight into and understanding of God's creation. Apparently, this creation had been completed according to mathematical principles. After the natural sciences separated from religious beliefs, the celebration of mathematics continued, with mathematics becoming nominated as the language of science.

In this chapter, we leave behind any such unexamined admiration of mathematics and look for the emergence of a critical conception of mathematics. Elsewhere in this volume, different conceptions of mathematics have been outlined, for example those of Godfrey Hardy, Nicolas Bourbaki, George Pólya, Ludwig Wittgenstein, L. E. J. Brouwer, and Hans Freudenthal. However, when it comes to the formulation of a critical conception of mathematics, it is not possible to provide any such name-related simplification.

In many common-sense interpretations, critique means pointing out flaws, weaknesses, and problematic issues. This forms part of a critical activity, but a critique can also point out strengths, advantages, and positive qualities. This is generally assumed when one talks about film critique, literary critique, art critique, and this also applies when one talks about a critical conception of mathematics. We will follow three routes towards a critical conception of mathematics: seeing mathematics as a multiplicity of actions, as a plurality of constructions, and as a critical resource.[1]

Mathematics as a multiplicity of actions

The extreme optimism that accompanied modernity assigned a crucial role to science and technology as being the true motors of the progress that would ensure economic welfare, a richer cultural life, and a

1 More routes lead to the formulation of a critical conception of mathematics, including those addressed by Houman Harouni in the next chapter. Feminism represents another such route, see Chapter 19 in this volume and also Leone Burton (1995) on a feminist epistemology of mathematics, and Gabriele Kaiser and Pat Rogers (1995) for providing a broader overview of the movement. Also, critical race theory (see Chapter 19) and related movements open for a critical conception of mathematics, as addressed in the work of Danny Martin and his colleagues (Martin, 2013, 2019; Martin & Gholson, 2012).

permanent state of peace. This optimism, that reached its peak by the turn of the nineteenth century was, however, not for everybody. It was based on ignoring the atrocities caused by colonialism and the brutal exploitations of workers. Such reservations being disregarded, the world was seen as enjoying steady progress. Science and technology could be celebrated, and the general admiration of mathematics seemed well-grounded.

The outbreak of the First World War called into question and undermined this narrow-minded optimism. In the most dramatic way, this war demonstrated that science and technology form an integral part of the machinery of war. Airplanes were constructed for military purposes, and chemical weapons used for the first time in history. It is true that for centuries mathematicians had worked on ballistics, but now it became greatly more obvious how mathematics contributes to the further development of war technologies.[2]

It would seem that the time had come where the blind admiration of mathematics could, and should, be questioned. But it was not. It took some more time before a proper critical conception of mathematics became formulated.

Fachkritik

Emerging during the late 1960s, the students' movement advanced a broad spectrum of political ideas, also about university education. Demands emerged for a new organisation of this education, not according to traditional disciplines. It should address social issues; it should be problem-oriented; it should be project-organised. The students should have a principal say in what to study, and how to study it. Professorial dominance should be broken, and the topics taught at the university had to be subjected to a profound critique.

In German this critique was referred to as a *Fachkritik*; in Danish we have a similar word *fagkritik*, but I have never seen any adequate English translation. *Fach* means school subject or discipline, and *Fachkritik* refers to a critique of subject matter issues. One can be specific and talk about

2 For discussion of relationships between mathematics and war, see Booss-Bavnbek and Høyrup (2003).

a *Fachkritik* in biology, physics, or in mathematics. The notion became broadly applied within the students' movement.

Fachkritik developed from a broad critique of positivism as a philosophy of science. Inspired by the work of the Vienna Circle, logical positivism claimed that science should ensure objectivity and neutrality, and that scientific investigations should be kept separated from political issues.[3] Logical positivism underwent a transformation from its programmatic formulations during the 1930s to, from the beginning of the 1950s, becoming a broadly assumed working philosophy of science. As part of this working philosophy, the normative claim about what science should do turned into the descriptive claim that science, as it is actually acted out at universities and research institutions, observes objectivity and neutrality. The *Fachkritik* reacted strongly to such an interpretation of science, claiming that positivism represents a dubious ideology rather than a proper philosophy of science.

One source of inspiration for a critique of positivism came from critical theory. In 1968, Jürgen Habermas published the first German version of *Knowledge and Human Interests*, where he presented the idea of knowledge-constituting interests. In other words, he argued that no knowledge exists that is entirely neutral. A technical interest guides natural science and technical disciplines; an interest in understanding guides the humanities; and an emancipatory interest guides the social sciences. This stipulation invited a strong critique of the social sciences to the extent they were organised in accordance with positivist principles. Assuming Habermas' (1971) terminology, positivism advocated a technical interest in general, also with respect to the social sciences. When guided by such an interest, the social sciences came to ally with oppressive forces. This observation provided a powerful departure for *Fachkritik* with respect to the social sciences.

Habermas associated a technical interest with the natural sciences, which did not motivate any *Fachkritik* related to these disciplines. Nevertheless, such a *Fachkritik* was developed profoundly, with other sources of inspiration. This observation applies also to the development of a *Fachkritik* of mathematics. An important inspiration came from

3 See Chapter 4 in this volume.

many critical investigations of mathematical modelling, on which we will concentrate in the following section.

During the same period, the notion of *kritischer Mathematikunterricht* started circulating; this is the German phrase for critical mathematics education. Initial ideas were presented by Peter Damerow, Ulla Elwitz, Christine Keitel, and Jürgen Zimmer (1974), and by Dieter Volk (1975). Soon after, Volk edited *Kritische Stichwörter zum Mathematikunterricht* (Volk, 1979), which provides a wide range of references and ideas, not only for a critical mathematics education, but for a critical conception of mathematics as well.[4] The development of critical conceptions of mathematics and of mathematics education includes much overlapping, but here I concentrate on the routes leading to a critical conception of mathematics.

Critique of mathematical modelling

Although the further formulation of a critical conception of mathematics took place in different contexts, I concentrate on what took place in Denmark. In 1972, Roskilde University Centre opened, organised according to priorities of the students' movement. In 1974, Aalborg University Centre opened with a similar profile; later it was renamed Aalborg University.

At these two universities, problem-orientation and project-organisation were implemented in all study programmes: sociology, biology, history, physics, mathematics, etc. Through problem-orientation the studies gained an interdisciplinary format. It was a period in which much educational innovation and experimentation took place, accompanied by the enthusiasm of both students and teachers.[5]

In mathematics many different problems became addressed through the students' project work. An approach often applied was to investigate real cases of mathematical modelling. Project groups investigated

4 During the same period, I started pondering what a critical mathematics education could mean in a Danish context, and in 1977 I had a PhD study approved with this ambition in mind.

5 For a presentation of how the mathematical study programmes at Aalborg University became problem-oriented and project-organised, see Vithal, Christiansen, and Skovsmose (1995). See also Jensen, Stentoft, and Ravn (2019), and Kolmos, Fink, and Krogh (2004).

such macro-economic models as the Annual Danish Aggregate Model (ADAM) applied by the Danish Ministry of Finance. Other projects addressed more theoretical economic models like the Goodwin Model that provides a possible interpretation of cyclic movements within a capitalist economy.[6] Project groups explored models applied by airlines for seat reservations, revealing that overbooking is not due to any systemic mistakes but to a carefully elaborated strategy for maximising profit.[7] The North Sea Model constructed by the Danish Institute for Fishery and Marine Research in order to maximise the fishing yield in the North Sea was examined in order to show how particular industrial interests became engraved in the mathematical structure of the model. One finds students' project reports addressing the Rasmussen Report, which is based on models that estimate an extremely low probability of a serious accident occuring at a nuclear power plant.[8]

All such project works contributed to the further development of a *Fachkritik* of mathematics. A recurring preoccupation was to identify possible political and economic interests embedded in mathematical models, thereby revealing that the postulate of mathematics-based objectivity and neutrality in such models is an illusion.

As an illustration of what such a critique might include, let us observe the book *Beskæftigelsesmodellen i SMEC III* [The Model of Employment in SMEC III], written by Mogens Niss and Kirsten Hermann (1982). Niss has been working at Roskilde University Centre from its opening and has been deeply engaged in its whole development and in supervising project works in mathematics. Hermann has worked as a secondary school mathematics teacher also with a deep concern about mathematical modelling.

1. The Simulation Model of the Economic Council (SMEC) was developed by Danish economists with the aim of advising the government and politicians about economic policy and its possible consequences. The SMEC exists in different versions,

6 For further comments on the Goodwin Model, see, for instance, Chapter 18 in Skovsmose (2014, pp. 263–280).

7 For a further discussion of such a model, see, for instance, Skovsmose (2005, pp. 79–82).

8 I supervised the group of students investigating both the North Sea Model and the Rasmussen Report. See Skovsmose (2023) for a more detailed discussion of the North Sea Model.

but Niss and Hermann concentrate on the third version and on the parts of the model dealing with employment. The *Beskæftigelsesmodel* in the SMEC III relates the level of unemployed to different economic factors, some of which can be influenced by political decisions. It is important to be able to specify as accurately as possible what could be the implications of such decisions before they are carried out. A function of the *Beskæftigelsesmodel* is to provide hypothetical reasoning to identify potential consequences of not-yet completed political actions.

2. As pointed out by Niss and Hermann, in order to develop a critical attitude towards the *Beskæftigelsesmodel*, one has not only to understand its mathematical components, but also to identify its assumptions. To construct a model, choices must be made, for instance which parameters to consider and how to integrate them into equations. There does not exist an economic reality as such waiting 'out there' to be described by a model. A macro-economic model is not just any representation of 'reality', rather it is an expression of an interpretation of economic activities and relationships. This interpretation can be guided by theoretical insights, economic priorities, political assumptions, and a range of particular interests.
3. For bringing together the whole structure of the *Beskæftigelsesmodel*, the Cobb-Douglas function of production plays a crucial role. According to this, we have $f = f(L, I)$ stating that the national product f is a function of two variables, namely the labour force L and capital investment I.[9] This function becomes used in the model-building process for identifying relevant parameters and for integrating them into equations. Simultaneously the Cobb-Douglas function creates

9 A first step in specifying this function is to stipulate that , where is an arbitrary constant. This stipulation is based on the idea that in order to increase the production by a factor , one needs to increase both the number of workers and the investment by the same factor . The function can be given particular algebraic expressions. One is , where is an arbitrary constant and is a constant between and . See also Chapter 13 in Skovsmose (2014, pp. 181–196) for comments about the SMEC III model.

a particular economic outlook by being a principal component of a classic liberal approach to economics. Through this function, an overall liberal outlook becomes installed in the way economic expertise advises the Danish government and politicians.

4. I see the critique of the *Beskæftigelsesmodel* as provided by Niss and Hermann as being exemplary in addressing not only the mathematical content of the model, but also assumptions incorporated in its mathematical structure. Such assumptions shape the space of possible recommendations that can be derived from the model. More generally, a mathematical model provides a particular description of a phenomenon, which reflects features of the model and of how it was constructed, rather than just features of what it was supposed to describe.

Critique of mathematics in action

A picture theory of language was presented by Ludwig Wittgenstein (2002) in the *Tractatus Logico-Philosophicus*, first published in a German-English edition in 1922. According to this theory, the principal function of language is to provide descriptions or pictures of reality. Furthermore, Wittgenstein claimed, at that time, that it is a formal language such as mathematics that can ensure such a picturing. An important development away from the picture theory of language became established through speech act theory. This theory highlights that language is not primarily descriptive, but performative. One does something through language. Speech act theory was anticipated by Wittgenstein (1997) in *Philosophical Investigations*, first published in 1953, two years after his death, and further developed by John Austin (1962) and John Searle (1969). The concept of speech acts was further developed in discourse theory, where the whole performative feature of language became addressed also in terms of its political ramifications.[10]

Inspired by speech act theory and discourse theory, one comes to recognise the performative aspect of mathematics. One can do things, not only with words, but also with mathematics. As one can do

10 See, for instance, Torfing (1999).

speech acting, one can do mathematics acting. By acknowledging the performative interpretation of mathematics, the scope of a critique of mathematics and of mathematical modelling gets a new profoundness. It is one thing to criticise mathematical picturing as being more or less reliable; it is quite a different thing to criticise mathematics-based actions. I aim to establish critique not only as a critique of mathematical representations but also as a critique of the actions that rely on them.[11] Here I will indicate what this could mean by referring to some examples.

Mathematics can form part of a fabrication of risks and crises.[12] Many processes get automatised by means of mathematical algorithms. This phenomenon can be observed in all kinds of economic transactions: when paying with a credit card at the supermarket, when selling and buying on the stock market. In order to operate on the stock market, one needs to make decisions, and do it fast: decisions about selling or buying, about when to do it, and about how much to trade. Any such decisions can be systematised and condensed into algorithms. This means that stock market decisions can become executed automatically. In 2006, a third of all stock market transactions in the European Union and the United States had the form of algorithmic trading.[13] That makes it possible for economic transactions to become accelerated. It can accelerate the whole situation out of balance, even with an economic crisis as a consequence. The economic crisis that took place in 2008 can be related to the mathematics that was brought into operation in the stock market.[14]

Automatic trading is just one example of mathematics brought into action in order to bring efficiency to a process. One finds automatisation in all kinds of processes, steering an aeroplane being one example. The automatic pilot can take over completely, but even when the real pilot is in control many manoeuvres are made automatically. The degree of automation gets more and more profound, and any such automatisation is made possible by a configuration of mathematical algorithms. Like any such configuration, unexpected implications can occur. As with

11 See, for instance, Skovsmose (2004, 2012, 2015).

12 For a discussion of mathematics and crises, see Skovsmose (2021). A reworked version of this text appears as Chapter 9 in Skovsmose (2023).

13 For a detailed presentation of algorithmic trading, see Johnson (2010).

14 For discussion of the relationship between mathematics and the economic crisis in 2008, see O'Neil (2016).

financial crashes, so also airplane crashes might have their explanation in some mathematical automatics going astray.[15]

Mathematics can form part of the fabrication of decisions. This encompasses all kinds of decisions, for instance decisions regarding medical issues and health care programmes. As an illustration of what this could imply, I refer to the efforts in calculating the value of a human life. Kathrin Hood (2017) points out that experts have spent 'over a century trying to develop a scientifically sound way to measure the economic value of human life' (p. 442). Whatever method the experts have arrived at, an extensive mathematical calculation is put into operation. We are still dealing with an example of mathematics brought into action.

Hood (2017) presents the following example: 'Every day, government analysts make calculations about how much human lives are worth compared to the cost of saving or prolonging them' (p. 442). One could think of the health system as forming an integral part of a humanitarian effort to save or prolong lives. However, it is also possible to look at the health care system from an economic perspective. One can think of it as making part of the government's investments. This leads one to consider to what extent the government is dealing with a good business. In order to clarify this, one needs to compare the amount of money spent on the system compared with the amount of money gained. This consideration makes it necessary to calculate the value of a saved or prolonged life. To complete such calculations, mathematics makes available a range of approaches. To me, they illustrate that mathematics fabricates decisions as well as overall perspectives that may guide decisions.[16]

Mathematics can also form part of the fabrication of possibilities. An important feature of technological development is technological imagination. Like sociological imagination, technological imagination

15 See Hawkins (2019), who raises the question whether the Boeing plane crashes can be related to automatisation.

16 The approach to identifying the value of a human life has to consider what the person would have been able to produce during the rest of his or her lifetime. A more recent approach is to consider life as a commodity, meaning that the value of a life should be identified with the price one is ready to pay for it. A recent development is found in paying attention to both the marginal costs and the marginal gains of saving a life. For a short discussion of such approaches, see Skovsmose (2021); a revised version of this text appears as Chapter 8 in Skovsmose (2023).

refers to the conception of things that do not, as yet, exist. For formulating a technological imagination, natural language might be relevant. However, natural languages have some limitations in articulating technical possibilities. Mathematics provides a different kind of language for this purpose. Whatever modern construction we might think of has been specified through a mathematical blueprint before it was constructed. This statement applies to drones, cell phones, tablets, and so on.

A mathematical conception makes it possible to identify radical new technological possibilities. As an example, one can think of the conception of the digital computer and of possibilities of digitalisation. An important step was taken by Alan Turing (1937), when he presented the abstract calculating device that later became referred to as the Turing machine. This is a mathematical conception of a computer. For doing any computing, one needs to represent an object. Within analogue computing this representation is somehow similar to the original object. In digital computing the representation is quite different: objects become represented by numbers. The whole idea of digitalisation cannot be formulated through natural language; only through mathematics can one recognise the possibility and power of digitalisation. In terms of industrial revolutions, the third one has been characterised as the digital revolution.[17] This revolution would not have been possible without mathematics.

I have restricted myself to three examples of mathematics-based fabrications; many more could be mentioned. Mathematics makes an integral part of automatisations at workplaces, and in the constructions of robots. One finds mathematics-based patterns of surveilling and controlling in all possible domains, face recognition in public places being just one example of such mathematics-based Big Brothering. Medical technologies become mathematics-based, and so do modern war machineries. Modern communication technologies are mathematics-based, and so are all security measures on the Internet. Mathematics-based fabrications can be found in all spheres of life, and with all kinds of social impact.

---

17 The First Industrial Revolution was characterised through the innovations of a range of new technologies. The Second Industrial Revolution refers to a phase of standardisation and automatisations of production processes.

Summary

A mathematical *Fachkritik* inspired a critical conception of mathematics. Initially such a critique concentrated on revealing that application of mathematics is not a neutral activity.

The critical conception of mathematics gained a new profoundness through a performative interpretation of mathematics. A critique of mathematics turned into a critique of mathematics-based actions, which I also refer to as mathematics-based fabrications. Such fabrications can concern any aspect of our life-worlds. There are no particular qualities associated to such fabrications due to the fact that they are mathematics-based. They can be interesting, reliable, questionable, cynical, risky, inefficient, misleading, accurate, disastrous, destructive, expensive, etc. Accordingly, a critique of mathematical performatives comes to address a range of socio-political and ethical issues.

Mathematics as a plurality of constructions

One can talk about engineering mathematics, street mathematics, applied mathematics, everyday mathematics, school mathematics, pure mathematics, any kind of Ethnomathematics. One may posit that behind this diversity there exists some kind of definitive mathematics. In fact, it is common to assume that so-called 'pure' mathematics represents *the* real mathematics.

An important idea developed along the second route towards a critical conception of mathematics is that a 'real mathematics' does not exist. Mathematics as manifest in its many varieties is a social construction, constructed in a diversity of ways. This means that 'mathematics' is an open and dynamic concept with a range of different interpretations, and new interpretations will continue to occur. There is no essence to be located within the notion of mathematics. The glorification of mathematics was based on the idea that mathematics is something unique and sublime, elevated above historical and social processes. That mathematics is a social construction removes a singularity from mathematics and its flavour of being divine.

The recognition of mathematics as a diversity of social constructions has different sources, and I will refer to three of them. First, to some philosophical observations concerning mathematics and grammar. Second, to historical observations showing that mathematics did not develop along any one-way route. Third, to ethnomathematical studies documenting the different cultural manifestations of mathematics.

A plurality of linguistic constructions

In 1939, Wittgenstein gave a series of lectures in Cambridge, where he elaborated on conceptions of mathematics.[18] As was his custom, it was for a rather closed group of people. Wittgenstein challenged such well-established conceptions of mathematics as Platonism and formalism. Turing – already well known for his presentation of the Turing machine (Turing, 1937) – joined the lectures, and he argued consistently for a formalist outlook. His presence was crucial to Wittgenstein, and once when Turing was not able to assist the lecture, Wittgenstein cancelled it.[19]

During the lectures, Wittgenstein argued against any uniform conception of mathematics, and through invented examples and thought experiments, he presented the prospect of seeing mathematics as a social construction. Furthermore, he made it possible to recognise the existence of a diversity of different forms of mathematics. In his lectures, Wittgenstein distanced himself from the conception of mathematics that he had presented in the *Tractatus*.

Wittgenstein was radical in his anti-Platonism. Rules play a crucial role in mathematics, but according to Wittgenstein mathematical rules are not based on any discoveries. There is no reality behind such rules. Mathematical rules are constructed; they have the nature of being conventions. Wittgenstein also confronted any formalism. According to formalism, a mathematical theory needs to be consistent, otherwise it has no place in mathematics. Turing insisted on this, but according to Wittgenstein consistency is not an essential requirement for a mathematical theory. What is considered consistent, and what not, depends on the rules that are put into operation – and rules could always be different. There do not exist any *a priori* rules for judging mathematical theories. We have to make do with social constructions.

18 The lectures have been collected edited by Cora Diamond and published as Wittgenstein (1989). See also Wittgenstein (1978).

19 For a mention of this episode, see Monk (1990).

One can compare mathematical rules with grammatical rules.[20] Grammatical rules become formulated and developed during history. There is nothing 'eternal' about grammatical rules. They are social constructions. Grammatical rules have some degree of permanency, and one can make grammatical mistakes by not observing the rules that are considered valid at the present period of time. Still, there is nothing ahistorical or eternal in grammatical rules.

In a similar way, mathematical rules can be interpreted as formed through historical processes, specifying ways of counting and using notions like number, point, line, and plane. There is nothing Platonic that brings validity to certain ways of using such notions. The conventions that guide their use are social constructions. Mathematical rules bring about mathematical truths, and such truths become social constructions as well. As with language, mathematics makes part of an ongoing social development. There is one more important observation to be made out of this metaphorical comparison between language and mathematics. There exist many different languages, guided by many different sets of grammatical rules. In the same way one can conceptualise the possibility of a diversity of mathematics, guided by different sets of rules.

Wittgenstein fiercely attacked dominant conceptions of mathematics. However, his philosophical critique can be taken further: it reveals that one can see mathematics as a plurality of social constructions with political coloration.[21] This observation is crucial for articulating a critical conception of mathematics.

A plurality of historical constructions

The Eurocentric presentation of the history of mathematics constitutes an integral part of the celebration of mathematics. Mathematics becomes articulated as a Western phenomenon, and simultaneously as a unique form of human knowledge. This Eurocentrism has been repeated again

---

20 Wittgenstein (1989) presents this idea in the following way: 'I have no right to want you to say that mathematical propositions are rules of grammar. I only have the right to say to you, "Investigate whether mathematical propositions are not rules of expression [...]"' (p. 55).

21 That mathematics is a social construction has also been highlighted by Ernest (1998) and Restivo (1992). See also Restivo, Bendegem, and Fisher (1993).

and again, for instance in many textbooks that include summaries of the history of mathematics.

A strong effort to show that the Eurocentric presentation is biased if not simply wrong, has been presented by George Gheverghese Joseph (2000), who exposes in detail the multi-cultural roots of mathematics. As a start, Joseph presents what he refers to as the 'classic' Eurocentric trajectory of mathematics. According to this trajectory, nothing of significance took place before the Greeks formulated mathematics in an axiomatic way and by doing so eliminated the empirical features in mathematical thinking. Then follows a 'Dark Age', where nothing happened in mathematics, except that Greek mathematics was carefully reproduced in the Arabic world. The 'Dark Age' became interrupted by the rediscovery of Greek philosophy, mathematics, and culture in general; and through the Renaissance and onwards mathematics developed in Europe and in 'her cultural dependencies'. Joseph describes a modified Eurocentric trajectory, according to which Mesopotamia and Egypt are acknowledged as important resources for Greek mathematics.

Joseph reveals any such Eurocentric trajectories, modified or not, as gross simplifications by outlining the very many cultural centres where mathematics developed during the so-called 'Dark Ages'. He provides a radically different picture of the historical development of mathematics than that outlined by any version of Eurocentrism. By also paying attention to what took place in India, China, Japan, Africa, and South America, Joseph documents the multiplicity of historical constructions of mathematics.[22]

Through his work, Joseph helps us to recognise that mathematics is not any unique and uniform phenomenon. It develops through complex historical processes, including many forms of interaction and communication combined with local discoveries and achievements. Mathematics appears in many different versions in different historical settings. We come to recognise that mathematics represents pluralities. The plurality of mathematics as pointed out by Wittgenstein gets an additional historical interpretation.

---

22 Raju (2007) argues that there are two streams of mathematics: a mathematics 'that was spiritual, anti-empirical, proof-oriented, and explicitly religious', emerging from Greece and Egypt; and a mathematics emerging from India via Arabs 'that was pro-empirical, and calculation-oriented, with practical objectives' (p. 413).

A specific issue addressed by Joseph, namely the development of calculus in India prior to its development in Europe and the transmission of that knowledge to Europe, has been analysed in depth by Chandra Kant Raju (2007).[23] Raju presents a careful analysis of the nature of knowledge transmission and the historical methods for evaluating evidence of such transmission. In addition to historical analysis, Raju argues for what he terms 'epistemological continuity' (p. 274), namely a conceptual interpretation of the available mathematical texts to seek signs of transmission of forms of thinking. Raju refers to the phenomenon of 'Hellenisation' which he describes as 'a simple trick by which a pure Greek origin was attributed to any incoming knowledge regarded as useful to Europeans' (p. 268).[24]

How could it be, then, that the Eurocentric version of the history of mathematics took such a dominant position? Important explanations are presented by Marin Bernal (1987) in *Black Athena: Afroasiatic Roots of Classical Civilization, Volume 1: The Fabrication of Ancient Greece, 1785–1985*. Let us take an extra look at the title and subtitle of the book. Bernal talks about the fabrication of ancient Greece, and he provides an account of that fabrication during the two-hundred-year period, 1785–1985. In fact, he starts his account much earlier than 1785. He covers the whole period when colonisations took place around the world, and where racism formed part of the Western outlook. By talking about 'fabrications' of Ancient Greece and not, say, about 'discoveries' of Ancient Greece, Bernal highlights that profound interpretations and re-interpretations of Antiquity took place during the colonial times. Such interpretations were guided by deep and extensive sets of preconceptions.

During the historical period explored by Bernal, it was difficult in the West to accept that Greek culture could have been influenced by the East. The whole interaction with the East as well as with Africa had to be played down in order to establish Greek culture as an integral part of Western culture. It was simply unthinkable that the White supreme European culture could have any roots in Africa or in the Orient.[25] Bernal shows

23 For a discussion of Kerala mathematics and its possible transmission to Europe, see Almeida and Joseph (2009) and Joseph (2011).

24 A particularly important example is the capture of Toledo and its library in 1085 and the subsequent translation of its books into Latin.

25 A particular version of racism took the form of 'orientalism', as coined by Said (1979).

how linguistic studies during the 1800s, not least at German universities, tried to explain away that Greek language demonstrates much influence from Eastern languages. Simultaneously, efforts were made to establish connections between Greek and German; such connections were in turn explained by a migration taking place thousands of years ago of people from the Caucasian region, with some groups moving into Greece while other groups continued into Germany. As Bernal ironically remarks, this migration seems to be the only one in history that did not leave behind a trace of broken pottery or other residues. Every kind of effort was made in order to fabricate the celebrated Greek culture as being a genuine Western culture.

The Westernisation of Greek culture also applied to the history of mathematics. When Greek culture was safely reinterpreted as being Western, Greek mathematics turned Western as well, and could be conceived of as a Western achievement. This process of 'Hellenisation', as referred to by Raju, highlights how the political power of the Christian church played a central role in this transformation.[26] Everything got together: a celebration of mathematics, a celebration of Greek culture, and a celebration of the West.

This whole worldview has to be revealed as being false, and Joseph, Raju, and others have made a huge effort to do so. The outlook is based on a wrong conception of the history of mathematics combined with layers of racism. To turn this explicit is a crucial component of establishing a critical conception of mathematics. The view of mathematics as a sublime discipline has developed together with processes of colonisation, the development of racism, and the formation of ideologies about the supremacy of European culture.

A brief discussion of the use of the expression 'Western mathematics' is warranted here: this phrase is often used by those aiming to critique precisely what they refer to in this way. However, the term troubles me for the following reason: if we look around the world today, there is nothing inherently Western about mathematics. Mathematics research occurs globally, mathematics is applied everywhere, and taught in schools as a discipline worldwide. If we go back in history, as Joseph and Raju have shown, mathematics was not exclusively 'Western' either.

26 See also Raju (2012).

So, when was mathematics fabricated as being Western? This occurred during the historical period addressed by Bernal, when colonialism and racism dominated the Western outlook, leading to a revision of the history of mathematics to fit this narrative.

A plurality of cultural constructions

More than any other research programme, Ethnomathematics has addressed the cultural plurality of mathematics.[27] In 1984, the ethnomathematical outlook was presented by Ubiratan D'Ambrosio in a plenary lecture at the International Congress on Mathematics Education (ICME-5) in Adelaide in Australia. Ethnomathematics refers to the mathematics of any cultural group. It could be Indigenous people, shoemakers, bank assistants, engineers, pure mathematicians. D'Ambrosio (2006) highlighted the importance of establishing this broader interpretation of Ethnomathematics.[28]

While Joseph provides an insight into the diversity of historical constructions of mathematics, Ethnomathematics reveals the plurality of cultural constructions of mathematics. This concretises what Wittgenstein indicated by seeing mathematics as a rule-following activity. Mathematical rules can be different like grammatical rules can be different. Mathematics can be different as languages can be different. Just as one can operate with a diversity of language games, one can operate with a diversity of mathematics.[29]

As an example of a recent contribution to Ethnomathematics, let us refer to Aldo Parra's (2018) study *Curupira's Walk: Prowling Ethnomathematics Theory through Decoloniality*. Curupira is a figure from Indigenous mythology, who walks with his feet pointing backwards, which makes it difficult to follow his route. Parra provides an extensive empirical study, and at the same time he contributes with an important conceptual development of Ethnomathematics.

27 See D'Ambrosio (1992, 2006).

28 For a recent overview of the field see Rosa, D'Ambrosio, Orey, Shirley, Alangui, Palhares, and Gavarrete (2016). Important examples of ethnomathematical studies are found in, for instance, Gerdes (2008, 2012) and Palhares (2008).

29 Explicit references to Wittgenstein's notion of language game have been made by Knijnik (2012, 2014, 2017).

Parra presents Ethnomathematics as studies of relationships. Let me illustrate what he means by that. Parra worked in a Nasa community, a group of Indigenous people living in Colombia. Among many things, he was interested in coming to understand their conception of space, and how they measured areas and distances. They had elaborated techniques for doing this, and Parra learned about these. However, he does not see research in Ethnomathematics as just identifying and describing culturally embedded notions and techniques. He sees Ethnomathematics as being concerned with relationships, which can be formed between such insights embedded in the Nasa culture and other forms of mathematics. Parra not only registered techniques applied by the Nasa community, but he also introduced alternative mathematical approaches; thus he showed how Google Maps functions.

Parra sees Ethnomathematics as 'composed of a series of contingent and purposefully constructed relations between mathematics and culture' (p. 13).[30] This could be mathematics from within a specific culture, but also mathematics that does not belong to that culture. Parra engaged himself in the educational programme developed by the Nasa community, and he also tried to contribute to this. He was not making studies *of* the Nasa people; he was doing studies together *with* them in an attempt to contribute to their environment and to meet their interests. Parra did not see himself as first of all an observer, but as a participant.

This way Parra adds a new aspect to the ethnomathematical research programme. Ethnomathematics is not first of all a study of some culturally embedded mathematical techniques and insights, it is as well a process of *involvement*. Being involved also means 'collaboration', 'participation', and 'sharing'. Such notions capture the way Parra acts out his approach to Ethnomathematics.

Milton Rosa and Daniel Clark Orey (2016) outline different dimensions of Ethnomathematics, one being the political. I find it extremely important to recognise this dimension, and Parra illustrates what this dimension might include by his involvement. Let me just indicate a different example of what involvement could mean. The Amazon Rainforest has been shrinking. It represents huge economic resources: trees can be cut and sold, and new farmland can be opened

30 In making the claim that Ethnomathematics has to do with relationships, Parra mentions that he is inspired by Alangui (2010).

up. The life conditions for the Indigenous people living there are threatened. Any ethnomathematical study engaged with Indigenous people in the Amazon needs to be involved in this drama. To try to operate as a descriptive observer means to become, not a neutral, but a cynical observer.[31]

Parra's study concerned the Nasa community. Other ethnomathematical studies address other communities. Taken together it becomes documented how different cultural contexts and practices give rise to different forms for mathematics. The ethnomathematical research programme presents mathematics as a plurality of cultural constructions. To recognise this is crucial for formulating a critical conception of mathematics.

Summary

Mathematics is formed through linguistic, historical, and cultural processes of construction, which bring about different versions of mathematics. Mathematics includes diversities. Mathematics is plural. This observation is crucial for formulating a critical conception of mathematics that does not assign any divine qualities to mathematics.

Above, I highlighted that mathematics-based actions can have any kind of qualities. This applies to any form of mathematics: engineering mathematics, street mathematics, applied mathematics, everyday mathematics, school mathematics, pure mathematics, any kind of Ethnomathematics. Whatever version of mathematics that is brought into action, also any version of Ethnomathematics, the result can be interesting, reliable, questionable, cynical, risky, inefficient, misleading, accurate, disastrous, expensive, etc. Mathematics, in whatever version we are dealing with, is in need of being critically addressed.

Mathematics as a critical resource

As already emphasised, critique not only means pointing out flaws, weaknesses, and problematic issues, but also strengths, advantages, and

31 The very idea of involvement also forms part of Knijnik's (1996) outlook. She points out resistance as being important for acting out the political dimension of Ethnomathematics.

positive qualities. In fact, a critique can point out any kind of qualities (positive or negative) of a phenomenon.

If we return to the *Fachkritik* of mathematics, as formulated during the 1970s and 1980s, one not only finds extensive questionings of mathematical modelling, but also efforts to identify models that could contribute to critical enterprises. This could be by documenting levels of economic inequalities, revealing different treatments of men and women, identifying dangers at the workplaces, showing implications of automatisation of production processes, and so on. In the initial period of the students' movements, much effort was made to ally with the general interests of workers. So, during that period, mathematics was also thought of as a possible critical resource. This possibility we will concentrate on now.

Questioning hegemonic ideologies

By publishing the article 'Critical Mathematics Education: An Application of Paulo Freire's Epistemology', Marilyn Frankenstein (1983) formulated the idea of critical mathematics education in the English-speaking context. She drew inspiration from the pedagogical ideas developed by Paulo Freire, combined with a profound critical perspective on mathematics.

Frankenstein pointed out the importance of addressing mathematics critically. This is due to the fact that forms of oppression can be masked by layers of numbers that establish an appearance of a 'necessity' of oppressive socio-economic structures. This point she formulates in the following way:

> A significant factor in the acceptance of this society's hegemonic ideologies is that people do not probe the mathematical mystifications that in advanced industrial society function as vital supports of these ideologies. (p. 327)

A pedagogy should address critically all forms of hegemonic structures, and Frankenstein highlights that mathematics also has a role to play in this critique:

> Critical mathematics education can challenge students to question these hegemonic ideologies by using statistics to reveal the contradictions (and lies) underneath the surface of these ideologies by providing

> learning experience where students and teachers are 'co-investigators' [...]. Further, critical mathematics education can link this questioning with action, both by illustrating how organized groups of peoples are using statistics in their struggles for social change and by providing information on such local groups as students may wish to join. (p. 329)

Frankenstein operates with the two main features of a critical conception of mathematics. She highlights that critique of mathematical mystifications that may function as vital support for hegemonic ideologies is essential. Simultaneously, she states that mathematics, in the form of critical mathematics education, can challenge students to question any patterns of explicit or implicit forms of oppression. Frankenstein is explicit in pointing out critical potentials of mathematics. She finds that critique is not only a reflective activity; questionings can be linked with actions. To Frankenstein critical mathematics education is not only a classroom practice; it makes part of a struggle for social justice.

In the book *Relearning Mathematics: A Different Third R – Radical Maths*, Frankenstein (1989) presents a richness of examples illustrating how a critical mathematics education can address oppressive structures. Frankenstein's work has inspired many, but I will focus here just on the work of Eric Gutstein (2006, 2016, 2018), who develops further the inspiration from Freire by showing what 'reading and writing the world' with mathematics could mean and how to combine educational activities with an activist approach.

The notions of reading and writing the world are inspired by Freire, who talked about 'reading the word' and 'reading the world'.[32] By 'reading the world', he refers to processes of interpreting social phenomena. It can be with respect to patterns of oppression that might be integrated in daily-life routines, and in this way concealed within practices that are taken for granted. By 'writing the world', Gutstein refers to processes of changing the world. This could be with respect to any kind of experienced social injustices. By seeing reading and writing the world as features of educational processes, education comes to

32 See Freire and Macedo (1987). Here Freire states: 'It is impossible to carry out my literacy work or to understand literacy [...] by divorcing the reading of the word from the reading of the world. Reading the word and learning how to write the word so one can later read it are preceded by learning how to *write the world*, that is, having the experience of changing the world and touching the world' (italics in the original).

include reflective as well as activist features. By talking about reading and writing the world with mathematics, Gutstein (2006) captures the critical potential of mathematics. It could be that mathematics forms part of hegemonic structures, as highlighted by Frankenstein, but simultaneously mathematics has a critical potential. It can be used for identifying forms of oppression and come to be a resource for political activism.

Media and racism

Racism is addressed by Frankenstein, Gutstein, and many others contributing to critical mathematics education.[33] Here I want to refer to the project *Media and Racism*, organised by Reginaldo Britto (2013, 2022) in a Brazilian context.[34]

Many statistics shows the degree of acted-out racism in Brazil. As an example, at a prominent university in São Paulo the students completing a degree in medicine are almost all White.[35] If we look at people put in jail the vast majority is Black, and the same goes for the soaring number of police killings – in 2019 in Rio de Janeiro, 1,810 people.[36] There are many such statistics that can be explored. In *Media and Racism* the focus was on making the students experience how their own observations could become expressed in numbers.[37] The project also illustrates what 'reading the world with mathematics' could mean.

The departure point for the students' investigations was the visibility of Black children in magazines circulating in Brazil. The students were divided into groups, and they were given the task of collecting photos of children shown in the magazines during a one-week period. A first observation was to see if the child was identifiable as Black or White (other racialised identities were not considered). Next, the students were asked to classify the environment in which the child was located as being either 'positive' or 'negative'. A positive environment could mean that the child was located in a wealthy and nice-looking setting, while

33 See, for instance, Davis and Jett (2019) and their chapter is this volume.
34 I have also presented Britto's projects in Chapter 2 in Skovsmose (2023).
35 See Silva and Skovsmose (2019).
36 See 'Rio violence' (2020).
37 During years, the project *Media and Racism* has been conducted by different groups of students, but in the following I concentrate on one occasion.

a negative environment could show poverty or violence. The process of classification called for a range of questions. In some cases, the classification of the child was not straightforward, and the classification of environments in being positive or negative also called for further discussions. Sometimes, the negative features were only hinted at.

Of the 41 photos that were collected by one of the groups, 36 presented White children, while 5 presented Black children. Of the White children 35 appeared in situations that were classified as positive, while only 1 appeared in a negative situation. Of the 5 photos of Black children, 3 appeared in positive situations while 2 appeared in negative. These observations became expressed through the notion of Degree of Visibility (DV).

The degree of visibility DV_i, where i refers to a particular ethnic group, is a number calculated as:

$$DV_i = \frac{\text{Number of photos with a person from the ethnic group i}}{\text{Number of photos with a person from any ethnic group}}$$

Thus, the visibilities of black children DV_b and of white children DV_w, as observed by the group in question, can be calculated as:

$$DV_b = 5/41 = 0.12$$

$$DV_w = 36/41 = 0.88$$

The numbers reflect the students' impressions: White children appear much more often in the magazines than Black children.

The DV_i provides the degree of visibility, whatever is positive or negative. However, it is possible to consider the quality of the environment, and the Degree of Negative Visibility of an ethnic group i, DNV_i, and of positive visibility, DPV_i, can be defined as:

$$DNV_i = \frac{\text{Number of appearances with negative content of the ethnic group i}}{\text{Number of appearances in total of the ethnic group i}}$$

$$DPV_i = \frac{\text{Number of appearances with positive content of the ethnic group i}}{\text{Number of appearances in total of the ethnic group i}}$$

The students could then calculate the following:

$$DNV_b = 2/5 = 0.40$$
$$DNV_w = 1/35 = 0.03$$
$$DPV_b = 3/5 = 0.60$$
$$DPV_w = 35/36 = 0.97$$

The point of making such calculations was to show that qualitative experiences can be expressed in numbers. However, such numbers are not the final words in the discussion; rather they provide a starting point for more profound explorations.

Any process of investigation calls for more investigations. We are dealing with open-ended processes. It would be natural for the students to ask, for example, if one could identify variations according to magazines. One could calculate the value of DV_i, DNV_i, and DPV_i, for Black and White children in different magazines. One could also investigate if there are changes over time. One could concentrate on photos from commercials, or on photos related to news. *Media and Racism* creates an opening for a range of further investigations. Through such activities, the potentials of operating with a notion like Degree of Visibility can become experienced by the students.

The approach used for showing the visibility of Black and White children can be used for showing the visibility of Black people and White people in any context. One can address visibility in different workplaces, different neighbourhoods, different educational settings, different political settings, different governments, etc. Furthermore, the approach can be used in relation to any classification of people in mind: women, men, immigrants, people with disabilities, etc. The very notion of Degree of Visibility and its associated calculations can be applied in many contexts. Furthermore, it is an approach whose mathematical part can be developed into a powerful tool for addressing any issue of representativity.[38]

38 For a further analysis of representativity, see Barros (2021). See also the presentation of the Bias Index in Skovsmose (2023).

Climate change

Many observations indicate that the climate of the planet is changing, and that the changes are caused by human beings. The changes are so profound that it is plausible to assume that we are entering a new historical period.[39]

Mathematics is intimately related to the discussion of climate change. A first observation is that it is impossible to talk more specifically about climate change without using mathematical modelling. In order to conceptualise climate change one needs to do forecasting. Weather forecasting is a common practice, which is now entirely based on mathematical modelling, in particular the application of dynamic system analysis. Weather forecasting has been practiced for centuries, and it existed before mathematics became applied. However, the forecasting with respect to climate changes cannot exist in any quantified form without mathematics. Mathematical climate models are tremendously complex.[40]

Through climate models, we might grasp the degrees of climate change taking place. We might also get an idea of what actions would prevent, or slow down, further changes. We are dealing with an example of *experimental forecasting*. Such forecasting is impossible without mathematics. A range of parameters forms part of a climate model, and the actual values of these parameters can be estimated through empirical observations. With changes in the values of such parameters, one might get a description of the present climate situation and how this will develop if no interventions are made. In experimental forecasting, one changes the value of some parameters in the model and observes how this will change the forecast. Through a systematic experimental forecasting one might ascertain the relevant initiative with which to respond. Experimental forecasting is crucial for formulating political recommendations for how to cope with climate changes.

Let us now return to the notions of reading and writing the world with mathematics. These expressions were used with reference to cases

---

39 Such a claim is made by Crutzen and Stoermer (2000), who refer to the Anthropocene as characterised by the fact that human beings are influencing the atmosphere of the earth.

40 See McKenzie (2007), Coiffier (2011), and Warner (2011).

of social injustices concerning, for instance, economic exploitations, sexism, and racism. It was also acknowledged that such readings and writings could be mis-readings and mis-writings. These possibilities are obvious when we think of reading and writing climate changes with mathematics. Any climate model might incorporate a range of assumptions, political priorities, industrial interests, also of the most dubious nature. Any mathematics-based reading and writing needs to be critically addressed.

In 'A Critical Mathematics Education for Climate Change: A Post-Normal Approach', Richard Barwell and Kjellrun Hiis Hauge (2021) discuss how climate change is addressed through mathematics.[41] Based on this discussion they provide recommendations for how to address problems concerning climate change in mathematics education. Barwell and Hauge present three groups of educational principles, which concern authenticity, participation, and reflections on mathematics.

With reference to authenticity, they recommend that we address problems concerning climate change that students find relevant in their lives; that students come to work with real data as much as possible; that the students' own ideas and values adopt a central role; and that students get the opportunity to engage in meaningful debate relating to climate change. With respect to participation, Barwell and Hauge recommend that students take part in the selection of problems, the mathematising of problems, the selection of data, the selection of mathematical tools, and the construction of models. Furthermore, they recommend that students actively participate in their communities and get engaged in public debates. With respect to reflections on mathematics, Barwell and Hauge recommend that students are afforded opportunities to reflect on the usefulness of mathematics, but also on the limits of mathematics.

These recommendations emerged through discussion of climate change, but they are relevant whenever we try to make use of the critical potentials of mathematics. I see these recommendations as applicable to any form of reading and writing the world with mathematics. They are crucial whenever one brings mathematics into action.

---

41 See also Barwell (2013).

Summary

The references presented in this section illustrate that mathematics can be mobilised to work for social justice. To acknowledge this is an important feature of a critical conception of mathematics. Naturally the point is not to claim that such efforts will be successful, only that they are possible.

Formulating a critical conception of mathematics presupposes that one addresses any form of mathematics brought into action and tries to show the different qualities such applications might have. Also, any mathematics that tends to reveal and document forms of social injustices requires critique.

Critical mathematics education

A critical conception of mathematics education is formed through many contributions. The formulation of this conception cannot be related to a few people. It appears as a collective achievement.

A critical conception of mathematics challenges any understanding of mathematics as being a sublime subject representing a unique and unquestionable form of human knowledge. Instead, mathematics becomes interpreted as a powerful structure when brought into action; as a social construction incorporating linguistic, historical, and cultural pluralities; and also as a possible resource for working for social justice.

I do not try to make any distinction between critical mathematics education and mathematics education for social justice. Both approaches draw on a critical conception of mathematics. They acknowledge that mathematics can be implicated in forms of oppression and exploitation, but also that mathematics includes potentials for reading and writing the world critically. How a critical mathematics education can be put into practice is naturally an open question, but numerous attempts and reflections have been presented and explored in the literature.[42]

---

42 See, for instance, Alrø, Ravn, and Valero (2010), Andersson and Barwell (2021), Avcı (2019), Bartell (2018), Ernest, Sriraman, and Ernest (2015), Greer, Mukhopadhyay, Powell, and Nelson-Barber (2009), Skovsmose (2011, 2014), Skovsmose and Greer (2012), and Wager and Stinson (2012).

References

Alangui, W. (2010). *Stone walls and water flows: Interrogating cultural practices and mathematics* [Doctoral dissertation, University of Auckland, Auckland, New Zealand].

Almeida, D. F., & Joseph, G. G. (2009). Kerala mathematics and its possible transmission to Europe. In P. Ernest, B. Greer, & B. Sriraman (Eds.), *Critical issues in mathematics education* (pp. 171–188). Information Age.

Alrø, H., Ravn, O., & Valero, P. (Eds.). (2010). *Critical mathematics education: Past, present, and future*. Sense. https://doi.org/10.1163/9789460911644

Andersson, A., & Barwell, R. (Eds.). (2021). *Applying critical mathematics education*. Brill. https://doi.org/10.1163/9789004465800

Austin, J. L. (1962). *How to do things with words*. Oxford University Press.

Avcı, B. (2019). *Critical mathematics education: Can democratic mathematics education survive under neoliberal regime?* Brill. https://doi.org/10.1163/9789004390232

Barros, D. (2021). *Lendo e escrevendo o mundo com a matemática em um movimento social: Discussões sobre representatividade e as lutas da comunidade LGBT+* [Reading and writing the world with mathematics in a social movement: Discussions on representation and the struggles of the LGBT+ community, Doctoral dissertation, São Paulo State University, Rio Claro, Brazil].

Bartell, T. G. (Ed.). (2018). *Towards equity and social justice in mathematics education*. Springer. https://doi.org/10.1007/978-3-319-92907-1

Barwell, R. (2013). The mathematical formatting of climate change: Critical mathematics education and post-normal science. *Research in Mathematics Education, 15*(1), 1–16. https://doi.org/10.1080/14794802.2012.756633

Barwell, R., & Hauge, K. H. (2021). A critical mathematics education for climate change: A post-normal approach. In A. Andersson & R. Barwell (Eds.), *Applying critical mathematics education* (pp. 166–184). Brill. https://doi.org/10.1163/9789004465800_008

Bernal, M. (1987). *Black Athena: Afroasiatic roots of classical civilization, Volume 1: The fabrication of Ancient Greece, 1785–1985*. Rutgers University Press.

Booss-Bavnbek, B., & Høyrup, J. (Eds.). (2003). *Mathematics and war*. Birkhäuser.

Britto, R. (2013). Educação matemática e democracia: Mídia e racismo [Mathematics education and democracy: Media and racism]. In *Anais do VII Congresso Iberoamericano de Educação Matemática* (pp. 3355–3362).

Britto, R. (2022). Media and racism. In M. G. Penteado & O. Skovsmose (Eds.), *Landscapes of investigation: Contributions to critical mathematics education* (pp. 39–56). Open Book Publishers. https://doi.org/10.11647/OBP.0316.03

Burton, L. (1995). Moving towards a feminist epistemology of mathematics. *Educational Studies in Mathematics, 28*(3), 275–291. https://doi.org/10.1007/BF01274177

Coiffier, J, (2011). *Fundamentals of numerical weather prediction*. Cambridge University Press. https://doi.org/10.1017/CBO9780511734458

Crutzen, P. J., & Stoermer, E. F. (2000). The 'anthropocene'. *Global Change Newsletter, 41*(17), 17–18.

D'Ambrosio, U. (1992). Ethnomathematics: A research programme on the history and philosophy of mathematics with pedagogical implications. *Notices of the American Mathematics Society, 39*, 1183–1185.

D'Ambrosio, U. (2006). *Ethnomathematics: Link between traditions and modernity*. Sense. https://doi.org/10.1163/9789460911415

Damerow, P., Elwitz, U., Keitel, C., & Zimmer, J. (1974). *Elementarmathematik: Lernen* für die Praxis [Elementary mathematics: Learning for practice]. Klett.

Davis, J., & Jett, C. C. (2019). *Critical race theory in mathematics education*. Routledge.

Ernest, P. (1998). *Social constructivism as a philosophy of mathematics*. State University of New York Press.

Ernest, P., Sriraman, B., & Ernest, N. (Eds.). (2015). *Critical mathematics education: Theory, praxis, and reality*. Information Age.

Frankenstein, M. (1983). Critical mathematics education: An application of Paulo Freire's epistemology. *Journal of Education, 165*(4), 315–339. https://doi.org/10.1177/002205748316500403

Frankenstein, M. (1989). *Relearning mathematics: A different third R—radical maths*. Free Association Books.

Freire, P., & Macedo, D. (1987). *Literacy: Reading the word and the world*. Bergin & Garvey.

Gerdes, P. (2008). Explorando poliedros do Nordeste de Moçambique [Exploring polyhedra from Northeast Mozambique]. In P. Palharas (Ed.), *Etnomatemática: Um olhar sobre a diversidade cultural e a aprendizagem matemática* (pp. 317–359). Edições Húmus.

Gerdes, P. (2012). *Otthava: Making baskets and doing geometry in the Makhuwa culture in the Northeast of Mozambique*. Lulu.

Greer, B., Mukhopadhyay, S., Powell, A. B., & Nelson-Barber, S. (Eds.). (2009). *Culturally responsive mathematics education*. Routledge.

Gutstein, E. (2006). *Reading and writing the world with mathematics: Toward a pedagogy for social justice*. Routledge.

Gutstein, E. (2016). 'Our issue, our people–Math as our weapon': Critical mathematics in a Chicago neighborhood high school. *Journal for Research in Mathematics Education*, *47*(5), 454–504. https://doi.org/10.5951/jresematheduc.47.5.0454

Gutstein, E. (2018). The struggle is pedagogical: Learning to teach critical mathematics. In P. Ernest (Ed.), *The philosophy of mathematics education today* (pp. 131–143). Springer. https://doi.org/10.1007/978-3-319-77760-3_8

Habermas, J. (1971). *Knowledge and human interests*. Beacon.

Hardy, G. H. (1967). *A mathematician's apology*. Cambridge University Press.

Hawkins, A. J. (2019, May 15). Deadly Boeing crashes raise questions about airplane automation. *The Verge*. https://www.theverge.com/2019/3/15/18267365/boeing-737-max-8-crash-autopilot-automation

Hood, K. (2017). The science of value: Economic expertise and the valuation of human life in US federal regulatory agencies. *Social Studies of Science*, *47*(4), 441–456. https://doi.org/10.1177/0306312717693465

Jensen A. A., Stentoft, D., & Ravn, O. (2019). Interdisciplinarity and problem-based learning in higher education: Research and perspectives form Aalborg University. Springer. https://doi.org/10.1007/978-3-030-18842-9

Johnson, B. (2010). *Algorithmic trading and DMA: An introduction to direct access trading strategies*. Myeloma.

Joseph, G. G. (2000). *The crest of the peacock: The non-European roots of mathematics*. Princeton University Press.

Joseph, G. G. (Ed.). (2011). *Kerala mathematics: History and its possible transmission to Europe*. B. R. Publishing Corporation.

Kaiser, G., & Rogers, P. (Eds.). (1995). *Equity in mathematics education: Influences of feminism and culture*. Falmer.

Knijnik. G. (1996). *Exclução e resistência: Educação matemática e legitimidade cultural* [Exclusion and resistance: Mathematics education and cultural legitimacy]. Artes Médicas.

Knijnik, G. (2012). Differentially positioned language games: Ethnomathematics from a philosophical perspective. *Educational Studies in Mathematics*, *80*(1–2), 87–100. https://doi.org/10.1007/s10649-012-9396-8

Knijnik, G. (2014). Juegos de lenguaje matemáticos de distintas formas de vida: Contribuciones de Wittgenstein y Foucault para pensar la educación matemática [Mathematical language games of different forms of life: Contributions by Wittgenstein and Foucault to thinking about mathematics education]. *Educación Matemática*, *25*, 146–161.

Knijnik, G. (2017). A ordem do discurso da matemática escolar e jogos de linguagem de outras formas de vida [The order of the discourse of school mathematics and language games of other forms of life]. *Perspectivas da Educação Matemática, 10*, 45–64.

Kolmos, A., Fink, F. K., & Krogh, L. (Eds.). (2004). *The Aalborg PBL model: Progress, diversity and challenges*. Aalborg University Press.

Martin, D. B. (2013). Race, racial projects, and mathematics education. *Journal for Research in Mathematics Education, 44*(1), 316–333. https://doi.org/10.5951/jresematheduc.44.1.0316

Martin. D. B. (2019). Equity, inclusion, and antiblackness in mathematics education. *Race Ethnicity and Education, 22*(4), 459–478. https://doi.org/10.1080/13613324.2019.1592833

Martin, D. B., & Gholson, M. (2012). On becoming and being a critical black scholar in mathematics education: The politics of race and identity. In O. Skovsmose & B. Greer (Eds.), *Opening the cage: Critique and politics of mathematics education* (pp. 203–222). Sense. https://doi.org/10.1007/978-94-6091-808-7_10

McKenzie, D. (2007). *Mathematics of climate change: A new discipline for an uncertain century*. Mathematical Sciences Research Institute, Berkeley, California. http://library.msri.org/msri/MathClimate.pdf

Monk, R. (1990). *Ludwig Wittgenstein: The duty of genius*. Jonathan Cape.

Niss, M., & Hermann, K. (1982). *Beskæftigelsesmodellen i SMEC III* [The employment model in SMEC III]. Nyt Nordisk Forlag Arnold Busck.

O'Neil, C. (2016). *Weapons of math destruction: How big data increase inequality and threatens democracy*. Broadway Books.

Parra, A. I. S. (2018). *Curupira's walk: Prowling ethnomathematics theory through decoloniality*. Aalborg Universitetsforlag.

Palhares. P. (Ed.). (2008). *Etnomatemática: Um olhar sobre a diversidade cultural e a aprendizagem matemática* [Ethnomathematics: A look at cultural diversity and mathematical learning]. Edições Húmus.

Raju, C. K. (2007). *Cultural foundations of mathematics: The nature of mathematical proof and the transmission of the calculus from India to Europe in the 16th c. CE*. PearsonLongman.

Raju, C. K. (2012). *Euclid and Jesus: How and why the church changed mathematics and Christianity across two religious wars*. Multiversity and Citizens International.

Restivo, S. (1992). *Mathematics in society and history*. Kluwer. https://doi.org/10.1007/978-94-011-2944-2

Restivo, S., Bendegem, J. P. van, & Fischer, R. (Eds.). (1993). *Math worlds: Philosophical and social studies of mathematics and mathematics education*. State University of New York Press.

Rio violence: Police killings reach record high in 2019. (2020, January 23). *BBC News*. https://www.bbc.com/news/world-latin-america-51220364

Rosa, M., D'Ambrosio, U., Orey, D. C., Shirley, L., Alangui, W. V., Palhares, P., & Gavarrete, M. E. (Eds.). (2016). *Current and future perspectives of ethnomathematics as a program*. Springer. https://doi.org/10.1007/978-3-319-30120-4

Rosa, M., & Orey, D. (2016). State of the art in Ethnomathematics. In M. Rosa, U. D'Ambrosio, D. C. Orey, L. Shirley, W. V. Alangui, P. Palhares & M. E. Gavarrete (Eds.), *Current and future perspectives of ethnomathematics as a program*. Springer. https://doi.org/10.1007/978-3-319-30120-4_3

Said, E. (1979). *Orientalism*. Vintage Books.

Searle, J. (1969). *Speech acts*. Cambridge University Press. https://doi.org/10.1017/CBO9781139173438

Silva, G. H. G., & Skovsmose, O. (2019). Affirmative actions in terms of special rights: Confronting structural violence in Brazilian higher education. *Power and Education, 11*(2), 204–220. https://doi.org/10.1177/1757743819837682

Skovsmose, O. (2004). Mathematics in action: A challenge for social theorising. *Philosophy of Mathematics Education Journal*, 18.

Skovsmose, O. (2005). *Travelling through education: Uncertainty, mathematics, responsibility*. Sense. https://doi.org/10.1163/9789087903626

Skovsmose, O. (2011). *An invitation to critical mathematics education*. Sense. https://doi.org/10.1007/978-94-6091-442-3

Skovsmose, O. (2012). Mathematics as discourse. *Boletim em Educação Matemática, 26*(43), 1–18.

Skovsmose, O. (2014). *Critique as uncertainty*. Information Age.

Skovsmose, O. (2015). (Ethno)mathematics as discourse. In C. Bergsten & B. Sriraman (Eds.), *Refractions of mathematics education: Festschrift for Eva Jablonka* (pp. 155–172). Information Age.

Skovsmose (2021). Mathematics and crises. *Educational Studies in Mathematics, 108*(12), 369–383. https://doi.org/10.1007/s10649-021-10037-0

Skovsmose, O. (2023). *Critical mathematics education*. Springer.

Skovsmose, O., & Greer, B. (Eds.). (2012). *Opening the cage: Critique and politics of mathematics education*. Sense. https://doi.org/10.1007/978-94-6091-808-7

Torfing, J. (1999). *New theories of discourse: Laclau, Mouffe, and Žižek*. Wiley-Blackwell.

Turing, A. M. (1937). On computable numbers, with an application to the Entscheidungsproblem. *Proceeding of the London Mathematical Society, s2–42*(1), 230–265. https://doi.org/10.1112/plms/s2-42.1.230

Vithal, R., Christiansen, I. M., & Skovsmose, O. (1995). Project work in university mathematics education: A Danish experience: Aalborg University. *Educational Studies in Mathematics*, *29*(2), 199–223. https://doi.org/10.1007/BF01274213

Volk, D. (1975). Plädoyer für einen problemorientierten Mathematikunterricht in emanzipatorischer Absicht [Plea for problem-oriented mathematics education with an emancipatory intention]. In M. Ewers (Ed.), *Naturwissenschaftliche Didaktik zwischen Kritik und Konstruktion* (pp. 203–234). Belz.

Volk, D. (Ed.). (1979). *Kritische Stichwörter zum Mathematikunterricht* [Critical keywords on mathematics education]. Fink.

Wager, A. A., & Stinson, D. W. (Eds.). (2012). *Teaching mathematics for social justice: Conversations with mathematics educators*. National Council of Mathematics Teachers.

Warner, T. T. (2011). *Numerical weather and climate prediction*. Cambridge University Press. https://doi.org/10.1017/CBO9780511763243

Wittgenstein, L. (1978). *Remarks on the foundations of mathematics*. Blackwell.

Wittgenstein, L. (1989). *Wittgenstein's lectures on the foundations of mathematics: Cambridge, 1939*. University of Chicago Press.

Wittgenstein, L. (1997). *Philosophical investigations*. Blackwell.

Wittgenstein, L. (2002). *Tractatus logico-philosophicus*. Routledge and Kegan Paul.

12. Art and anti-mathematics

Houman Harouni

Scattered across history and cultures, we encounter instances of people trying to limit or reject the expansion and application of mathematics. These actions, which we can refer to as "anti-mathematics", are particularly common among artists of the modern era. This chapter tries to decipher, through a close reading of a large group of examples, the different motivations and desires that give rise to anti-mathematics across different contexts. The author argues that such actions are attempts at shielding particular ways of life from the encroachment of forces (economic, philosophical, and administrative) that use mathematics as their main instrument. In art, the pain and confusion caused by the uses of mathematics can be hurled back at those uses and expose their underlying violence. Anti-mathematics, however, does not only expose. It always creates new zones, new approaches, new products for thinking and life. The author finally connects these historical examples with the experience of children in contemporary schools and suggests that a study of anti-mathematics might be the key to developing an autonomous and rational relationship to the irrationality of mathematized reason.

Let it not be four: Anti-mathematics and science

"Every prayer", Ivan Turgenev wrote in 1881, "reduces itself to this: Great God, grant that twice two be not four" (Turgenev, 2015, p. 102). Almost eighty years earlier, in Germany, Novalis had written that "miracles, as facts contrary to nature, are anti-mathematical" (Novalis, 2021, p. 289). Both sayings open, almost immediately, onto a familiar battlefield: the one between the hard rationality of science on one side and the softened allure of religion, occultism, and romanticism on the other. It is a sentiment that energized many in the European eighteenth

©2024 Houman Harouni, CC BY-NC 4.0 https://doi.org/10.11647/OBP.0407.12

and nineteenth centuries, as they stood amid the transformation of the world through a power whose surest instrument, beside gunpowder, was mathematics. Almost seventy years earlier, in 1813, Lord Byron had written to his wife-to-be, the mathematician Anne Milbanke: "I know that two and two make four—& should be glad to prove it too if I could—though I must say if by any sort of process I could convert 2 & 2 into *five* it would give me much greater pleasure" (Byron, 1899, p. 404). In all of these pronouncements on the nature of 2+2 we can hear two ideas, stated simultaneously. The first is a desire for something anti-mathematical, something that breaks the bounds of certainty. The second is a confession that reality has already become fully subject to mathematical models and explanations. 2+2 was not merely a mathematical sentence, but a weapon that could be used to promote or forestall social transformation. Only a few years before and not far from where Turgenev wrote 'The Prayer', the revolutionary Mikhail Bakunin had used the formula to dismiss deism, perhaps the last rationalist attempt at a religious perspective, as "a philosophical vinegar sauce of the most opposed systems [...] accompanied, of course, by an ignorance, as contemptuous as it is complete, of natural science, and proving just as two times two make five, the existence of a personal god" (Bakunin, 1910, p. 63).

For most of history, whenever a person has been called "anti-mathematical", or labeled as someone who wants 2+2 to equal a number other than four, the connotation has been purely negative. It has been an epithet reserved for the ignorant, the occultist, or the dimwitted.[1]

1 To relieve the text of the burden of multiple citations, I offer a few historical examples in this footnote. Fauvel-Gouraud (1845, p. 86) applies the term antimathematical to a young person who is too slow to learn numbers. Similar uses are frequent, for example in Sonnenschein (1889, p. 577): "even the dullest and most antimathematically minded boy can hardly fail to understand". Medical scientists could use the term to disparage colleagues who, in their opinion, refused to get with the times and apply hard science to their craft (e.g., Young, 1813, p. 603). The astronomer Heaviside (1893, p. 309) applies it to those who do not know enough mathematics to understand basic physics. In education, the term was used at least once (by White, 1919, p. 29), to refer to those "who give much encouragement to the movement against mathematics as a required subject in the high school and who try to persuade our present and future teachers and our school officials that mathematical training does not have sufficient value to justify requiring it in the high school". It is important to note that White does not cite any of his opponents, and probably could not do so, because arguments against

The connotation of being a loser in the game of reality is particularly strong whenever the term "anti-mathematical" is applied to a person or a movement, an adjective purporting to describe a way of being that is out of step with truth itself.[2] Almost no one has ever applied the term to themselves.[3] The only serious exception to the rule happens to be extremely common: I am referring to the many students who proclaim or admit, without self-irony, that they "hate mathematics". We will have a chance, before the end, to discuss the relationship between this cry of exasperation and the more self-assured statements by those like Byron and Turgenev.

It is a defeated territory, defenseless and open to all forms of trespass, that we set out to explore in this chapter. But, for all that, it is not unpopulated or eventless. Many figures, many chains of occurrence pass through here.

We should suspect that there is more to the prayer that wishes to unmoor the laws of mathematics than mere occultism. After all, religion itself is more than the opiate of masses: Karl Marx saw in religion "the sigh of the oppressed creature, the heart of a heartless world, and the

---

teaching mathematics in schools were not published in scholarly journals of the time (see also Barber, 1990, pp. 103–105).

2 In philosophy, August Comte (1876) was probably the first to use the term, applying it to Plato and his followers, whom he put in opposition to Aristotle: "Archimedes and even Hipparchus intellectually emanated from Aristotle, as did Leibnitz, and even Newton, from Descartes. The other schools, not excepting those that made the greatest noise, never shared in the great scientific discoveries the reaction of which on methods of reasoning was thoroughly repugnant to them. The bent of Plato's talent—his pompous inscription notwithstanding—was just as emphatically anti-geometrical as, on the contrary, the character of Aristotle's genius was mathematical" (p. 266). For Comte, Plato is a loser in the game of positive science. It would take more than a century before a philosopher, Alexander Koyré, would use the term neutrally. Incidentally, Koyré applied the characteristic to Aristotle: "Aristotle's physics", he wrote, "is based on sense perception, and for that reason it is resolutely anti-mathematical" (1966, p. 207). In this latter instance there is no hidden, negative connotation. Koyré admired Aristotle and was simply describing what he saw as an aspect of the old master's approach to physics.

3 I can locate only two instances. In both, the term is used jokingly. In Germany, in the 1890s, a group of engineering professors campaigned against the excessive teaching of mathematics in their schools, and they might have referred to themselves as "the anti-mathematical movement" (see Hansson, 2018). The other instance is the mathematician Florentin Smarandache, who, in his pamphlet *Aftermath and Antimath* (2012), promotes absurdist plays on word problems and other mathematical questions.

soul of soulless conditions" (1970, p. 11). When Fyodor Dostoevsky has his mouthpiece in *Notes from the Underground* protest "Two times two is four is no longer life, gentlemen, but the beginning of death!", he might, on one plane of thought, be militating for that obsolete religious faith that he held on to for most of his painful life. On another level, however, he is arguing on behalf of something new and yet-to-come, a creative project that recedes in the face of numerical certainty. "Gentlemen", he says, "what sort of will of one's own can there be if it comes to tables and arithmetic, and the only thing going is two times two is four?" (Dostoevsky, 2011, p. 30).

It is only by looking at the positive, creative aspects of the above examples that we can perceive the real differences between the speakers. Byron, for example, who would "be glad" to partake of the work of proofs, and who respects in his correspondent, Anne Milbanke, her work on mathematics, does not wish merely to uproot numbers, but to search far enough until a zone of chaos can present itself. This is how his letter continues:

> The only part [of my mathematical education] I remember which gave me much delight were those theorems (is that the word?) in which after ringing the changes upon A, B and C, D etc., I at last came to "which is absurd"—"which is impossible" and at this point I have always arrived and I fear always shall through life—very fortunate if can continue to stop there. (Byron, 1899, p. 404)

Here Byron, a man who looked for liberty within the discipline of the military or of the strictest rhyme schemes, has greater affinity with a figure like Goethe—also accused of being "anti-mathematical" (e.g., in Read, 1898, p. 216)—who explored within scientific logic itself for pathways that could not be reduced to logic. He is far from, to look at another example, Turgenev's spiritualized nihilism, the acquiescence to living in a world absolutely bereft of miracles, while nonetheless believing in them. For Turgenev, the only solution, as the ending of 'Prayer' suggests, is a defiant resignation:

> And if they set about confuting him [the believer] in the name of truth, he has but to repeat the famous question: "What is truth?"
>
> And so let us eat, drink, and be merry—and say our prayers. (Turgenev, 2015, p. 104)

This kind of quietism, in turn, is alien to Dostoevsky, whose lonely characters often rage against the logic of their surroundings—money, morals, measures—by violent means, doomed to exhaustion and disgust.

In each of these cases, the artist is carrying an opposition to the uses of mathematics on behalf of a *form of life*. The more a society rationalizes its terrain, the more strictly it defines (i.e., cuts from the infinite) the world by dividing it into manageable zones—then the more likely for various forms of life to become threatened. It is then not unlikely for certain people to take on the burden of illuminating chaos in their actions, in their mode of existence. We do not have to think of this as a modern phenomenon. Highly traditional, small societies, with strict rules of conduct and a willingness to let the unknown be unknown, give rise to the figure of the shaman with his or her chaotic and magical relationship to social norms. It is in very rare instances, such as early medieval Islam, where tremendous diversity and curiosity are allowed to live side-by-side with imposed order, in such a way that the task of engaging with chaos does not become a specialized activity. There, polymaths proliferate, so that an Omar Khayyam can both write the most precise treatises on cubic equations, the theory of parallels, and astronomic calculations, and also compose spiritual poems that render all precision and prediction, other than death, ineffective. At the same time, Khayyam's doubts as a spiritualist were not exiled from his work on mathematics, which, by doubting the assumptions of the old masters, produced the first inkling of a non-Euclidean geometry (see Smith, 1935). The Islamic Golden Age is a long procession of such complexities: Avicenna, Al-Biruni, Ibn Firnas, Al-Farabi. Everywhere chaos and order mingle, so that the same person, in a single train of thought, moves from science to mysticism to poetry in patterns that define the vastly varying cosmologies of that all-too-brief period.

A major distinction of the modern era is the solidification of science as a separate realm, for which consistent patterns of movement can be devised. At the most decisive moment of this new era, that is at the moment when *Philosophiæ Naturalis Principia Mathematica* obliterated the need for philosophy in understanding nature, it was still possible for its author, a devout Christian and an alchemist, to think, in all seriousness, that divine intervention might be necessary to keep the machinery of the universe running without a hitch (Newton, 1952, p. 402). But Isaac Newton did not work this hypothesis into the *Principia*. The axiomatic logic had finally become so tightly bound and so extensible that it left no room for such metaphysical conjectures. Philosophy and art now stood outside the periphery of science. Science did not need them, but they could not ignore it. Of the three (the scientist, the philosopher, and the artist), it was the artist who received most clearly the task of dealing with chaos and uncertainty as they impact the senses. We can say with Gilles Deleuze and Félix Guattari that the "artist brings back from the chaos *varieties* that [...] set up a being of the sensory, a being of sensation, on an anorganic plane of composition that is able to restore the infinite" (1994, pp. 202–203). However, we need to restore to the statement the historical background that Deleuze and Guattari have ignored. What they describe is the artist as the product of a specific, historical division of labor.[4] In other words, "restoring the infinite" is neither the artist's sole vocation, nor is it solely the artist's burden. Anyone who feels the overwhelming power of the absolutely-defined and the irrevocably-measured also has the opportunity to pose counter-measures to that power.

This is why we are on more secure ground when we think of anti-mathematics as a series of actions (rather than a way of being, as implied by the term "anti-mathematical") which are scattered across epochs and cultures. Their apparent goal is to either limit the encroachment—the expansion—of mathematics into a way of life, or to encroach on the realm set apart by mathematics. The motivating purposes of these acts, however, are open only to speculation. The evidence for any definitive statement is lacking. Almost everywhere, the actors have held back from disclosing, or even exploring, their own motives at length. Perhaps the opposing force has always remained too powerful, held too overwhelming a claim to truth. Whoever speaks against this force, speaks in the self-doubting voice reserved for prayer.

4 Deleuze and Guattari contrast *varieties*, as the order-producing result of the arts in relation to the infinite, with *variations* and *variables*, which they view as the instruments of philosophy and science, respectively. One has only to consider the early days of a science—e.g., psychology—to see to what extent such distinctions are inapplicable: for example, "ego," "the self," and "the I" are all varieties of variations on a variable that did not yet have a clear definition.

Beyond a prayer: The larger terrain of anti-mathematics

Not all acts of anti-mathematics are the work of artists per se. Various philosophers have also put up a resistance to the science. We have Zeno of Elea's paradoxes, which might have been devised to curb the ambitions of the Pythagoreans (Matson, 2001), Augustine of Hippo's highly influential exhortations in *De Doctrina Christiana* (1995, p. 123) that the clergy limit their study of mathematics to the most rudimentary topics, Blaise Pascal's expressions of horror at the idea of a rationalized universe (see Zakai, 2010), and Martin Heidegger's (1969) vilification of "calculative thought", to name only a few examples. Elsewhere, I study these instances in detail (Harouni, forthcoming) and so will only pause here to point out two important theses. First, all of the above actors recognize the legitimate claim of mathematics to encroach on what we can call their "domains of interest". In fact, all of them had a relatively strong command of the mathematics of their time, and, just as in the statement "two plus two does not equal four", some also use numerical reasoning to bolster their ideas (e.g., Zeno's paradoxes). Second, in all cases, the opposition, in contrast to what Charles Wolfe (2017) claims regarding his seventeenth-century examples, is not based on some scientific scepticism regarding the utility of mathematics.[5] It is purely ideological. The problem is never the instruments of arithmetic and geometry in and of themselves. It is what an opposing ideology is trying to do with those instruments.

It is important to remember that mathematics is not only the instrument of an explanatory (scientific) power. People shape it into a tool for a wide range of activities—for example, commerce, administration, and construction (Harouni, 2015b)—and in turn humanity is shaped by these uses. There are forms of anti-mathematics that do not aim their

5 One of the most systematic studies of anti-mathematics to date is that of Schliesser (2017, 2011) and, following him, Wolfe's (2017). They call the phenomenon "antimathematicism" (i.e., a stance, not an act, as I have formulated) which, for them, arose in reaction to what Schliesser calls "Newton's Challenge"—the possibility of mathematizing all science. "Antimathematicists" are those who try to limit the utility of mathematics based on doubts regarding its universal applicability. The definition is, on the one hand, extremely limited in its historical scope, and on the other, far too expansive to be of use: all conscientious statisticians, for example, who try to limit the implications of their studies, suddenly turn out to be antimathematicists.

opposition at a scientific perspective at all, but at certain organizations of social life. Nothing in the examples I have given so far prepares us to understand, for example, the taboo against the counting of human beings that we find among the ancient Israelites (Park, 2013), various communities in Africa (Githuku, 2001), and the Quechua in South America (Urton, 1997). In each case, the taboo appears alongside an administrative or cultural system that obsessively reckons people and their possessions (an opprobrium against counting livestock sometimes accompanies the one against counting humans). It is as if the culture, knowing that it must surrender all its members as units in a giant scheme of reckoning refereed by kings, empires, or avaricious men, tries at the last moment to warn itself of the ultimate consequences. The Torah establishes the ban on counting humans just at the moment that Yahweh demands a military census from Moses (Exodus 30:11–16). Such taboos are not merely outdated superstitions. Contemporary culture still carries forms of aversion to the infringement of numbers on certain aspects of life: A calculating mindset in the context of family or romantic love disgusts us (Belk, 2005). To demand that you be paid back, in equal monetary terms, for a gift you gave out of love does not belong in the harmonious sphere of marriage, but in the explosive zone of divorce. The certainty of mathematical reasoning in the realm of exchange or administration is not in itself a problem; but when extended into other realms, it can pose a formidable challenge to those forms of life that must shirk it to survive.

The nineteenth century, my point of departure in this chapter, marks the era in which money economy and state administration (*statistics*, the science of the state) finally overtook nearly every arena of life. It also marks the beginning of a widespread awareness of the consequences of such a takeover. Medieval Europe, particularly among its aristocracy, had harbored an opposition to monetary relations and, with them, to the craft that made monetary exchange possible: arithmetic (see Davis, 1960; Harkness, 2007). But this was an issue for the high-handed fringes of society who had access to surplus resources, and to those who built a living around the movement of this surplus through loans and luxuries. The serf, the priest, the craftsman, or the soldier, placed in immutable social positions and exploited or rewarded according to set formula, had very little to do with these concerns. We must travel a long way to arrive at a system that promotes the total fungibility of labor, objects, and values in the manner that, for example, plagued the thinking of Charles Dickens. His books are populated by characters who are, on the one hand, obsessed by money, as it determines their movement through society (upward or downward—Dickens had experienced both in his life), and who, on the other hand, are incapable of comprehending its value when they have it in hand. The incomprehension is essential to the humanity of these characters—the upstart, Pip, for example, in *Great Expectations*—but they are not heroes fit for their own adventure if they do not give in to the desires that their universe dictates (in the same novel, the saintly blacksmith, Joe, is not a hero, but a pole that attracts or repels young Pip).

Dickens could see the same push and pull in the operations of the state and its agents. To reduce individuals to statistical units both sheds light on social problems and, at the same time, annihilates the individuals within those problems. This is made ham-fistedly clear in a passage from *Hard Times*. I quote it at length, because it concerns the character with whom this discussion will end: the child who comes to say that he/she hates mathematics. Here we see her in the person of little Sissy, who has come home after receiving a scolding at school and is describing the ordeal to her benefactor:

> 'Then Mr. M'Choakumchild said he would try me again. And he said, "This schoolroom is an immense town, and in it there are a million of inhabitants, and only five-and-twenty are starved to death in the streets, in the course of a year. What is your remark on that proportion?". And my remark was—for I couldn't think of a better one—that I thought it must be just as hard upon those who were starved, whether the others were a million, or a million million. And that was wrong, too.'
>
> [Louisa] 'Of course it was.'
>
> 'Then Mr. M'Choakumchild said he would try me once more. And he said, "Here are the stutterings—"'
>
> 'Statistics,' said Louisa.
>
> 'Yes, Miss Louisa—they always remind me of stutterings, and that's another of my mistakes—of accidents upon the sea. And I find (Mr. M'Choakumchild said) that in a given time a hundred thousand persons went to sea on long voyages, and only five hundred of them were drowned or burnt to death. What is the percentage? And I said, Miss;' here Sissy fairly sobbed as confessing with extreme contrition to her greatest error; 'I said it was nothing.'

> 'Nothing, Sissy?'
>
> 'Nothing, Miss—to the relations and friends of the people who were killed. I shall never learn,' said Sissy. 'And the worst of all is, that although my poor father wished me so much to learn, and although I am so anxious to learn, because he wished me to, I am afraid I don't like it.' (Dickens, 1854, p. 69)

Sissy in the above passage acts as an artist. The sentences she crafts rise, supposedly, from working-class experience, and they twist the utilitarian logic of the schoolmaster (the brutal Mr. Gradgrind, with his motto "the Facts, sir; nothing but Facts") to express what bourgeois calculations stifle. But these words are an obvious idealization of what real working-class children can usually articulate in schools. Sissy's tears, her pain, and her confusion are closer to reality than her words. "I am afraid I don't like it" is anti-mathematics in its most defeated form. A contemporary expression of the sentiment is the exasperated cry of Detective McNulty in the most Dickensian of all-American television series,[6] *The Wire*: "Fuck the fucking numbers already! The fucking numbers destroyed this fucking department" (Simon & Burns, 2008). Each season of *The Wire* centers on a social institution that, under the pressure of calculated costs and benefits, has lost its capacity to serve its purpose. Almost every time someone makes a calculation in *The Wire*, it is an act that affirms the supremacy of the economic factor. In one scene (Simon & Burns, 2002), a young girl comes to her older brother to ask for help with a math homework problem. The children live in a dilapidated squat, and the older brother, a very low-level drug dealer, is the breadwinner. It is a simple word problem, about the number of people on a bus after such and such number step in or out at various stations. The girl cannot answer it, and so the brother, frustrated, restates the question in terms of street drug dealing:

> Wallace: Damn Sarah, look! Close your eyes. You working the ground stash. Twenty tall pinks. Two fiends come up to you and ask for two each and another one cops three. Then Bodie hands you off ten more. But some white guy rolls up in a car, waves you down, and pays for eight. How many vials you got left?

---

6 See Joy DeLyria and Sean Micheal Robinson's (2011) clever essay that brings out the relationship between *The Wire* and Dickens's critique of social institutions by presenting the television series as a serialized, nineteenth-century novel.

Sarah: Fifteen.

Wallace: How the fuck you able to keep the count right and not be able to do the word problem then?

Sarah: Count be wrong, they'll fuck you up.

Educational scholars have seen in this very scene (e.g., Dixon-Román, 2014), and ethnographic data like it (e.g., Mesquita et al., 2011), the antidote to working-class children's resistance to learning school mathematics. If only someone had the wherewithal, Ezekiel Dixon-Román (2014) muses, to take advantage of these "deviantly marked cultural repertoires" (i.e., child labor in the extreme violence of the drug economy) in order "to pedagogically mediate [the children's] textbook learning experience," then "the academic mathematics" would "function effectively within the particularities of marginalized communities" (p. 20).

Charitable approaches, like the one described above, do not offer the marginalized child a way into the academic system. Rather, they deprive them of the last vestiges of dignity with which their indignation at the system had equipped them. The children do not receive Sissy's ability to see through the uses and abuses of numbers. Rather, they are lulled (the educationalist hopes) into total capitulation to that all-encompassing economic system within which, if "count be wrong, they'll fuck you up."

Sissy's soul and feelings

In art, the pain and confusion caused by the uses of mathematics can be hurled back at those uses. The action often relies on an extrapolation of what mathematical models leave out or override. This is, in part, similar to what critical social science tries to do with economic and statistical data (see Harouni, 2015b). Marx's theory of value, for example, walks back the calculations of commodity exchange until one reaches the element that buying and selling had obscured—that is human labor. In fact, there are entire arenas of art that cannot be distinguished from politically motivated data visualization. The German artist K. P. Brehmer, in 1972, rigged a West German flag so that the sizes of the three colors—black, red, and gold—were determined by the distribution of wealth in the country (Figure 12.1).

Fig. 12.1 K. P. Brehmer, *Korrektur der Nationalfarben, Gemessen an der Vermögensverteilung* (*Version I*) [Correction of the National Colours, Measured by Distribution of Wealth (Version I)], 1972, Collection Alexander Schröder, Berlin. Exhibition photo from *KP Brehmer. Real Capital-Production*, Raven Row, 2014. Photograph by Marcus J. Leith. Reprinted with permission.

The extrapolation here is still thoroughly within the realm of mathematical sense-making and indistinguishable from social science. We can compare the flag to another of Brehmer's works, the series titled *The Soul and Feelings of a Worker*, where the artist renders the incalculable elements of the worker's life (Sissy's tears and Sissy's imagination) as precise but nonsensical geometrical constructions arranged on graph paper (Figure 12.2).

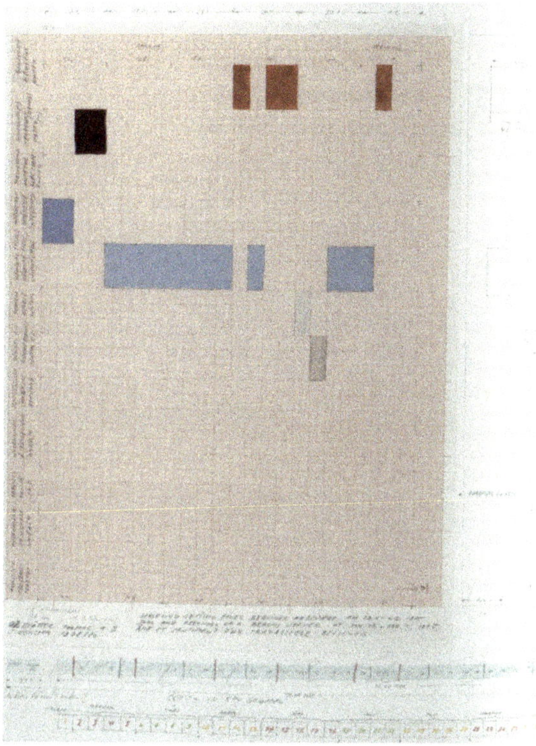

Fig. 12.2 K. P. Brehmer, *Seel und Gefühl eines Arbeiters* [Soul and Feelings of a Worker] 1980, Photo from K. P. Brehmer, *Wie mich die Schlange sieht*. Daadgalerie. 1986. Reprinted with permission.

The obvious and intentional failure of Brehmer's graph to capture "the soul and feelings" illuminates the irreducibility of the inner life of the worker to the products of calculation. Of course, so much of the external life of labor revolves around hard numberings that even social or governmental movements with the expressed aim of improving that life have to return, again and again, to those calculative reductions. By placing itself within this tension, Brehmer's graph becomes an act of anti-mathematics. In one sense, it is part of a tradition that tries to oppose the administrative function of mathematics through rendering it senseless. The Slovakian artists, Stano Filko, Alex Mlynárčik, and Zita Kostrová, in their 1965 project, HAPPSOC (a mix of "happy," "happening," and "socialism") created census data that pushed against the communist regime's obsession with representations of its own achievements. "One

Danube... six cemeteries", one piece of data claimed (Hoptman & Pospiszyl, 2002, p. 86). Zhang Huan's 1995 performance piece in which a group of naked artists piled on top of each other to "Add One Meter to an Anonymous Mountain" can also be seen as a ridicule of Chinese governmental propaganda that constantly recounted the roads, dams, schools, and hospitals built in a year. But, in another sense, Brehmer's piece differs from these other works in that it contains also a call for taking into account, in the mathematical sense of the phrase, the non-productive aspects of a worker's life.

Fig. 12.3 Zhang Huan, *To Add One Meter to an Anonymous Mountain* (1995). Reprinted with permission.

There is an intellectual and emotional sophistication in each example we have studied so far—a sophistication born of residing in the tension of a dilemma: Mathematics has become one of the most powerful instruments of definition and measurement, and, as such, it calls to the artist as a topic of investigation that, nonetheless, cannot be treated artistically. Or, to put it another way, each artist must acknowledge the "unreasonable effectiveness of mathematics" (Wigner, 1960) in science and administration, even try to incorporate some of that power, while giving voice to powers that are rejected by it.

That sophistication may not need to have any impact on the world beyond its own expression. But I would like to suggest at least one potential extension of its impact. Consider Gustave Flaubert's famous "Age of the Captain" problem, which he invented for his younger sister:

> Since you are now studying geometry and trigonometry, I will give you a problem. A ship sails the ocean. It left Boston with a cargo of wool. It grosses 200 tons. It is bound for Le Havre. The mainmast is broken, the cabin boy is on deck, there are 12 passengers aboard, the wind is blowing East-North-East, the clock points to a quarter past three in the afternoon. It is the month of May. How old is the captain? (Flaubert, 1893, p. 39)

The nonsensical little story mounts a small opposition to the way mathematics was, and continues to be, taught in schools. Different versions of it have been used to that effect ever since its composition (see Verschaffel et al., 2000). Its oppositional force, however, only extends to what happens in schools and schoolbooks. It critiques the use of fictional word problems that reduce the complexity of a context such as a merchant ship to a few cliched situations within which a mathematical calculation can find an example. Works like Dickens's, Brehmer's, Huan's, and HAPPSOC, however, go further. They challenge the social systems that made mercantile-administrative mathematics (Harouni, 2015b), with its constant emphasis on calculation and reduction of objects to values, a staple of the modern education system (Harouni, 2015a). They do not stay with the age of the captain, but hint at a challenge to the movement of wool from Boston to Europe in 1841 (the year Flaubert wrote his word problem), and the structures that make such a movement possible or necessary.

We find a stronger version of this challenge in the works of the Chilean poet (and physicist) Nicanor Parra, who simultaneously disrupts our conception of school mathematics and the social activities that give rise to it. The piece titled "Mission Accomplished" (Parra, 2004, pp. 59–61) begins and ends as follows:

trees planted	17
children begotten	6
works published	7
sum total	30

[...]

European capitals	548
lice and fleas	333333333
Apollo 16	1
sum total	49

regular kisses	48
" with tongue	17
" at the mirror	1
" luxury	4
" Metro Goldwyn Mayer	3
sum total	548

tears	0
drops of blood	0
sum total	0

In form, Parra (probably unwittingly) returns us to the earliest word problems on record—the little stories that Babylonian scribes created to teach their craft and worldview to the youth who, one day, would take over the duties of accounting and organizing labor on behalf of the state:[7]

649,539	barley-corns
72,171	ears of barley
8,091	ants
891	birds
99	people
730,791	

7 For a thorough discussion of the historical occurrences of this Babylonian problem see Friberg (2005), and for an analysis of its relationship to forms of labor see Harouni (2015a, 2015b). I have simplified the representation of the original text, which is written on two sides of a tablet and in two numerical systems.

In order for a sum to be gained, all these different objects (ants, birds, corn, people) must lose their identity, even as genera and species, until nothing but an evenly distributable identity—i.e. pure value for accounting—remains to them (see Chapter 5 in this volume). It is such losses and gains that Parra comingles in his writing, acknowledging and discarding both: all kisses, or injuries, or works accomplished, can indeed be summed up—but the final results may be wrong, and in the end both profit and loss will return to that grounding of blood and tears wherein they both draw and lose their meanings. That Parra does not divorce this image from its historical background can be seen in his riff on the 2+2 motif, in the poem "Watch Out for the Gospel of the Times" (p. 5):

> 2 parallel lines that always intersect
> create a perfect marriage
> a river that flows against its own current
> never arrives at a happy end
> everything is permitted
> absolute freedom of movement
> that is, without leaving the cage
> 2+2 doesn't make 4:
> once it made 4 but
> today nothing is known in this regard

For students in today's math-obsessed schools, a study of anti-mathematics will at least elevate their sense of opposition to the school subject from defeatism to a critical, historical, and artistic stance. It might even help free them from the designs of all those educationalists who conspire to spoon-feed them, day after day and for more than a decade, a set of skills that, should we accept Jacques Ranciére's (1991) historical examples, one could pick up with a few months of interested and diligent study, paced according to one's own needs or desires.

New spaces

The examples I have gathered so far seem to stand at a distance from the work of mathematics itself. They comment on it, impact its role in social or private life, but do not touch it directly. So-called pure, theoretical mathematics might prove immune to the influence of anti-mathematics. As a strictly reasoned discourse, it forces all opposition to express itself in strictly reasoned terms, at which point it has already proven itself victorious. This is what it did to Zeno's paradoxes, which, in the beginning, might have been posed against the influence of the Pythagoreans on philosophy. The paradoxes were posed in mathematical terms—for example, "that which is in locomotion must arrive at the half-way stage before it arrives at the goal [and so will never reach it]" (Aristotle, 1984, 239b11). Eventually, the paradoxes became part of mathematics itself, rather than any reaction against it. Their arguments fueled not less, but more mathematical exploration and power (Salmon, 2001).

But, as various activities draw on mathematical power to expand their reach, the strict reasoning of mathematics loosens, sometimes to such an extent that it becomes merely a mask for the chaotic desires that try to impose themselves on life. In these instances, anti-mathematics can expose the unreason that masquerades as reason. The false equations and orders break apart under the power of an opposition that through transposition of mathematical terms, expresses their limits. To return to Dickens's *Hard Times:*

> 'I am almost ashamed,' said Sissy, with reluctance. 'But today, for instance, Mr. M'Choakumchild was explaining to us about Natural Prosperity'.
>
> 'National, I think it must have been,' observed Louisa.
>
> 'Yes, it was. But isn't it the same?' she timidly asked.
>
> 'You had better say, National, as he said so,' returned Louisa, with her dry reserve.
>
> 'National Prosperity. And he said, "Now, this schoolroom is a Nation. And in this nation, there are fifty millions of money. Isn't this a prosperous nation? Girl number twenty, isn't this a prosperous nation, and a'n't you in a thriving state?"'
>
> 'What did you say?' asked Louisa.
>
> 'Miss Louisa, I said I didn't know. I thought I couldn't know whether it was a prosperous nation or not, and whether I was in a thriving state or not, unless I knew who had got the money, and whether any of it was mine. But that had nothing to do with it. It was not in the figures at all', said Sissy, wiping her eyes.
>
> 'That was a great mistake of yours', observed Louisa.
>
> 'Yes, Miss Louisa, I know it was, now'. (Dickens, 1854, p. 68)

The sum total of available currency in a country might, and only might, be a tightly reasoned fact; but that reason does not extend to a discussion of "national prosperity," a term whose meaning is decided subjectively. Sissy shatters the veneer of objectivity under which the term parades as a fact.

Anti-mathematics, however, does not only expose. It always creates new zones, new approaches, new products for thinking-life. We can see this more clearly if we look at a plastic and three-dimensional example. In his architectural designs, Friedensreich Hundertwasser mounted an opposition to the straight line. According to his own words, he was "against rationalism in architecture" (1958). The straight line, he said, is "something cowardly drawn with a rule, without thought or feeling," and "any design undertaken with the straight line will be stillborn. Today we are witnessing the triumph of rationalist knowhow and yet, at the same time, we find ourselves confronted with emptiness. An esthetic void, desert of uniformity, criminal sterility, loss of creative power" (as quoted in Peitgen & Richter, 1986, p. v).

In his own buildings, lines undulate and spiral. The floors and stairs are uneven, forcing the inhabitant to think and feel with every step: "An uneven floor is melody to the feet" (quoted in Karberg et al., 1995). The anti-mathematical force of these statements and actions does not destroy mathematics as such: Hundertwasser still needs numbers to erect his buildings. Rather, it disrupts the march of standardization and anonymization that, in modernism, cloaks itself in reasonability—efficiency paraded as justice.

> The apartment-house tenant must have the freedom to lean out of his window and as far as his arms can reach transform the exterior of his dwelling space. And he must be allowed to take a long brush and as far as his arms can reach paint everything pink, so that from far away, from the street, everyone can see: there lives a human who distinguishes himself from his neighbors, the pent-up livestock! (Hundertwasser, 1958)

Hundertwasser himself could not clearly grasp the implications of his own work. He confused false reason with reason itself, and in his writings, such as the *Mouldiness Manifesto against Rationalism in Architecture* (1958), he proposed that it was rationality itself that must be fought. The work, nonetheless, surpasses the words. In his buildings a new rationality survives the onslaught of the fiendish powers that can only rest happy by turning people into objects living within objects.

The students who know as much anti-mathematics as they do mathematics will not betray the latter with the former. Rather, it is only such students who, rendering the limits of mathematics discernable to themselves, return the science to its proper rationality.

References

Aristotle. (1984). Physics. In W. D. Ross (Trans.), *The complete works of Aristotle*. Princeton University Press.

Augustine. (1995). *De doctrina Christiana* [On Christian Doctrine]. Clarendon.

Bakunin, M. A. (1910). *God and the state*. Freedom.

Barber, B. (1990). *Social studies of science*. Transaction.

Belk, R. W. (2005). Exchange taboos from an interpretive perspective. *Journal of Consumer Psychology, 15*(1), 16–21. https://doi.org/10.1207/s15327663jcp1501_3

Byron, G. G. B. (1899). *The works of Lord Byron*. J. Murray.

Comte, A. (1876). *System of positive polity*. Longmans, Green, & Company.

Davis, N. Z. (1960). Sixteenth-century French arithmetics on the business life. *Journal of the History of Ideas, 21*(1), 18–48.

Deleuze, G., & Guattari, F. (1994). *What is philosophy?* Columbia University Press.

DeLyria, J., & Robinson, S. M. (2011, March 23). 'When it's not your turn': The quintessentially Victorian vision of Ogden's 'The Wire'. *Hooded Utilitarian*. https://www.hoodedutilitarian.com/2011/03/when-its-not-your-turn-the-quintessentially-victorian-vision-of-ogdens-the-wire

Dickens, C. (1854). *Hard times: For these times*. Bradbury & Evans.

Dixon-Román, E. J. (2014). Deviance as pedagogy: From nondominant cultural capital to deviantly marked cultural repertoires. *Teachers College Record, 116*(8), 1–30. https://doi.org/10.1177/016146811411600802

Dostoevsky, F. (2011). *Notes from underground*. Knopf Doubleday.

Fauvel-Gouraud, F. (1845). *Phreno-mnemotechny: Or, the art of memory*. Wiley and Putnam.

Flaubert, G. (1893). *Correspondance*. G. Charpentier.

Friberg, J. (2005). *Unexpected links between Egyptian and Babylonian mathematics*. World Scientific.

Githuku, S. (2001). Taboos on counting. In M. Getui, K. Holter, & V. Zinkuratire (Eds.), *Interpreting the Old Testament in Africa: Papers from the International Symposium on Africa and the Old Testament in Nairobi, October 1999* (pp. 113–118). Peter Lang.

Hansson, S. O. (2018). The rise and fall of the anti-mathematical movement. In S. O. Hansson (Ed.), *Technology and mathematics: Philosophical and historical investigations* (pp. 305–323). Springer. https://doi.org/10.1007/978-3-319-93779-3

Harkness, D. E. (2007). *The Jewel House: Elizabethan London and the scientific revolution*. Yale University Press.

Harouni, H. (2015a). Reframing the discussion on word problems: A political economy. *For the Leaning of Mathematics, 35*(2), 27–32.

Harouni, H. (2015b). Toward a political economy of mathematics education. *Harvard Educational Review, 85*(1), 50–74.

Harouni, H. (forthcoming). *Anti-mathematics: A programmatic study*.

Heaviside, O. (1893). *Electromagnetic theory*. 'The Electrician' Printing and Publishing Company.

Heidegger, M. (1969). *Discourse on thinking*. Harper & Row.

Hoptman, L. J., & Pospiszyl, T. (2002). *Primary documents: A sourcebook for Eastern and Central European art since the 1950s*. Museum of Modern Art. Distributed by the MIT Press.

Hundertwasser, F. (1958). *Mouldiness manifesto against rationalism in architecture*. Hundertwasser Archive. https://www.hundertwasser.at/english/texts/philo_verschimmelungsmanifest.php

Karberg, R., Jalving, A., & Jalving, C. (1995). Hundertwasser: The phenomenon. In *Hundertwasser, Arken* (pp. 15–43). Museum of Modern Art.

Koyré, A. (1966). Études *d'histoire de la pensée scientifique* [Studies in the history of scientific thought]. Presses Univeritaires de France.

Marx, K. (1970). *A contribution to the critique of Hegel's Philosophy of Right*. Cambridge University Press.

Matson, W. (2001). Zeno moves! In A. Preus (Ed.), *Essays in Ancient Greek philosophy VI: Before Plato* (pp. 87–108). State University of New York Press.

Mesquita, M., Restivo, S., & D'Ambrosio, U. (2011). *Asphalt children and city streets: A life, a city and a case study of history, culture, and ethnomathematics in São Paulo*. Sense.

Newton, I. (1952). *Opticks: Or a treatise of the reflections, refractions, inflections & colours of light*. Dover.

Novalis (2021). Mathematical fragments. *Symphilosophie International Journal of Philosophical Romanticism, 3*, 273–290.

Park, S.-M. S. (2013). Census and censure: Sacred threshing floors and counting taboos in 2 Samuel 24. *Horizons in Biblical Theology, 35*(1), 21–41.

Parra, N. (2004). *Antipoems: How to look better & feel great*. New Directions.

Peitgens, H. O., & Richter, P. H. (1986). *The beauty of fractals: Images of complex dynamical systems*. Springer.

Ranciére, J. (1991) *The ignorant schoolmaster: Five lessons in intellectual emancipation*.

Stanford University Press.

Read, C. (1898). *Logic, deductive and inductive*. Simpkin, Marshall, Hamilton, Kent, & Co.

Salmon, N. (2001). The limits of human mathematics. *Philosophical Perspectives, 13*, 93-117. https://doi.org/10.1111/0029-4624.35.s15.5

Schliesser, E. (2011). Newton's challenge to philosophy: A programmatic essay. *The Journal of the International Society for the History of Philosophy of Science, 1*(1), 101–128.

Schliesser, E. (2017). *Adam Smith: Systematic philosopher and public thinker*. Oxford University Press.

Simon, D., & Burns, E. (Directors). (2002). Lessons (Season 1, Episode 8) [Television series episode]. In *The Wire*. Blown Deadline Productions; HBO Entertainment.

Simon, D., & Burns, E. (Directors). (2008). Not for attribution (Season 5, Episode 3) [Television series episode]. In *The Wire*. Blown Deadline Productions; HBO Entertainment.

Smarandache, F. (2012). *Aftermath & antimath*. Zip.

Smith, D. E. (1935). Euclid, Omar Khayyam, and Saccheri. *Scripta Mathematica, 3*(1), 5–10.

Sonnenschein, S., et al. (1889, June 5). A first Euclid. *Bookseller: A Newspaper of British and Foreign Literature*, 577.

Turgenev, I. S. (2015). *The novels of Ivan Turgenev: Dream tales and prose poems*. BiblioBazaar.

Urton, G. (1997). *The social life of numbers: A Quechua ontology of numbers and philosophy of arithmetic*. University of Texas Press.

Verschaffel, L., Greer, B., & De Corte, E. (2000). *Making sense of word problems*. Swets & Zeitlinger.

White, E. C. (1919). Mathematics and anti-mathematics. *School Science and Mathematics, 19*(1), 20–37.

Wigner, E. (1960). The unreasonable effectiveness of mathematics in the natural sciences. *Communications on Pure and Applied Mathematics, 13*(1), 1–14. https://doi.org/10.1002/cpa.3160130102

Wolfe, C. T. (2017). Vital anti-mathematicism and the ontology of the emerging life sciences: From Mandeville to Diderot. *Synthese, 196*(9), 3633–3654. https://doi.org/10.1007/s11229-017-1350-y

Young, T. (1813). *An introduction to medical literature, including a system of practical nosology, intended as a guide to students, and an assistant to practitioners*. Underwood and Blacks.

Zakai, A. (2010). *Jonathan Edwards's philosophy of nature: The re-enchantment of the world in the age of scientific reasoning*. T & T Clark.

PART 2

13. How children, under instruction, develop mathematical understanding

Brian Greer

The relationship between the development and institutionalisation of mathematical understanding across millennia and its development for an individual child is the starting-point for this chapter. Greatly influenced by the writings of Hans Freudenthal, a position is taken in opposition to the theory propounded by Jean Piaget. The counterposition emphasises that a child can only be said to acquire any but the most elementary mathematics under more or less formal instruction and other forms of social and cultural interactions. The perennial debate about the relative weights that should be afforded in school mathematics to procedural competence and deep understanding is also related to the historical development of mathematics, particularly in relation to conceptual restructuring. This relationship is illustrated by the progressive enrichments of what is meant by 'number' and the basic arithmetical operations. The expansion of mathematical modelling from physical phenomena to the complexity of human interactions remains to be adequately addressed in school mathematics. And the question 'What is mathematics education for?' should be constantly revisited.

Introduction

How humanity collectively has created and systematised mathematics as a discipline is sketched in Chapter 2 of this volume. This chapter is, likewise, necessarily extremely selective. The vast literature on theory and research related to the teaching and learning of mathematics (e.g., Lerman, 2020) is minimally touched upon. The focus is restricted

©2024 Brian Greer, CC BY-NC 4.0 https://doi.org/10.11647/OBP.0407.13

largely to the context of formal schooling (not including the tertiary level) in advanced industrial countries. The fascinating relationship between mathematics and language is barely touched upon. Many of the assertions made are offered as hopefully provocative (in the best sense of the word) speculation.

Building on Chapter 2, I attempt to elucidate the complex relationships between the development of mathematics as a project of humanity and the development of mathematics as a project for a contemporary school pupil and his/her teachers and others in social/cultural/political contexts. To provide an overview, the following key points will serve as an advance organiser:

1. *Millennia versus years.* Many have pointed to the immense challenge that is implied by expecting children to learn in a few years mathematics that took the combined intellectual efforts of humankind millennia to develop. Insofar as the individual development of mathematics happens, it happens under instruction in schools, and in other milieux, with the benefit of resources created and systematised during history, refined by evolutionary processes. Any precise correspondence between the two projects is simplistic.

2. *School mathematics should be democratic.* The number of people who become academic or professional users of formal or technical mathematics is small in relation to essentially 100% of children who attend formal schools where such exist and spend a lot of time in mathematics classes. Accordingly, it seems reasonable to recommend that mathematics education should be designed to serve the bulk of the population, while by no means neglecting cultivation of the next cadre of mathematical specialists. In pursuing this ideal, the guidance of mathematicians is obviously necessary, but far from sufficient, and sometimes obstructive.

3. *Curricular issues.* School mathematics curriculum has been showing signs of rigor mortis for decades, characterised as it is by inertia, slowness to incorporate new content and resources, dominated by the twin fetishes of algebra and calculus, permeated by premature formalisation, failure to address

the nurturing of critical attitudes towards, and agency about applying, mathematics.

4. *Intellectual rights.* Children should be accorded intellectual rights, including the right to sense-make, to receive teaching that is developmentally appropriate and conducive to understanding, and that is relevant to issues important to them, their future as adults, and their communities and cultures.
5. *The dynamic balance between homogenisation and diversity.* Within academic mathematics there is constant interplay between what Ian Hacking (2014, p. 13) calls 'unification and diversification'. In contrast to the diversity of manifestations of mathematics within cultural practices, applications and work, and everyday life, school mathematics is becoming locally and globally more homogenised, in particular due to the convergence of curricula and standardised testing.
6. *Two faces of mathematics.* Mathematics may be thought of as having two faces. On the one hand, there is the formal apparatus of pure mathematics; on the other, there is the use of mathematics in modelling aspects of reality, including physical phenomena and, increasingly for some time, phenomena involving the complexities of human life.
7. *Two places of mathematics.* Children learn mathematics beyond school, whether under some form of instruction (for example, by parents or community members) or through their own creativity when interacting with their environments, and by absorbing manifestations of mathematics within their cultures. Much more could be done to articulate the learning that occurs in the two places.

An overarching question is: 'What is mathematics education *for*?'. This cannot be separated from a consideration of the ethical responsibilities of mathematicians and mathematical educators. Mathematics education, like mathematics itself, is embedded in historical, cultural, social, and political – in short, human – contexts. And the challenge is to embrace the possibility that things can be different.

Under instruction, for understanding

> Teaching is one of the immense social influences that can affect a child, but its effects can be out of proportion to any other kind of social influence once the first beginnings of a child's life are past. In it once again knowledge builds on knowledge, but the form of experience that makes it possible is really quite unlike those forms of experience that come the individual's way when teaching is not involved. (Hamlyn, 1978, p. 144)

In my opinion, David Hamlyn's point is particularly true when it comes to mathematics teaching/learning. As argued further below, it makes no sense to posit that any child could formulate much, for example, about fractions and operations on them without instruction from others, and without a collective representational system. The *origins* of mathematical cognitive activity may be traced, as Jean Piaget has it, to reflection on actions on the physical environment, but how far can that take one? Likewise, the neuroscientists – in their study of how people develop constructs about number, in particular – seem to exhibit the same form of what might be called the foundationalist fallacy, namely that the development of any complex, multi-levelled edifice of understanding can be analysed by focusing on its beginnings.

In Chapter 2, the notion of universities and other institutions/sites as constituting constructed environments for the doing of mathematics was introduced, and the same, of course, goes for mathematics classes within schools. Following Jean Lave (1992) and many others, it will be emphasised that children learn much more than technical mathematical content in such classes. They may learn or be taught, in some general sense, to think mathematically (if they are lucky), for example to become solvers of mathematical problems in the tradition of George Pólya. They are much less likely to be taught – though it is argued below that they should be – how mathematics is embedded in human contexts; worse, they may be inculcated into harmful beliefs about the dehumanising power of numbers and equations. Unfortunately, for too many, their recollections of school mathematics are suffused with alienation and perceived irrelevance. It will also be suggested that school mathematics is instrumental in forming lasting and consequential facets of an individual's worldview, in particular relating to a simplistic view of mathematical modelling.

Researchers of mathematical teaching and learning devise their own particular constructed environments, as when, for example, an experimenter sits with a child and presents a Piagetian conservation task. Such activity represents one form of the general problem of understanding the Other. Assessment may be viewed through a similar lens, as an activity involving communication. As emphasised below, the term 'assessment' needs to be differentiated in relation to very different activity systems, from its use by the state as an instrument of control to its embeddedness as an integral part of teaching and learning. And, in general, a major issue with testing arises when the test item refers, at least on the surface, to 'real-life' scenarios, since the reactions of students, and indeed the evaluations of their responses, are affected by the degree to which the reality sketched in the item lies within the life experience of the person being tested, evaluating the test, or using the evaluations to inform their teaching.

The other emphasis implied by the title for this section reflects an aspiration that mathematics education produce 'understanding' as contrasted with superficial competence in 'pawing at symbols' (by analogy with Paulo Freire's canine metaphor of 'barking at text', which he contrasted to reading in what he considered the full sense). 'Understanding' is not so easy to define, but it is not difficult to exemplify, particularly in its absence (examples are given below).

Beginnings and continuations

> Children acquire number in the stream of their physical and mental activities, which makes it difficult for researchers to find out how this happens in detail. (Freudenthal, 1991, p. 6)

Schools have existed for a long time, but not always, and there are still societies in which 'our' form of schooling is not practiced. However, discussion here is limited to the familiar forms of schooling, to the interplay between learning in and out of school, and, in particular, to 'the poor permeability of the membrane separating classroom and school experience from life experience' (Freudenthal, 1991, p. 5).

Children develop and are taught by others before they go to school and once they are at school they continue to learn in out-of-school contexts. Consider a subset of what a five-year-old child starting school might know about *uses* of the number 5 (beyond being able to count to 5): her age is 5, perhaps represented by 5 candles on a birthday cake; everyone (essentially) has 5 fingers on each hand and 5 toes on each foot ('digits'); he may be familiar with a single coin representing the same value as 5 coins each representing 1 with the same unit; 5 spots in a pattern on a die or playing card; she may live in a house numbered 5 (between numbers 3 and 7), or travel in a bus with that number; and know that there are 5 days in the school week, 5/5 represents the 5th of May, 5.05 is five minutes past five o'clock (with the minute hand pointing to 1, representing 5), and on and on and on...

Many mathematicians (e.g., Schoenfeld, Chapter 14, this volume) report childhood insights; I can do likewise. While playing with some cardboard boxes (age five?) I found I had two boxes of different sizes, neither of which would fit inside the other. That struck me as odd, until a simple thought experiment involving a roughly cubical box and a long thin one elucidated it for me. Or take my various encounters with probability. As a child growing up some seventy years ago in a small seaside town, I had plenty of opportunities for gambling and so developed some intuitive understanding of probabilistic events (and an innoculation against gambling). Years later I was introduced to probability theory at school; later at college it was characterised as a branch of measure theory; more recently, I have written about it in relation to socio-cultural issues. So, it is possible to see what Piaget is getting at when he talks about mathematics originating in reflections on our actions. But there is a vast chasm between that and the mathematical content that even a ten-year-old is expected to engage with in school.

After a relatively short time in school, the contextual and phenomenological richness and spontaneous thinking of the child are liable to be inhibited. Further, as the child progresses through school, the disconnect referred to by Hans Freudenthal (cited above) may be strengthened through the norms of the mathematics classroom. Consider the following observation:

> It was a lesson under the heading of 'ratio and proportion' and the teacher told me that she wanted to approach the mathematical concepts in a practical way. So she offered [...] [a scenario involving mixing paints to reproduce a particular colour]. The problem seemed quite clear and pupils started to calculate using proportional relationships. But there

> was one boy who said: 'My father is a painter and so I know that, if we just do it by calculating, the colour of the room will not look like the sample. We cannot calculate as we did, it is a wrong method!'. In my imagination I foresaw a fascinating discussion starting about the use of simplified mathematical models in social practice and their limited value in more complex problems [...] but the teacher answered: 'Sorry, my dear, we are doing ratio and proportion'. (Keitel, 1989, p. 7)

Constructed environments of school mathematics

In Chapter 2, I introduced the idea of 'constructed environment' in relation to the doing of mathematics, and the same applies to the learning/teaching of school mathematics. As Lave (1992, p. 81) put it, schooling 'is a site of specialized everyday activity – not a privileged site where universal knowledge is transmitted'.

While being mindful of the distorting lenses of contemporary framings, more or less similarly organised schools have been around in many cultures for a very long time. If you stop and think about it, there is something very artificial about 'children spending large amounts of time in formal schools where their activity is separated from the daily life of the rest of the community and mediated by technologies of literacy and numeracy as well as specialized uses of language' (Cole, 2005, p. 195).

Many of the issues are exemplified very clearly in the ways in which 'word problems' (or 'story problems') are presented in school mathematics (Lave, 1992; Verschaffel, Greer, & De Corte, 2000). Children learn that there is a 'Word problem game' (Verschaffel et al., Chapter 5) whose rules include ignoring what the child knows of reality. A striking example is the following statement by a teacher in the course of a discussion with a student's mother:

> Of course, we all know that nowadays a loaf of bread costs considerably more than 21.5 francs. But after all, that's not what students have to worry about when doing algebra problems. It's the construction and execution of the mathematical expression that counts, *all the rest is décor*. (Van der Spiegel, personal communication (1997), cited in Verschaffel, Greer, & De Corte, p. 57, emphasis added)

More generally, the notion of the didactical contract between teacher and students (Brousseau, 1997) is a useful construct for describing the mutual norms that are progressively created, often implicitly, governing interactions in mathematics classrooms. By contrast with the typically implicit nature of the didactical contract, Paul Cobb (e.g, Yackel & Cobb, 1996) advocated for explicit promotion of what he termed 'sociomathematical norms'.

During their very considerable amount of time in mathematics classes, children form images about the nature of mathematics. Too often they infer from what they are exposed to that mathematics historically was the intellectual achievement of predominantly White males. They come to believe that low marks on mathematics tests are an indication of stupidity and a deserved lack of access to educational and economic opportunities. They are much less likely to form a critical disposition or sense of agency in relation to uses of mathematics. Likewise, they form images of the nature and purposes of mathematics education. For many of them, and repeatedly, when they ask the reasonable question 'Why do we have to do this?' they get the answer 'Because it will be useful later'.

Students also learn, too often the hard way, about how society constructs success and failure (Varenne & McDermott, 2018), in particular through testing. That instrument is particularly powerful in relation to attaching numbers to mathematical performance, against the background of the unreasonable political effectiveness of 'mathematics'. In the United States, a racially coded message is sent by the pervasive use of the term 'achievement gaps' when test score gaps are being referred to. Most generally, mathematics classrooms constitute constructed environments within which children learn how to fit within state systems.

Constructed environments of research on mathematical cognition

Whenever a researcher who is not the teacher engages with a child in order to try to understand that child's mathematical learning or thinking, it constitutes another very special kind of constructed environment. Here I am restricting discussion to the scenarios in which a researcher comes from outside the school and engages, typically for a short time, with students one by one. Rather than cursorily survey this vast field, I draw attention to specific aspects.

Typically, in such research, the child is asked to address a task designed by the theorist/researcher, often involving customised equipment or representations. The vast range of experiments carried out by Piaget's team constitute a familiar example, and will serve to make important points, especially through the critique of Freudenthal (1973, Appendix I, pp. 662–677). The typical experiment constitutes a very particular kind of social interaction; by analogy with the notion of a didactical contract, the idea of an experimental contract may be invoked. It is necessary to consider how the children in this situation construe what is going on, why they are there, what is required of them. In my experience and observation, such considerations are often minimally addressed by experimenters.

Aligned with this framework, Freudenthal (1973) very strongly criticised Piaget's insufficient attention to the language used and whether or how the child understands it; a parallel may be drawn with the role of communication in assessment (see below). Experiments on conservation, for example, are particularly open to this kind of scrutiny and a range of experiments has shown that altering the experimental contract or the nature of the communication in apparently minimal ways can have a marked effect on the responses (see, e.g., Donaldson, 1978).

Most seriously, we may ask the general question: How does the experimenter/theorist know that the design and presentation of the task, and the children's responses, constitute an appropriate test of the constructs embedded in the theory? Is it possible that the experimental tasks and communications, consciously or unconsciously, are designed to support rather than test the theory? That is an extremely serious charge that deserves to be taken seriously. For example:

> By a suggestive design of the experiments it is achieved that the subjects reconstruct a landscape according to the Piaget theory of multiplication of relations, that is by means of a Cartesian coordinate system. (Freudenthal, 1973, p. 669)

To return to the point made in the opening quotation, in relation to conservation of volume, I would be much more impressed by a report of two children being poured lemonade from identical bottles into differently shaped glasses and one of them objecting that it was unfair.

'Assessment'

> If you want to sort people, make them run a race; if you want to see if kids can 'do it', then give them adequate time to 'do it'. (Stage, 2007, p. 358)

For a long time, I have found it problematic to use the same word to refer to two very different families of practices. One family, the focus of this section, involves producing measures that allow students, and groups of students, to be measured and ranked, often with high stakes attached. Another family has to do with interacting with the student in order to form conjectures about the student's understanding, cognition, beliefs, and so on; as such, it is an integral part of teaching/learning.

Following these introductory remarks, I consider just three from the vast range of relevant aspects: the analysis of assessment in terms of communication; the diversity of social realities in relation to attempts to include mathematical modelling in test items; political issues in the uses of testing for purposes of the state.

Assessment as communication

Concentrating on the communicational functions of assessment affords pointed contrasts between the two activity systems distinguished above. In general, assessment involves: communication to the student about a specific competence to be demonstrated through a particular task; action by the student in an attempt to demonstrate the required competence insofar as they understand it; some form of evaluation of that attempt; and communication of the interpretation of that evaluation to the teacher and others. In these terms, a standardised written or computer-administered test may be seen as extremely impoverished in terms of communication at every stage, particularly when there is no opportunity for clarification through subsequent iterations of communication. 'In short, the typical written assessment is closed in terms of time, in terms of information, in terms of activity, in terms of social interaction, in terms of communication' (Verschaffel, et al., 2000, p. 72).

To take a simple contrasting example, a teacher may ask a student to subtract 17 from 24 and the student might give the answer '13'. The teacher may conjecture that the student has exhibited the 'subtract the smaller from the larger within any column' misconception and ask further questions to test this conjecture. Finally, the communicative act of evaluation may consist of much more than a simple statement that the student's answer was wrong, but be combined with an explanation of why, and of how that misconception can arise.

Assessment and modelling

Test items often resemble word problems in presenting a description of a real-world situation that the assessed is expected to interpret and model mathematically. In the absence of open communication, it then often happens that the model depends on the life experience of the generalised modeller as well as that of the testee, as in the following example discussed by William Tate (1995, p. 440):

> It costs $1.50 each way to ride the bus between home and work. A weekly pass is $16.00. Which is the better deal, paying the daily fare or buying the weekly pass?

It should be obvious that assumptions made (e.g., that work occurs five days a week) will affect a person's interpretation and response and that the person's form of life will influence the assumptions made. There is no 'right answer' and if it is assumed that there is, and there is no opportunity for clarification, the item is accordingly unfit for assessment. In particular, as Tate (1995, p. 440) pointed out, 'the underpinnings of school mathematics, assessment, and pedagogy are more often closely aligned with the idealised experience of the White middle class'. More generally, testing may be seen as an instrument of cultural violence, as when 'test-score gaps' are mislabelled as 'achievement gaps' with no qualifications as to how achievement is defined.

Testing as an instrument of the state

Above all, 'assessment' is a political issue. Episodes of recent history within the United States in terms of clashes between political, corporate, and educational goals are analysed by Alan Schoenfeld (Chapter 14, this volume), a battle-scarred veteran of many campaigns. A key point that he makes is that another communicative function of testing is to convey to teachers and students what is expected, encapsulated in the acronym 'WYTIWYG' (What You Test Is What You Get). Schoenfeld illustrates from his experience how the ways teachers teach are liable to be distorted under the pressure of upcoming high-stakes tests. The politics of global testing are analysed by Paola Valero and Lisa Björklund Boistrup (Chapter 15, this volume) and Mark Wolfmeyer (Chapter 16, this volume).

The goal of understanding

> Most people have been taught mathematics as a set of rules of processing – an agreeable experience when they have learned to master them, and a disagreeable one if they have failed. (Freudenthal, 1991, p. 3)

Later, Freudenthal argues that elementary arithmetic cannot be learned *other than through insight*, but as the school student progresses to more advanced mathematics, 'the learner's insight tends to be superseded by the teacher's, the textbook writer's, and finally by that of the adult mathematician' (Freudenthal, 1991, p. 112).

The section title expresses an aspiration that mathematics education should produce 'understanding', something that is difficult to define but easy to illustrate, particularly in its absence. A simple example comes from *Productive Thinking* (Wertheimer & Wertheimer, 1982/1945). Children were asked (p. 130) to find what number the following expression is equal to:

$$\frac{274 + 274 + 274 + 274 + 274}{5}$$

A child who correctly computes a repeated addition, or multiplication, followed by a division, demonstrates computational fluency, but such a performance surely suggests a lack of understanding. The authors related his surprise that, while most of the bright students he asked 'enjoyed the joke' (p. 112), 'a number of children who were especially good at arithmetic [...] were entirely blind' (p. 113).

As a second example, I posed this question to future elementary school teachers studying slope as represented on graphs:

> A candle, initially 24 cm high, is burning *down at the rate of 3 cm per minute*. If you plot the graph of height of the candle (in cm) against time (in minutes), what will be the slope of the line?

Most of the students demonstrated competence by plotting the line and calculating the slope; hardly any showed understanding by pointing to the answer (-3) given in the italicised part of the question.

A distinction may be drawn between 'internal' understanding and 'external' understanding. The former refers to making connections, noticing and exploiting structure, within 'disembedded' (Donaldson, 1978) mathematics, as in the example from the Wertheimers' book cited above. The second example is about articulating procedures (plotting points and calculating slope) as opposed to understanding that slope of a straight line corresponds to a constant rate of change in some variable.

The relationship between procedural competence and conceptual understanding is central in discussions on mathematics and mathematics education. The problem, as suggested in the opening quotation, arises when procedural competence dominates (as it does in communicationally impoverished forms of testing).

Learning from history

> We know nearly nothing about how thinking develops in individuals, but we can learn a great deal from the development of mankind. (Freudenthal, 1991, p. 48)

> It is with children that we have the best chance of studying the development of logical knowledge, mathematical knowledge, physical knowledge, and so forth. (Piaget, 1970, pp. 13-14)

The first obvious comment is that these quotations illustrate the chasm between the positions of Piaget and Freudenthal, as reflected below. Studying the history of mathematics is extremely difficult to do for many reasons, and it has often been done very poorly, as Jens Høyrup, for one, has made clear (Greer, 2021). The central question in this section of the chapter is: 'What guidance for mathematics education can be derived through studying the history of mathematics?'. Many have pointed out that children in school are expected in a few years to come to grips with mathematics that took humankind millennia to develop:

> School is seen as a magical shortcut that allows ideas arduously developed by humanity over thousands of years to be transmitted in a few years to a random human being. (Hofstadter & Sander, 2013, p. 391)

To the extent that those ideas can be transmitted, how is it possible? The short answer is that it is achieved predominantly through instruction in a constructed environment.

The position taken here is that any kind of simplistic version of 'ontogeny recapitulates phylogeny' (Gould, 1985) is untenable. The complexity of the interactions between biological and cultural evolution must be addressed (Cole, 2005). Some general comments on Piaget's treatment of mathematics within his theory of genetic epistemology are followed by a specific critique of the supposed correspondence he claimed between the 'mother structures' of Bourbaki and constructs within his theory of cognitive development. A contrasting position is based largely on Freudenthal's (1991) conception of 'guided reinvention' and James Kaput's (1994) conception of 'applied phylogeny'.

Among the facets of the cultural environments in which children grow up is the panoply of representations, both formally introduced within mathematics classes and encountered in the environment in general. Assimilation/accommodation of existing, collectively sanctioned, representations is a very different matter from the original slow development, with evolutionary selection, of those representations. Another glaringly obvious historical observation, evident through a cursory glance through Florian Cajori's (1928) painstaking work, is that notations and representations are contingent, arbitrary, underdesigned – whether that will ever change is doubtful.

Finally, in this section, to illustrate some of the issues, I take a look at a particular content area, that of negative numbers.

Simplistic parallelism: Piaget and Bourbaki

> The fundamental hypothesis of genetic epistemology is that there is a parallelism between the progress made in the logical and rational organization of knowledge and the corresponding formative psychological processes.
>
> (Piaget, 1970, p. 4)

In this section, I first offer some general assertions about the significance of Piaget for the study of mathematical cognition and the practice of mathematical education, then I specifically discuss Piaget's claim that the mother structures of Bourbaki correspond to empirically supported developmental cognitive structures.

Piaget identified himself primarily as a genetic epistemologist, not a developmental psychologist. He became the latter in service of the former, arguing, as cited earlier, that since little is knowable about the origins of human thinking, the best recourse is to study children. Moreover:

> The defining feature of Piaget's approach [...] is that the stages and mechanisms that he postulates are not psychological, or historical (so he is not 'reporting' an accidental parallel between the two), but rather, epistemological – this is how knowledge is inherently constructed. (Kaput, 1994, p. 84)

Rather than attempt a systematic critique of this position – an enormous undertaking – I merely state my conviction that I find it unconvincing, unless it is reduced to the banal statement that knowledge grows through developmental processes which can be described in such general terms as to fit both domains. I would even conjecture that having framed his position on intellectual-aesthetic grounds, Piaget devoted much of his life as an experimentalist to confirming it.

Further, while in the spirit of making controversial statements, I will suggest that in his emphasis on adaptation, initially with respect to the physical environment and originating in his first experimental investigations as a biologist studying adaptation of molluscs, Piaget extended, in a kind of metaphorical way, to the other environments that I have labelled cultural and constructed (educational). In contrast, it has been argued that

> the contemporary study of the role of culture in human development is hampered by the continued failure of behavioral scientists to take seriously the co-evolution of phylogenetic and cultural-historical change in shaping processes of developmental change during ontogeny. (Cole, 2007, p. 236)

Piaget's monumental oeuvre is of great importance, especially against the backdrop of behaviourism in the earlier part of the twentieth century (space does not allow consideration of the part played by Lev Vygotsky and other Russian psychologists in theorising the social, collective complexities of education, nor the intimate involvement of Russian mathematicians in school mathematics). At a macro-level, Piaget revealed the complexity of children's thinking; however, there are several criticisms that are particularly relevant to mathematics education. Of these, perhaps the most important is that, in postulating an explanation in terms of adaptation to environment, he understated the differences between physical, social, political, and constructed environments. Having spent a reasonable amount of time studying his work, and doing related research on children's cognition, I, like others, find the claim unconvincing that people, whether historically and collectively, or contemporarily and individually, construct mathematics through reflection on their interactions with the environment:

> The epistemological approach which starts from the position of the individual alone is so wrong. The fact that such an approach fits in with the biological approach which similarly considers the individual organism in relation to its environment equally shows the inappropriateness of that as a model on which to construe the growth of knowledge and cognitive development generally. (Hamlyn, 1978, p. 59)

Piaget believed that the systematisation of (some parts of) mathematics by Bourbaki, in particular their postulation of three 'mother structures' constituted a striking confirmation of his position. It is my impression that the Bourbaki group was willing to collude in this belief as it strengthened their own case to be central to a network of structuralist ideas. The strongest critique of the supposed relationship was made by Freudenthal, who wrote:

> Poor Piaget! He did not fare much better than Kant, who had barely consecrated Euclidean space as a 'pure intuition' when non-Euclidean geometry was discovered! Piaget is not a mathematician, so he could not know how unreliable mathematical system builders are. [...] Mathematics is never finished – anyone who *worships* a certain system of mathematics should take heed of this advice. (Freudenthal, 1973, p. 46, emphasis added)

Piaget (1970) did acknowledge the emergence of category theory as a systemic reformulation, but without suggesting how that affected his correspondence hypothesis.

Another mathematician who tangled with Piaget was René Thom, a topologist known for his development of catastrophe theory. In particular, Thom argued that Piaget's position on geometry was gravely wrong, as did Freudenthal: 'it is a serious mistake if, to justify a particular kind of didactics, people tell you Piaget proved Euclidean geometry to start psychologically with Cartesian coordinates' (1973, p. 669).

To close, a particular issue that has puzzled me for nearly forty years is the ontological status of 'cognitive structures' as something that a child 'has' (Jeeves & Greer, 1983, pp. 65–69). In those pages, we used a lengthy quotation from Feldman and Toulmin (1976, p. 426), which seems to go to the heart of the issue:

> Nowhere, it seems, are the differences between the problems involved in formally representing a theory and the problems in empirically testing it so difficult to keep separate as in the area of cognition. Just because the theoretical system in question can plausibly be represented as corresponding to some mental system in the mind of an actual child, we may be led to conclude that the formalism of the theoretical system must be directly represented by an isomorphic formalism in the mind of the child... In this way, ontological reality is assigned to the hypothetical mental structures of the theory simply on the basis of the formal expressions by which they are represented in the theory.

Guided reinvention, applied phylogeny

> Children should repeat the learning process of mankind, not as it factually took place but rather as it would have done if people in the past had known a bit more of what we know now. (Freudenthal, 1991, p. 48)

The qualification within the statement is crucial; Freudenthal's vision of reinvention was with strategic instruction, guided by what Kaput (1994, p. 83) termed 'applied phylogeny'. Kaput introduced this term with appropriate warnings about the cognitive appeal of 'ontogeny recapitulates phylogeny' including 'the differences between a collective historical enterprise and an individual's learning' and 'the irregularity of historical developments'. To use an obvious example, nobody would suggest that children should be taught the Roman way of labelling natural numbers before the decimal system.

By way of example of 'repeating the learning process of mankind', consider multiplication and division of positive real numbers. In ancient Mesopotamia (as extensively documented by Høyrup) and in Peru (Urton, 1997), for example, the operations were linked polysemously to cultural practices. For the Quechua of Peru, multiplication and procreation were closely connected (and remember that, in the Bible, people are told to 'go forth and multiply'). To put it another way, multiplication and division can be used to model many classes of situation. In particular, there is a marked contrast between 'asymmetrical' situations – in which the quantities multiplied are clearly distinguishable as multiplicand and multiplier – and 'symmetrical' situations, such as rectangular area, in which they play equivalent roles. As a consequence, in the former case, there are two distinct forms of division, but not in the latter (Greer, 1992).

Contrast the above with the formal treatment of the operations. From a Bourbakian perspective, they are applied in decontextualised computation and organised within groups and other disembedded structures. In schools, arguably a great deal too much emphasis is on computation and formal properties such as commutativity, treated abstractly and not in relation to situations modelled, within which its nature varies greatly – sometimes addition and multiplication are trivially commutative, sometimes not. Similar comments apply to the statements that addition and subtraction, multiplication and division are inverse operations.

The contrast between an abstract structure, such as a group, and diverse instances of it, such as transformation groups in geometry, was characterised by Freudenthal (1991, p. 20) in terms of 'rich' and 'poor' structures. Groups, historically, were encountered as rich structures in multiple different contexts and only axiomatised relatively recently. In Freudenthal's vision for teaching mathematics, the axiomatisation should come as the culmination of a long process – starting with the axioms or the poor/pure structure was, in his view, a gross pedagogical error, a 'didactical inversion'. Thus (p. 29) 'the didactically recommendable direction will be the same as that in which mathematics arises, that is, from rich to poor'.

In Freudenthal's vision, also, he emphasised changing the view of mathematics learning as accumulating content and 'neatly tailored abilites, the mastery of which can be tested "objectively" (as they call it) – that is, by computers' (1991, p. 49) to experiencing important mathematical activities:

> The learner should reinvent mathematising rather than mathematics; abstracting rather than abstractions; schematising rather than schemes; formalising rather than formulas; algorithmising rather than algorithms; verbalising rather than language...

Cajori (1898) wrote that: 'The history of mathematics may be instructive as well as agreeable; it may not only remind us of what we have, but also teach us how to increase our store'. Citing Augustus De Morgan, he continued: 'The early history of the mind of men with regard to mathematics leads us to point out our own errors; and in this respect it is well to pay attention to the history of mathematics'. This principle lies at the heart of Kaput's notion of applied phenology. Painstaking research in the historical record can identify cognitive obstacles and the ways in which they were, often after considerable time, resolved. That can then guide the teaching of children, in line with the quotation at the start of this section.

Material representations

> A class in arithmetic [...] will find it astonishing that it should have taken so long to invent a notation which they themselves can now learn in a month.
>
> (Cajori, 1928, p. 3)

The importance of representations in the growth of mathematics historically is discussed in Chapter 2. There are, of course, huge differences between, on the one hand, the invention by mathematicians of representations in the service of the mathematics symbiotically being created and systematised, and, on the other, the presentation to children of evolutionarily stable representations. Is it any wonder that the presentation of the products of such long-drawn-out efforts as off-the-shelf resources for children to use is rife with complications? For example, mathematicians are prone to regard the graphical representations of functions in the Cartesian plane as perspicuous, yet a mass of empirical evidence exists to show that misinterpretations are extremely common and difficult to dislodge.

In passing, a point that is obvious to anyone reading Cajori's (1928) painstaking survey is that mathematical vocabulary, notations, and representational conventions are created very arbitrarily (an example being the conventional use in algebra of a, b, c as parameters and x, y, z as variables). Why do children (in English, as in many other European languages, but not German) have to deal with 'isosceles' rather than the Anglo-Saxon 'twesided' used by Robert Recorde in the sixteenth century (Cajori, 1922)? For a discipline whose exponents pride themselves on their rationality, the representations used in mathematics are surprisingly user-unfriendly.

One way in which design *has* been prominent is in the conscious development of material teaching/learning resources termed 'manipulatives'. The earlier history of these in (some parts of) European mathematics education is well covered in De Bock and Vanpaemel (2019). Reflecting a distinction that is very clear for computer representations (Kaput, 1992), these are representational resources which children can use for recording, but also for acting upon. For teaching arithmetic, examples include Cuisenaire rods and the multibase arithmetic blocks designed by Zoltán Dienes.

The prominence of manipulatives has declined. One reason is that their pedagogical effectiveness has been called into question. Rather like the problems in trying to turn Pólya's heuristics into classroom gold the issue is that you cannot understand how to exploit a heuristic such as 'think of a related problem' unless you know a great deal already about what a related problem looks like. Likewise, those who can understand how a manipulative relates to the mathematics it is designed to illuminate have little need for the manipulative. Conversely, manipulatives may be of limited effectiveness for those who do not understand the connections (like the child who told Kath Hart that 'bricks is bricks and sums is sums'). With respect to the multibase arithmetic blocks:

> Children who already understood base and place value, even if only intuitively, could see the connections between written numerals and these blocks [...] But children who could not do these problems without the blocks didn't have a clue about how to do them with the blocks [...] They found the blocks [...] as disconnected from realty, mysterious, arbitrary, and capricious as the numbers that these blocks were supposed to bring to life. (Holt, 1982, pp. 281–219)

New representational windows

> [...] information technology will have its greatest impact in transforming the meaning of what it means to learn and use mathematics by providing access to new forms of representation as well as providing simultaneous access to multiple, linked representations. (Kaput, 1986, p. 1)

I seem to remember Benoit Mandelbrot, speaking in 1992 saying that 'the computer has put the eye back in mathematics'. In evolutionary perspective, the computer age represents a new epoch in the creation of representational resources, with consequent massive implications for cognition (Kaput & Shaffer, 2002). Examples follow.

One of the first such revolutionary visions exploiting developments in information technology (IT) was the language Logo, designed by Seymour Papert. It rivals the Turing machine in terms of the simplicity of its primitives. Inspired by the young Papert's fascination with gears, the basic mechanism (literally embodied in drawing machines called 'turtles' which drew geometrical configurations controlled by the language) was an axle with equally sized wheels on the ends, which turn at the same speed either in the same direction, thus moving the turtle forward a stipulated distance, or in opposite directions, rotating it through a stipulated angle. The second element in generativity is the programming language in which the user can define and name routines; the names are then appended to the language. Logo produced a way of conceiving planar geometry very differently from Euclidean geometry. A circle, for example, is approximated to any desired degree of precision, as a regular polygon.

Another geometrical system, more closely linked with traditional geometry is Geometric Supposer (GS) (see also Cabri, and 3-D versions). Within GS, constructions can be defined similar to those of Euclidean geometry and recorded as *procedures* rather than *drawings*. The theorem that if you construct any quadrilateral and join the midpoints of the four sides you get a parallelogram takes on a different feel when you can grab it by a vertex and make the whole construction waltz on the screen. Among many other wonderful creations may be mentioned Fathom, which affords exploration of probability and statistical modelling.

Kaput himself designed SimCalc and other software as resources for teaching calculus exploiting the kinds of features that he analysed (Kaput, 1992). And STELLA makes dynamic system modelling accessible to high school students (Fisher, 2021).

Reading such work in the 1990s, a reader might well have thought: 'Just think what they'll be able to do in schools thirty years from now'.

They would be greatly disappointed. There are many reasons for this, some of which are to do with the IT industry seeing more profit in other kinds of product than in tackling the complexity of teaching mathematics. Another major reason is the failure to grasp the need for teacher training and to provide the necessary support. For example, Papert's vision of Logo as a mathematical world in which children could learn by themselves was arguably overoptimistic, and it progressively became clear that its effectiveness could only be realised under the guidance of skilled teaching.

A historical example: Directed numbers

3 – 8 is an impossibility, it requires you to take from 3 more than there is in 3, which is absurd. (De Morgan, 1810/1931)

Minus times minus makes a plus.
The reason for this we need not discuss. (Attributed to W. H. Auden)

The case of directed numbers may be taken as paradigmatic for considering how a study of the history of mathematics might inform contemporary teaching. De Morgan was an eminent mathematician but balked at an arithmetical expression that quite young children today are expected to take in their stride. He was right if the only interpretation of $n - m$ (where n and m are whole numbers) is the removal (in some sense) of m countable entities from a set of n. (And he did acknowledge the *algebraic* interpretation of $n - m$ when $m > n$.)

It could be argued that there is a fundamental epistemic shift illustrated here, from $n - m$ as a direct representation of a situation (taking objects away from a set of objects) to $n - m$ as a mathematical expression that can be used to model many situations – such as bank balances (did De Morgan never get into debt?) or scales with a zero such as those for measuring temperature or altitude relative to sea level. (When children later are being taught multidigit subtraction, e.g. 43–18, the teacher may say something like 'you can't take 8 from 3, so you borrow 10 and take 8 from 13'). From a purely structural point of view, the expansion of the positive whole numbers to all integers means that subtraction is closed over the set of whole numbers, which form a group under addition, with identity element 0.

Ademio Damazio (2001, p. 209), on the basis of classroom observations of children being taught about directed integers, concluded that 'the students did not overcome the concept of number as an ability to count concrete objects instead of as abstract objects that can be operated independently'. Well, why would you expect them to achieve such a feat over a series of twenty class lessons, what De Morgan, after a full mathematical education and career, failed to achieve? But try telling that to curriculum developers!

On the basis of observations of a teacher, Damazio (2001, p. 208) commented that 'the teacher ceases to evidence relevant aspects for concept formation. You can do that if you are concentrating on calculative fluency alone. In particular, the notion of a relative zero (as a reference point) as opposed to that of absolute zero [...] is the foundation of the concept of relative whole numbers'.

The case of multiplying and dividing negative numbers is much more complex than addition and subtraction and took even longer to resolve to the satisfaction of rigour-demanding European mathematicians (with false starts over centuries along the way, and eventual survival of what works). Within this context, the shift to modelling is even clearer. How can the plausibility of the rule Auden was told not to discuss (see above) be communicated to a child? There are a number of general approaches:

- *Patterns*. A two-dimensional table can be constructed with 0, 1, 2, 3, ... along each axis and the products in the body of the table. Extending back along each axis to –1, –2, –3, ... and following the patterns makes the rules for multiplying directed numbers at least somewhat plausible (for an excellent discussion, see Sawyer, 1964, pp. 297–300) and a similar exercise can be carried out graphically (pp. 300–309). (Such patterns could be thought of as 'localised structures', partial reflections of the structures of Bourbaki and the like.)
- *Modelling linear change over time*. Suppose you are walking on steps at a constant rate of n steps per minute, not up ($+n$),

but down ($-n$). Then t minutes earlier ($-t$) you were nt steps higher than you are now.

- *The algebraic/geometrical approach of the Babylonians.* Consider the expression $(x - a)(y - b)$. It is straightforward if $x > a$ and $y > b$ and easily verified through examples that its expansion as $xy + x(-b) + (-a)y + (-a)(-b)$ 'works' if interpreted as $xy - xb - ay + ab$. And there is a geometrical counterpart.
- In Greer (2005), I cited a hilarious formal 'proof' written by mathematicians for sixth-grade students and I cannot resist reproducing the start of it here:

> First, if a number a satisfies $b + a = 0$; then a is $-b$. That is how $(-b)$ is defined, as the additive inverse of b. Second, $N \times 0 = 0$ for any number N because the area of a rectangle with one side zero is zero [...] Third:
> $$0 = (-1) \times 0 = (-1) \times (1 + (-1) = (-1) \times (1) + (-1) \times (-1)$$
> (California Department of Education, 2000, p. 144)

(How I like mathematicians to speak of 'a rectangle with one side zero'!).

- *A student's pragmatism.* I asked students in a general college mathematics class to (a) say if they believed (–1) x (–1) = +1 and (b) explain why. The answers were, as you might expect, mainly appeals to authority of some kind. However one student wrote that he believed it because every time he had operated according to that belief in a test he had gained marks, and conversely.

I would be prepared to argue for what would generally be considered a radical solution, namely to postpone treatment of multiplication and division of directed numbers until college, at which point it could be treated with the formal and informal thoroughness it warrants with students who have more relevant experience. I can cite one prominent mathematician who wrote that 'the multiplication of negative numbers (like the addition of fractions) can and should be postponed' (Hilton, 1984, p. 8). Whenever it is introduced, it damned well ought to be discussed.

What is mathematics education for?

Introduction

> Mathematics as an aim in itself [...] is an important aspect, although of less concern to us here, since our subject of mathematics education embraces a much larger group than only future professionals of whom once again only a small minority choose mathematics in itself. (Freudenthal, 1991, p. 3)

In the above quotation, Freudenthal expresses clearly a theme that is omnipresent in the following discussion. Think of a pyramid representing the population of those who spend a lot of time in mathematics classes at school – in most of the world, essentially everyone. A very small section at the top then corresponds to those who will constitute the next cadre of mathematics researchers and tertiary level teachers of mathematics. A larger section below that represents those who use significant amounts of mathematics in their work – scientists, engineers, (some kinds of) social scientists and, generally, certain specialists within most fields (though there is considerable research showing that architects, for example, may use little of the formal mathematics of which they have been required to show mastery (Hacker, 2016)). The remaining bulk of the pyramid represents everyone else. Quite simply put, the thrust of this section of the chapter is to argue that the pyramid should be inverted, so to speak, so that school mathematics much more deliberately reflects the needs of the majority; to put it provocatively, school mathematics education is too important to leave to mathematicians who are primarily invested in perpetuation of their subspecies.

As a start, I pick up on earlier discussion of how formal mathematics influences curriculum, a particular case being the impact of Bourbaki. While the overt influence of Bourbaki has waned, its ghost still haunts mathematics classrooms (along with those of its extended family) in terms of premature formalisation. The Common Core State Standards used in the United States may be taken as a representative contemporary curricular design in terms of content that could be termed 'Bourbaki light' – a framework based on progressive mathematical structuration with premature formalisation rather than pedagogical and developmental considerations, and with scant attention to long-term pre-emptive planning or to the critical points at which conceptual change must be carefully nurtured.

Accordingly, a counter-position is presented whereby curricular design is fundamentally respectful of the child's capacity for understanding and accumulated experience at any point. The first point to be made is that given the vast amount of recorded and systematised mathematics, the selection problem (already mentioned in relation to Bourbaki) rears its head. In the face of what might be considered the (somewhat) reasonably reactionary nature of curricula, a number of radical alternative design principles are proposed, in particular aimed at making school mathematics useful to the adults that students become, rather than being a preparation for a small elite. As part of that argument, I contend below for substantial reductions in the level of formalisation of content and framing (which would also help teachers). These proposals are also linked with the proposal to move the centre of gravity, substantially, away from technical mastery and towards understanding – which, it is argued, would benefit also the elite who become advanced students of mathematics (and, again, teachers).

Given the degree to which mathematics formats our lives (to use Skovsmose's term), those who frame school mathematics now have a responsibility to include instruction about modelling, its purposes, and its limitations. These aspects arise sharply, and very early on in elementary school, in the context of 'the bizarre genre of word problems', a locus within which quite young children could be taught to distinguish between modelling that is precise, modelling that is a more or less good approximation, and modelling that is plain wrong.

All of these considerations build to the argument that the relationships between mathematics and the social sciences be re-examined (see Chapter 9, this volume). Particularly important aspects include:

- The nature and purposes of mathematical modelling.
- Talking with students about mathematics, what it is *for*, how it is taught/learnt, the cognitive obstacles, its history, and its political ramifications.
- Mathematics in relation to aspects of life important to the students, their families and communities.

Premature formalisation

The influence of Bourbaki on mathematics and mathematics education in the twentieth century is discussed in Chapter 2. While the overt influence of Bourbaki and other formally-intensive statements by mathematicians has waned, its lingering influence can be seen in the perseverance of premature formalisation.

By way of an example, the Common Core State Standards for Mathematics (CCSSM) developed in the United States may be taken as a reasonably representative curricular framework in terms of specifying mathematical content. It is critically flawed in many ways, in particular in its failure to offer pedagogical guidance. In his masterly comparative analysis of national curricula in fourteen countries, Geoffrey Howson (1994, p. 26) made the crystal-clear point that 'a curriculum cannot be considered in isolation from the teaching force which must implement it'. I argue below that the dominance of mathematicians in its framing illustrates the harmful effects that mathematicians can have on school mathematics education.

Unlike some manifestations of New Math of the 1960s (which I remember living through), set theory is not proposed as the starting point for children's learning of mathematics, thus avoiding the absurdity of confusing the foundations of mathematics education with the foundations of mathematics as traditionally presented by philosophers; nevertheless, premature formalisation is pervasive. The ghost of behaviourism lurks, in that the framework is very much presented in terms of an incremental progression on a superficial metric of complexity, logical in the sense of the adult mathematician's retrospection, not in terms of children's ability to understand.

As an example, consider the extension of multiplication and division beyond the natural numbers. This is what we read:

> **Apply and extend previous understandings of multiplication and division to divide fractions by fractions.**
>
> Interpret and compute quotients of fractions, and solve word problems involving division of fractions by fractions, e.g. by using visual fraction models and equations to represent the problem. For example, create a story context for $(2/3) \div (3/4)$ and use a visual fraction model to show the quotient; use the relationship between multiplication and division to explain that $(2/3) \div (3/4) = 8/9$ because $3/4$ of $8/9$ is $2/3$. (In

> general, $(a/b) \div (c/d) = ad/bc$. How much chocolate will each person get if 3 people share ½ lb. of chocolate equally? How many ¾-cup servings are in 2/3 of a cup of yogurt? How wide is a rectangular strip of land with length ¾ mi and area ½ square mi? (CCSSM, p. 42, for Grade 6)

If you are not laughing, you have not been paying attention. It is hard to expunge the image of someone embarrassedly saying to a guest 'I am so sorry, I can only offer you 8/9 of a serving of yoghurt'. Another compelling image is of the unfortunate person tasked with devising a believable story to go with $(2/3) \div (3/4) = 8/9$ for a sixth grader.

Indeed, the ways in which fractions are treated are indicative of the problems I am trying to elucidate. Here are some of the issues:

- Essentially nobody apart from children in school needs to compute something like 4/7 + 5/11. People like carpenters and engineers, who make things that work, use decimals or binary fractions. What would be lost by following their example, restricting instruction to the few fractions and uses of fractions that people generally find useful (as approximations, for example)? The loss for formalists would be the lack of the formal closure of the positive rationals under the four arithmetical operations.
- Not unrelated is the common observation that fractions often constitute the first wall of incomprehension in mathematics class. A *Peanuts* cartoon depicts a young child responding to her teacher's enquiry 'Do you have any questions (about fractions)?' by asking 'Do you hate us?'.[1]
- Typically, mathematics educators see fractions as having multiple aspects, embedded in the complex conceptual field of multiplicative structures (Vergnaud, 2009). On the other hand, Hung-His Wu (1999) objects to this position on various grounds, appealing to the mathematician's dogma that mathematics is formal, abstract, simple, precisely defined. For example, the student is expected to 'understand a rational number as a point on the number line' (CCSSM, p. 43). How can

1 See Charles Schulz (November 7, 1991), *GoComics*, www.gocomics.com/peanuts/1991/11/07

> a number be a point? A full treatment of Wu's position would require a book (which I may yet write). Mathematicians tend to approach fractions in terms of computational properties and embeddedness in formal structures, such as groups, and to try to project that perspective onto learners.

CCSSM (pp. 6–8) has a very short section on 'Standards for mathematical practice' namely:

1. Make sense of problems and persevere in solving them
2. Reason abstractly and quantitatively
3. Construct viable arguments and critique the reasoning of others
4. Model with mathematics
5. Use appropriate tools strategically
6. Attend to precision
7. Look for and make use of structure
8. Look for and express regularity in repeated reasoning

Together with a brief characterisation of each, with examples. As such, this is a fine list, but the teachers who plunge into CCSSM expecting enlightenment on how to cultivate such practices in their classrooms will find little. In particular, I find the treatment of Modelling (Standard 4) decidedly undernuanced and overly limited to a straightforward encoding of a described situation into a precise mathematical formulation.

Arguably, premature formalisation represents the most pervasive and harmful influence from mathematicians on mathematics-as-school-subject. For the sake of making an argument, let me list positions that can be found in the writings of mathematicians (sometimes bordering on caricature):

- The attitude that if you define everything with precision, build everything up logically, step by step, and they still don't get it, it's their fault.
- Mathematics-as-school-subject exists primarily for the preparation of the next cadre of research mathematicians.

Schools should teach a great deal of calculus, in particular, so that university mathematics instruction can hit the ground running.

- In lamenting what he perceives as a schism between mathematicians and mathematics educators, Michael Fried (2014, p. 4) expressed nostalgia for the time when 'asking about the distinction between mathematics and mathematics education would have been like asking about the distinction between mathematics and geometry'.
- In addressing the mathematical education of children, mathematicians tend to project their own images (or 'pictures', see Freudenthal, 1991, p. 131) of mathematics onto children.

Overall, it can be argued that mathematicians (of course, with many exceptions, have had harmful effects in multiple ways:

- Perpetuation of the Graecocentric narrative of the history of academic mathematics through a combination of laziness and ideology. In my opinion a line can be drawn between this and the manifestations of racism in contemporary classrooms.
- Denigration of the mathematical practices of those who make things that work (Chapter 17, this volume).
- A tendency to assume that being good at mathematics is not only a necessary but also a sufficient condition for being good at teaching mathematics.
- Alienation from, and perceived irrelevance of, mathematics combined with a propensity for intimidation.
- Failure to sew the seeds of criticality and agency in future citizens.
- Misuses of mathematics in the service of states.

As mentioned above, throughout history there have been exceptions. I will mention some personal heroes. Ubiratan D'Ambrosio brought to our field the necessary radically different kind of thinking that began to liberate Eurocentric anthropology and psychology from their imperial and colonialist roots. The influences of Hans Freudenthal and Jim Kaput on my thinking must be obvious in this chapter. (I vividly remember the latter commenting on a paper and drawing on the blackboard a large, amorphous creature, representing mathematics education. He then meticulously drew one toenail and commented 'We spend too much time analysing toenails on the creature when we should be analysing the creature'.) Reuben Hersh was one of the leaders in a radical reformulation of what philosophy of mathematics might be about, communicated accessibly what mathematicians do when they do mathematics, and illuminated the pervasive diversity within academic mathematics. Alan Schoenfeld, and his career, speak for themselves in his chapter in this volume.

Rethinking curriculum

Here I am using curriculum in the sense of a plan for the contents and sequencing of school mathematics. As a starting-point, remember the metaphor of inverting the pyramid, introduced above. Taking that position has heavy implications for content. Above all, combined with a shift in the centre of gravity from mere competence to understanding and problem solving, and attention to premature formalism, there could be a drastic reduction in the amount of 'technical' mathematics, including, as argued above, work with fractions and multiplying negative numbers (Hilton, 1984). Of course, there are protests against such a position. One such argument, that I find ill-founded, is that it hurts those children who are mathematically gifted. By way of counterarguments, I would characterise such giftedness as partly a cultural construct heavily loaded with connotations of cultural capital and that, in these days when masterclasses can be put online, enrichment is easily provided for those children who should (in whatever sense) have it (with careful provision to ensure such facilities are equitably accessible). As for students arriving at university with less technical knowledge under their belts, maybe the university teachers need to up their game. And, bearing in mind the adage that if you have four hours to chop wood, you should spend the first two hours sharpening the axe, if they arrive better able to 'think mathematically' (and enjoying mathematics rather than, at best, being rewarded by competence) that may be more than ample compensation.

As argued at various points, and see further below, in modern times mathematical modelling needs to be taken seriously, with a lot more attention to the socially and politically situated processes of assumption-making, simplification, mathematisation, interpretation, and communication of results. The historical alignment of mathematics with the physical sciences is discussed in Chapter 8, together with suggestions that this traditional alignment be reconsidered. The extension of mathematical modelling to social phenomena is reflected in the prominence of the use of mathematical techniques in social sciences such as experimental psychology (emphasis on measurement, psychometrics, and statistics).

Then there are what I think of as the rights of the child. As far as cognition goes, foremost of these is the right to sense-making. As far as identity goes, there are cultural rights, including access to an accurate (as far as possible) and balanced history of the development of academic mathematics, as well as an appreciation of the 'funds of knowledge', which is 'based on a simple premise: People are competent, they have knowledge, and their life experiences have given them that knowledge' (Gonzalez, Moll, & Amanti, 2005, p. ix). And then there are the multiple aspects of equitable treatment, educationally and personally.

Curriculum developers, in my opinion, can show a remarkable ability to fail to learn from history; this reflects, and partly explains, the stultifying inertia that characterises school mathematics, the slowness to embrace new content and resources. Lessons from the failure of the variety of attempts made under the banner of New Math have not been sufficiently absorbed. Paralleling trends in assessment, curriculum designers appear increasingly to benchmark against productions in other countries. This can lead to the error of importing a particular resource without the cultural embedding that makes it effective in its original milieu. It also encourages convergence (almost in a mathematical sense) with consequent implications for homogenisation.

If we look back as recently as the 1990s, at that time there were significant advances in assessment, even to the point of creating optimism (see review in De Corte, Greer, & Verschaffel, 1996, pp. 530–534). That has largely disappeared – by way of example, one only has to look at the fate of Smarter Balanced Assessment Consortium as narrated by Schoenfeld (Chapter 14, this volume). In a similar way, as described above, the potential of computers as reviewed by Kaput (1992) has yet to be adequately realised in classrooms.

And there are what might be called emotional rights. There is no reason why elementary mathematics instruction should not be an intellectual playground. There is no reason why mathematics teachers should tend to authoritarianism, but the subject certainly provides multiple opportunities for any such tendency. Mathematicians who love mathematics express sympathy for those who are alienated by it (or more accurately what they have been confronted with), but might spend more time thinking about whether they need to show intellectual empathy for the children who lack the facility with mathematics that they themselves typically enjoyed when young.

Teachers have rights, too, but that's another story.

Coherent long-term design

> Calculus might be regarded as a web of ideas that should be approached gradually, from elementary school onward, in a longitudinally coherent school mathematics curriculum. (Kaput, 1994, p. 78)

Kaput was talking specifically about calculus – which he suggests should be reconceptualised as 'the mathematics of change' (p. 152) and not necessarily built on the traditional foundation of algebra (pp. 77–78) – but the point applies equally to any major branch of mathematics. To give another example, instead of the framing 'the transition from arithmetic to algebra', the inherently algebraic nature of arithmetic may be recognised, and there are plenty of pedagogical moves to do just that.

Here I make what I see as vital points about curricular design being long-term, coherent, forward-looking, and mindful of conceptual obstacles and pedagogical dilemmas. At a very concrete level, a century ago, Edward Thorndike (1922) observed that children's mathematical conceptualisations are significantly framed by the examples to which they are exposed. A narrow range can result in a narrow understanding.

The lingering effects of behaviourism in folk pedagogy include a belief in the obviousness of the principle of monotonic and incremental movement along a simple/complex dimension, and the short-termism engendered by the desire to maximise scores on the next test. Efraim Fischbein pointed out the consequent dangers:

> From the educational point of view there is an important problem to be considered by curricula writers and by teachers. A certain interpretation of a concept or an operation may be initially very useful in the teaching process as a result of its intuitive qualities (concreteness, behavioral meaning etc.). But as a result of the primacy effect that first model may become so rigidly attached to the respective concept that it may become impossible to get rid of it later on. The initial model may become an obstacle which can hinder the passage to a higher-order interpretation – more general and more abstract – of the same concept. (Fischbein, 1987, p. 198)

A similar warning was expressed by De Morgan in relation to number and arithmetic:

> If we could at once take the most general view of numbers, and give the beginner the extended notions which he may afterwards attain, the mathematics would present comparatively few impediments. *But the constitution of our minds will not permit this.* (De Morgan, 1831/1910, p. 33, emphasis added)

Accordingly, it is essential to identify points at which conceptual change is difficult (and here history can be an indispensable guide) and then look for bridging resources. A clear example is the extension of multiplication and division beyond the positive integers to positive rational numbers, particularly those less than 1. The ramifications of this extension have been very extensively researched, in particular how 'multiplication makes bigger, division makes smaller' is no longer valid (here Thorndike's precept is particularly relevant). The consequent difficulties can be ameliorated, as discussed in Greer (1994), by pre-emptively including examples of multiplication and division by numbers less than 1 as early as possible and by the use of bridging example sets and representations. The point was well made by Cajori (1898):

> That, in the historical development, multiplication and division should have been considered primarily in connection with integers, is very natural. The same course must be adopted in teaching the young. First come the easy but restricted meanings of multiplication and division, applicable to whole numbers. *In due time, the successful teacher causes students to see the necessity of modifying and broadening the meanings assigned to the terms.* (p. 183, emphasis added)

As an overarching principle, the above considerations should be discussed with students. Fischbein, for example, has recommended telling students about historical examples showing that conceptual change is difficult for mathematicians too.

Many other examples come to mind of the consequences of lack of forethought. To a mathematician, it is obvious that when performing calculations on numbers that are measures of quantities, e.g., multiplying speed by time to get distance, the operation is invariant in relation to the numbers; for a child this principle is very far from evident, as shown by considerable research. Again, to a mathematician, multiplication is commutative, but in certain contexts it isn't, in the sense illustrated by the following example. To find the distance travelled by something at a constant speed of 0.75 miles per hour for 3 hours is instantly recognised as being found by multiplying the two numbers, but if it is 3 miles at 0.75 miles per hour, not so (many children will say the answer can be found by dividing 3 by 0.75, plausibly because they realise that the answer will be less than 3 and 'multiplication makes bigger, division makes smaller').

Modelling: From unreasonable effectiveness to reasonable ineffectiveness

> The unreasonable effectiveness of mathematics. (Wigner, 1960, title)

> The reasonable ineffectiveness of research on mathematics education. (Kilpatrick, 1981, title)

Eugene Wigner's seminal article addressed the question of how it is possible, for example, to predict through mathematics the movements of celestial bodies. Jeremy Kilpatrick, in relation to mathematics education, pointed out that the answer to 'How can we send a man to the moon, but cannot improve mathematics education?' is that the first is, however complex, a technical problem, while the second is a human problem. A similar contrast is evident in moving from the modelling of physical phenomena to the modelling of phenomena involving humans. Further, given the pervasiveness of what Skovsmose has characterised and analysed for decades as the formatting of our lives through mathematical modelling, it is important that the curriculum address the associated complexities.

The simple schematic for modelling in terms of mathematisation, development of implications, evaluation, and revision needs to be elaborated to include the following aspects:

- Considerations of who is doing the modelling and for what purpose; there is massive added complexity, particularly at the stages of mathematising the situation, including, in particular, what assumptions are made.
- Limitations imposed by technology and techniques available – these diminish with computing developments, but are still an issue for students.
- Evaluation of the outcomes of manipulating the model are also subject to the motivations of the modellers.
- Communication and dissemination of the results, especially some sense of their fragility (reasonable ineffectiveness); motivations of the modellers are also central at this point.

The foregoing considerations have massive implications for what should be taught – not just modelling in the sense of examples of how it is done, but questions of why. An extreme (in my view) counterargument was put, with admirable clarity, by André Toom (1999). His position was that, rather than viewing word problems as having anything to do with applications, the purpose of including such problems is simply to help teach pure mathematics and students to quickly learn the rules. The issue is pinpointed in the remarkable amount of discussion about the single equation $2 + 2 = 4$. As Houman Harouni (see Chapter 12, this volume) outlines, this discussion can become very rarified; I find the explanation by Reuben Hersh (1997, p. 16) straightforward and convincing, that the equation has a double meaning, as a statement of arithmetic, and as a description of what happens when 2 stable entities are put together, without interaction, with 2 other stable entities. To elucidate slightly, $2 + 2 = 4$ may afford a precise model – if I go to the bakery for donuts and my wife has said to get two for her, and two for myself, I could be in trouble if I come back with a number of donuts other than four. Or it could be totally wrong as a model – if your doctor says you can drink 2 pints a night and a second opinion confirms that recommendation, that is not a licence to drink 4 pints a night.

In Verschaffel, Greer, and De Corte (2000), we argued that what are called 'word problems' or, especially in the United States, 'story problems' could, indeed should, be treated as simple exercises in modelling. There is no reason why, through such problems, young children should not be introduced to the core insight that models may be exact, approximate, or plain wrong, and that it is possible to discriminate among those cases. Giyoo Hatano (1997) argued that the cost of increasing the demands on students by having them learn about complexities is too high; the position taken in this chapter that the cost of *not* doing so is also too high. In extension of this line of thinking, I would argue that mathematics education inculcates simplistic thinking. Children are taught the rules of the word problem game, foremost of which is that when you enter the mathematics classroom, you can ignore what you know about the real world and enter:

> A strange world in which you can tell the age of the captain by counting the animals on his ship, where runners do not get tired, and where water gets hotter when you add it to other water. (Back cover of Verschaffel, Greer, van Dooren, & Mukhopadhyay, 2009)

Early school mathematics can be seen as foundational in establishing not just the beginnings of understanding and competence, but also epistemological biases beneath the surface of mathematical content and techniques, including, in particular:

- The implicit assumption that essentially anything can be measured on a single dimension, and therefore individuals and groups can be measured in terms of that variable. The case of IQ provides a particularly clear and consequential example.
- The idea that real-world situations can be modelled unproblematically in terms of mathematical structures and operations and that once numbers and models have been specified, they cannot be disputed.

It does not have to be like that

Many issues about the development of mathematics-as-discipline by humankind were raised in Chapter 2. Likewise, there are fundamental questions in considering mathematics-as-school-subject:

- What are the relationships between the development of mathematics by humanity over millennia and the growth of mathematical understanding in an individual? How can a child be expected to come to grips with conceptual networks that took the combined intellectual resources of humankind millennia to create? How can this challenge be addressed within constructed environments?
- Why is school mathematics so alienating, and unused/ unusable for so many (including highly intelligent people), problematic even for those who succeed, and loved by only a few (Hersh & John-Steiner, 2011)?
- Why do states/societies ask children to endure such stupidity?
- Why is systematic design, illuminated by study of the past, conspicuously lacking?

Above all, we should always return to the basic questions 'What is mathematics education *for*?'. Why could it not be different, and in what ways? Some thirty years ago, I was asked to say briefly what I had learned about mathematics education. I responded: 'For too many people, school mathematics is a personally and intellectually negative experience. It does not have to be like that'. That remains a good summary of how I feel.

References

Brousseau, G. (1997). *Theory of didactical situations in mathematics.* Kluwer.

Cajori, F. (1898). *A history of elementary mathematics.* Macmillan.

Cajori, F. (1922). Robert Recorde. *The Mathematics Teacher, 15*(5), 294–302.

Cajori, F. (1928). *A history of mathematical notations.* Open Court.

California Department of Education (2000). *Mathematics framework for California public schools: Kindergarten through Grade 12*. California Department of Education.

Cole, M. (2005). Cross-cultural and historical perspectives on the developmental consequences of education. *Human Development, 48*, 195–216.

Cole, M. (2007). Phylogeny and cultural history in ontogeny. *Journal of Physiology, 101*, 236–246.

Common Core State Standards for Mathematics. http://www.corestandards.org/Math

Damazio, A. (2001). Mathematical cognition in the classroom: A cultural-historical approach. In M. Hedegaaard (Ed.), *Learning in classrooms* (pp. 191–210). Aarhus University Press.

De Bock, D., & Vanpaemel, G. (2019). *Rods, sets, and arrows: The rise and fall of modern mathematics in Belgium*. Springer.

De Morgan, A. (1910). *Study and difficulties of mathematics*. Open Court. (Original work published 1831)

Donaldson, M. (1978). *Children's minds*. Croom Helm.

Feldman, C. F., & Toulmin, S. (1976). *Logic and the theory of mind*. University of Nebraska Press.

Fischbein, E. (1987). *Intuition in science and mathematics*. Reidel.

Fisher, D. (2021). Global understanding of complex systems problems can start in pre-college education. In F. K. S. Leung, G. A. Stillman, G. Kaiser, & K. L. Wong (Eds.), *Mathematical modeling education in East and West* (pp. 35–44). Springer.

Freudenthal, H. (1973). *Mathematics as an educational task*. Reidel.

Freudenthal, H. (1991). *Revisiting mathematics education*. Kluwer.

Fried, M. N. (2014). Mathematics & mathematics education: Searching for common ground. In M. N. Fried & T. Dreyfus (Eds.), *Mathematics and mathematics education: Searching for common ground* (pp. 1–22). Springer.

Gonzalez, N., Moll, L. C., & Amanti, C. (Eds.). (2005). *Funds of knowledge: Theorizing practices in households, communities, and classrooms*. Erlbaum.

Gould , S. J. (1985). *Ontogeny and phylogeny*. Harvard University Press.

Greer, B. (1992). Multiplication and division as models of situations. In D. Grouws (Ed.), *Handbook of research on mathematics education* (pp. 276–295). Macmillan.

Greer, B. (1994). Extending the meaning of multiplication and division. In G. Harel & J. Confrey (Eds.), *The development of multiplicative reasoning in the learning of mathematics* (pp. 61–85). SUNY.

Greer, B. (2005). Ambition, distraction, uglification, and derision. *Scientiae Paedogogica Experimentalis, 42*, 295–310.

Greer, B. (2021). Learning from history: Jens Høyrup on mathematics, education, and society. In D. Kollosche (Ed.), *Exploring new ways to connect: Proceedings of the Eleventh International Mathematics Education and Society Conference* (Vol. 2, pp. 487–496). Tredition. https://doi.org/10.5281/zenodo.5414119

Hacker, A. (2016). *The math myth and other STEM delusions.* New Press.

Hacking, I. (2014). *Why is there philosophy of mathematics at all?* Cambridge University Press.

Hamlyn, D. W. (1978). *Experience and the growth of understanding.* Routledge and Kegan Paul.

Hatano, G. (1997). Cost and benefit of modeling activity. *Learning and Instruction, 7*, 383–387.

Hersh, R. (1997). *What is mathematics, really?* Oxford University Press.

Hersh, R., & John-Steiner, V. (2011). *Loving and hating mathematics.* Princeton University Press.

Hilton, P. (1984). Current trends in mathematics and future trends in mathematics education. *For the Learning of Mathematics, 4*(1), 2–8.

Hofstadter, D., & Sander, E. (2013). *Surfaces and essences.* Basic Books.

Holt, J. (1982). *How children fail.* Dover.

Howson, G. (1994). *International curricula in mathematics.* The Mathematical Association.

Høyrup, J. (1994). *In measure, number, and weight.* SUNY.

Jeeves, M. A., & Greer, B. (1983). *Analysis of structural learning.* Academic Press.

Kaput, J. J. (1986). *Information technology and mathematics: Opening new representational windows.* Available at: https://files.eric.ed.gov/fulltext/ED297950.pdf

Kaput, J. J. (1992). Technology and mathematics education. In: D. A. Grouws (Ed.), *Handbook of research on mathematics teaching and learning* (pp. 515–556). Macmillan.

Kaput, J. (1994). Democratizing access to calculus: New routes to old roots. In: A. H. Schoenfeld (Ed.), *Mathematical thinking and problem solving* (pp. 77–156). Lawrence Erlbaum.

Kaput, J. J. (1998) Representations, inscriptions, descriptions and learning: A kaleidoscope of windows. *Journal of Mathematical Behavior*, *17*(2), 283–301.

Kaput, J. J., & Shaffer, D. W. (2002). On the development of human representational competence from an evolutionary point of view. In K. Gravemeijer, R. Lehrer, B. van Oers, & L. Verschaffel (Eds.), *Symbolizing, modelling and tool use in mathematics* (pp. 277–293). Kluwer.

Keitel, C. (1989). Mathematics education and psychology. *For the Learning of Mathematics*, *9*(1), 7–13.

Kilpatrick, J. (1981). The reasonable ineffectiveness of research in mathematics education. *For the Learning of Mathematics*, 2(2), 22–29.

Lave, J. (1992). Word problems: A microcosm of theories of learning. In P. Light & G. Butterworth (Eds.), *Context and cognition: Ways of learning and knowing* (pp. 74–92). Harvester Wheatsheaf.

Piaget, J. (1970). *Genetic epistemology.* Columbia University Press.

Sawyer, W. W. (1964). *Vision in elementary mathematics.* Pelican.

Stage, E. K. (2007). Perspectives on state assessments in California: What you release is what teachers get. In A. H. Schoenfeld (Ed.), *Assessing mathematical proficiency* (pp. 357–363). Cambridge University Press.

Tate, W. F. (1995). School mathematics and African American students: Thinking seriously about opportunity-to-learn standards. *Educational Administraction Quarterly*, *31*, 424–448.

Thorndike E. L. (1922). *The psychology of arithmetic.* Macmillan.

Toom, A. (1999). Word problems: Applications or mental manipulatives. *For the Learning of Mathematics*, *19*, 36–38.

Urton, G. (1997). *The social life of numbers: A Quechua ontology of numbers and philosophy of arithmetic.* University of Texas Press.

Van den Heuvel-Panhuizen, M. (2004). Shifts in understanding: The didactical use of models in mathematics education. In H.-W. Henn & W. Blum (Eds.), *ICMI Study 14: Applications and modelling in mathematics education. Pre-conference volume* (pp. 97–102). Universität Dortmund.

Varenne, H., & McDermott, R. (2018). *Successful failure: The school America builds.* Routledge.

Vergnaud, G. (2009). The theory of conceptual fields. *Human Development*, *52*(2), 83–94.

Verschaffel, L., Greer, B., & De Corte, E. (2000). *Making sense of word problems.* Swets & Zeitlinger.

Verschaffel, L., Greer, B., Van Dooren, W., & Mukhopadhyay, S. (Eds.). (2009). *Words and worlds: Modeling verbal descriptions of situations.* Sense.

Wertheimer, M. (1959). *Productive thinking*. Harper. (Original work published 1945)

Wigner, E. (1960). The unreasonable effectiveness of mathematics in the natural sciences. *Communications in Pure and Applied Mathematics*, *13*, 1–14.

Wu, H. (1999). *Some remarks on the teaching of fractions in elementary school.* https://math.berkeley.edu/~wu/fractions2.pdf

Yackel, E., & Cobb, P. (1996). Sociomathematical norms, argumentation, and autonomy in mathematics. *Journal for Research in Mathematics Education*, *27*, 458–477.

14. Rethinking mathematics education

Alan Schoenfeld

I am now concluding my fiftieth year as a professional mathematics educator. That benchmark provides an opportunity to reflect on the emergence of ideas and understandings over the past five decades, and the persistence of challenges that the field continues to face. To quote from the opening page of A Tale of Two Cities, *"it was the best of times, it was the worst of times, it was the age of wisdom, it was the age of foolishness, it was the epoch of belief, it was the epoch of incredulity, it was the season of light, it was the season of darkness, it was the spring of hope, it was the winter of despair." On the one hand, our intellectual advances have been extraordinary. We understand thinking, teaching, and learning in ways that transcend previous understandings. In this chapter, we take a chronological tour through such discoveries—the nature of problem solving, of teaching, of powerful learning environments. On the other hand, both social progress and institutional progress have been hard to come by. Schools and classrooms reflect the structural and racial ills of American society; mathematics instruction, while potentially meaningful and useful in people's lives, has little to do with the kinds of sense-making it could support. If anything, school mathematics' distance from meaningful issues in people's lives serves to reify current structures rather than to problematize and challenge them. The chapter concludes with a proposal to address this state of affairs.*

Introduction

I had the good fortune to fall in love with mathematics as a child and to spend the early part of my career as a mathematician. Then, intrigued by George Pólya's ideas about problem solving, I turned to mathematics education. The challenge as I understood it was, can we understand

©2024 Alan Schoenfeld, CC BY-NC 4.0 https://doi.org/10.11647/OBP.0407.14

enough about mathematical thinking and problem solving to help students get good at it? Wouldn't it be great if increasing numbers of students could experience the power and beauty of mathematics, and even come to feel about it the way I do?[1]

From my current vantage point, the picture is far more complex. Over the years I have come to see mathematics as highly political, in the sense of realpolitik. It has become increasingly clear, as I have worked to help build a rigorous research base for productive change, that those of us who concern ourselves with the improvement of mathematics teaching and learning have far less to say about the enterprise than we might; that school math does little good in the "real world"; and that huge numbers of students are systematically excluded from participation in mathematics. Such realizations crystallized amidst the onset of COVID and the increase in racial tensions across the United States due to the murder of George Floyd. Schooling has been massively disrupted, with the concomitant exacerbation of already significant racial inequities. Yet nowhere have there been calls for re-thinking what is possible or appropriate—standards and testing remain unchallenged and meaningless concerns about "learning loss" predominate. Societal preoccupations and plausible academic goals conflict in uncomfortable ways.

This chapter provides a political/intellectual narrative, ultimately raising questions regarding the character of appropriate goals for mathematics education and how one might think about attaining them. It tells the larger, political story of my experiences as a researcher and developer as I have pursued deeper understandings of the nature of mathematical thinking, teaching, and learning. The narrative takes a turn at the end, as I reflect on my mathematics-related experiences in recent years. I still love mathematics for its beauty and power, but I am deeply concerned about its non-use (except for those who have a professional need for it) in real-world contexts that matter. In my personal life I have made significant use of K-12 mathematics over the past few years—but in very different ways than the current goals of K-12 mathematics would suggest or support. It is time to rethink the rhetoric and reality of mathematics education. As I reflect on my real-world mathematical

1 Of course, what anyone takes pleasure in is a matter of taste. But we can imagine all students having opportunities to experience mathematics (or art, or sports, or literature) in ways that open up the potential for such pleasure.

thinking during times of COVID, on the systematic exclusion of students from the mathematical pipeline, and on what I have found it important to emphasize in my ongoing problem-solving courses, I think significant change is necessary. The question is, what should the goals of mathematics instruction be and how might we attain them? This chapter concludes with a proposal for change. That proposal is unlikely to gain traction, but perhaps something good can come from the issues it raises.[2]

I am going to tell this tale as a chronological narrative, in that it reflects what might be called my sentimental education, from a young naïf to an older and perhaps somewhat wiser scholar-activist who has many of the same goals he had in his youth—to help students experience mathematics in ways that enable them to become mathematical sense-makers, experiencing its power and beauty—but is more cognizant both of the social obstacles that impede progress and of the failures of extant curricula.

The narrative begins with a focus on mathematical thinking, a brief recap of key ideas from my problem-solving work. Once I understood what it is to be a powerful mathematical problem solver, I focused on understanding teaching, and then learning environments. Increasingly, as well, I focused on things that can make a difference—curriculum, assessment, and professional development.[3] Second, there is my evolving political awareness. This was, of course, nascent early on; it was clear many decades ago that the nature of statewide standards and assessments shaped what was possible in the classroom. But the political nature of the standards process was not apparent to the young me—the "math wars" came after the first sets of standards were released. Similarly, the politics of professional development only became apparent to me when I engaged up close. These issues, in turn, paled in significance when society's utter disregard for teachers became apparent during the onset of the COVID pandemic. Amid the chaos of 2020, there was good reason to re-think the purposes and impacts of education; but there was

2 The issues that unfold in my narrative are sometimes grounded in the culture of the United States and sometimes general. Experiences within the socio-political context of the U.S. may or may not have analogs in other nations, but aspects of mathematical thinking are in large measure universal. My goals for mathematics instruction are thus a hybrid of the two.

3 This is somewhat oversimplified, of course. These issues overlap, substantively and chronologically—e.g., my first assessment project began in 1991.

little work to do so. Increased income inequality and the blatancy of racial violence make the challenges we face that much more salient.

As has always been the case, my teaching and thinking are deeply intertwined; such issues make their way into my course on mathematical thinking and problem solving, and experiences in the course shape my thinking about what matters. It goes without saying that I write from a position of privilege, and that my history and my perspective have been shaped accordingly.

In the beginning

I've always loved mathematics. When I was a kid, my entertainments were mathematical: trips to the library brought back books like George Gamow's (1947) *1, 2, 3, ... Infinity* and even on recreational outings I would find myself doing something like estimating the number of windows in a (large, New York City) housing development. Although my working-class parents desperately wanted me to become a doctor, I loaded up on math courses as an undergrad, and when I told the mathematics chair at Queens College[4] that I wanted to change from my pre-med chemistry major to mathematics, he asked "What took you so long?".

There are many reasons to love mathematics. Part of what makes mathematics so special is that it's not arbitrary; you can figure things out. I have no idea at what point my first conscious mathematical discovery was. Perhaps it was the observation that every time I added two odd numbers, the result *had to* be even. Perhaps it was something else. Whatever it was, it was magical. And, it was mine – I'd figured it out, I *owned* it! This wasn't somebody's rule, which I had to memorize; this was something I'd figured out, and I understood why it was true. In other fields I had to memorize things. Where did Ohm's law come from,

4 It's worth noting that aspects of the social compact were in place when I was an undergraduate. Education was considered a public good. Tuition and fees at Queens College (Part of the City University of New York) for New York City residents were $32 per semester. All the way into the 1990s, tuition and fees at the University of California were under $2000. Then, in a massive shift, politicians came to consider higher education to be a *private* good—people with college degrees earn more money over their lifetimes—and tuition fees began to skyrocket. The result is the massive student debt that current graduates suffer from—a distinctly U.S. phenomenon.

for example? Or biological taxonomies, or multiple and conflicting interpretations of Hamlet? In math, things made sense.

To me, that meant that math was fundamentally democratic. It was open for discovery. And, assuming you showed (proved) something correctly, it was true—period. Nobody could argue it away; no authority could declare otherwise. What fun, what power! I had the sense, long ago (see, e.g., Schoenfeld, 1994) that much of mathematics could be learned via sensemaking and problem solving. It seemed to me that most curricular content could emerge as the result of well-structured investigations rather than being imposed from on high. And, I had no idea that there could be anything political about mathematics. If anyone could do it and own it, how could it be political? (I remember a chat with some Spanish colleagues back in the late 1980s, when one of them claimed that mathematics was inherently political. I was incredulous—in fact, his statement didn't really "compute".)

Graduate school and early professional life: The late 60s and early 70s

Some time before I earned my Ph.D. in mathematics, I ran across Pólya's (1945/1957) *How to Solve It*. I read the book with fascination. Page after page, Pólya described methods of problem solving. As I read through the book, my smile got wider. If I was doing the things Pólya said mathematicians do, then I must be a real mathematician! But then I wondered, why hadn't I been introduced to these methods? Was mathematics a secret guild, where the price of entry was figuring such things out for yourself? (In a sense, the answer is yes; but it's more complex than that.)

In any case, doing math was fun. So was teaching. After earning my Ph.D., I taught for two years as a lecturer at University of California, Davis. That was my first introduction to academic politics: I was advised by my senior colleagues that I was spending far too much time with my students and that if I wanted to have a successful academic career I should limit my office hours and either close the door or go home to prove theorems. I very much enjoyed my teaching and earned high teaching evaluations; but I was told that that could be seen as a kiss of death among my department colleagues. The choice framed for me was, am I a researcher or a teacher?

At the time, I'd been reflecting (as a total amateur) on my teaching and had written about useful classroom techniques (Schoenfeld, 1977). I spoke with a biologist friend who was involved in educational efforts at Berkeley. She suggested that I chat with Fred Reif, a physicist who chaired an interdisciplinary group called SESAME (Search for Excellence in Science and Mathematics Education) at Berkeley. Fred convinced me that there was a future to cognitive science and education, so I took a postdoc at Berkeley. Basically, I did so on the basis of an informal expected value computation. On the one hand, I loved mathematics and I wasn't bad at it. But the odds that I'd do something transformative in mathematics were very slim – the pioneers of the previous few centuries were hard acts to follow, and the field itself had existed for two thousand years. By contrast, mathematics education was in its infancy. *Educational Studies in Mathematics* first appeared in 1968, *Journal for Research in Mathematics Education* in 1970. When I did my postdoc at Berkeley from 1975–1978, the field of cognitive science didn't really exist. (The first issue of the journal *Cognitive Science* appeared in 1977.) So, there were opportunities to participate in the growth of the field from the very beginning, bringing together my love of mathematics and my wish to go deeper into understanding mathematical thinking and teaching. In addition, I'd always felt somewhat guilty being a professional mathematician. Being paid for producing theorems felt like being paid for doing crossword puzzles. It was fun, but to what benefit? If research on problem solving made it more accessible, then there was a potential payoff for students in terms of teaching and learning. I was, of course, totally naïve about what it takes to have an impact on school systems. But, the opportunity to shape the emergence of a new field, to combine my love of mathematics with my love of exploring thinking and learning, and, if I was lucky, to have some influence on practice, was irresistible.

The early years: Problem-solving research and development, 1975–1990

For a number of reasons, I began my research on mathematical problem solving at the college level. I thought about working with doctoral students (but did they really need my help?) or on niche areas like the Putnam exam (but to tell the truth, I wasn't great at that kind of problem). I thought that Pólya's problem solving strategies were pretty sophisticated, so college students (rather than secondary students) were probably the right audience; I started with upper division Berkeley students, just to be safe.

What I found very quickly was that my students—among the best and the brightest—were woefully unfamiliar with even the most basic problem-solving strategies. They were smart, they were creative... and they had gotten as far as they had because they were very good at mastering the mathematics they were instructed to master. There was, not only for these students, but in general, an unspoken didactical contract: their teacher will establish the context and show the students what they are responsible for. Homework assignments will stretch the students a bit, but they are largely repetitive. Tests will, with the possible exception of problems designed to "reveal the A students," reward students who have done their homework. Although they were referring to K-12 curricula, Glenda Lappan and Elizabeth Phillips (2009) tapped into at least a K-14 universal when they referred to the dominant mode of instruction as 'demonstrate and practice'.

The net result was that students had little or no experience with problem solving, or what John Mason, Leone Burton, and Kaye Stacey (1982) called 'thinking mathematically'. I came to realize that my students were fundamentally deprived, in mathematical terms. I moved my problem-solving courses down to the lower division level, so that my students—whether intending math majors or not—could at least experience a dose of mathematical thinking. If they planned to go on in mathematics, they should at least have a sense of what mathematical sensemaking looks like. And if they didn't, then there was at least as much reason to give them a sense that mathematics could be interesting and exciting. There's enough mathophobia in the world as it is.

The core aspects of my work on mathematical problem solving are fully documented (see, e.g., Schoenfeld 1985, 1992), so I'll summarize them briefly. Then I'll discuss what I found along the way. The central work on problem solving evolved over a decade or so. I first focused on problem solving strategies, or 'heuristics', as identified by Pólya (1954, 1957, 1981). The key insight was that Pólya was right about the strategies—mathematicians do use them, having picked them up, idiosyncratically, from their experience. (Rough paraphrase: a technique used twice becomes a strategy.) But, the grain size of Pólya's descriptions was wrong: a strategy such as 'exploit an easier related problem' was unteachable on its own terms because it actually consists of at least a dozen different sub-strategies for identifying easier related problems and exploiting their solutions. My research showed that the sub-strategies could be learned and that when students learned enough of the sub-strategies, they could implement 'the strategy'. (Rough analogy: if you learn to cook a range of vegetables, and starches, and a variety of meats, then you can put together a complete balanced meal.)

Interestingly, solving the sub-strategy problem created a new problem. Pólya had identified perhaps two dozen major heuristic strategies, a manageable number. But if learning each strategy entailed learning a dozen sub-strategies, then the challenge jumped by an order of magnitude. The difficulty isn't simply a *learning* challenge, although mastering hundreds of techniques rather than dozens certainly ups the ante; it's a *management* challenge. How in the world do you decide which technique to use, when you have hundreds at your disposal? (Rough analogy: if I give you a key ring with a dozen keys, the odds are you'll be able to open a door within a reasonable amount of time. You can try them all if need be. But if I give you a key ring with hundreds of keys, the odds of your success diminish substantially.) That led to the issue of metacognition, more specifically the issue of monitoring and self-regulation. The bottom line is that self-monitoring can also be learned. With appropriate attention to reflecting on progress during problem solving, students can get good at it.

I was interested in what helped students succeed and what caused them to fail. There was no good reason to ask students to solve problems for which they didn't have the relevant knowledge, so I chose problems for which the students had the appropriate backgrounds. At the time, plane geometry was a required high school course, so I could be confident that the students knew the basics. I gave my first-year college students a simple geometry construction problem—which, despite their knowledge, they all approached empirically. I pursued the issue for some years, ultimately having the students prove results that solved the construction problem just before I gave them the construction problem to work on. Amazingly, the students ignored what they had just proved and made conjectures that contradicted it. Those findings led to the study of mathematical beliefs and their origins. And that pursuit led me into the schools, where I observed both what was taught and why. My experiences in schools led me to consider a series of structural issues, starting with the role of curricula and assessment in shaping students' learning.

Given that I had uncovered the challenge of unproductive beliefs in geometry, I started by sitting in on high school geometry classes. In the 1980s, New York was one of three major states (the other two being California and Texas) that had state-wide testing, along with state-supported curricula designed or selected in concert with the tests. What soon became apparent were the ways in which testing deformed instructional practice. The New York State Regents exams had a very specific format, with 10 points awarded for 'solving' (i.e., reproducing the proof of) each of two proof problems out of a dozen or so 'required' proofs. What happened of course was that students memorized all the proofs, for a guaranteed 20 points out of 100 on the exam. The test also had one 'construction problem'. Students could earn two points for producing a sequence of lines and arcs on the page that looked just like one of the 'required' constructions.

The way that instruction was organized in the school that I observed made the power of the exam very clear. Although geometric constructions were discussed about half-way through the text, the math department reorganized instruction so that constructions were taught just before the statewide exam. The rationale was simple: since students were intended to memorize the constructions and carry them out precisely, it was unwise to have too much time pass between memorizing and test-taking. Indeed, one of the most memorable quotes from instruction that year came from the teacher, shortly before students were to take a unit test on constructions: "You'll have to know all your constructions cold so that you don't spend a lot of time thinking about them." The emphasis was on speed and accuracy, tailored to test performance. What mattered when producing constructions was that the arcs on the page looked good, and that they were reasonably accurate.

A range of research findings included those observations (see, e.g., Schoenfeld 1988, 1989). These findings were not about any particular teacher; they were general. Hugh Burkhardt's acronym WYTIWYG (What You Test Is What You Get) accurately summarized the influence of high stakes assessments. That hasn't changed. Course texts were, and still are, tied to assessments. The literature has long indicated that teachers follow texts with great fidelity.

It should be stressed that the teachers in this and other studies were uniformly well intentioned—they were doing what they thought was in the best interests of their students. But, test pressures are enormous. That was the case even before the No Child Left Behind Act, and it remains so. I have had National Board Certified teachers tell me that they would try out the ideas in our professional development program for one year, but if their students' test scores dropped by even one point, they would leave. I have seen an equity-focused teacher who built a summer program based on ideas related to growth mindset that was designed to help prepare 'low-performing' students build confidence and agency completely forsake those ideas during the regular school year because there wasn't time for such things in a curriculum aimed at the high stakes state exams.

The point here is that by the early 1990s mathematics education researchers had a good idea of what mattered in mathematical performance. Understanding content—having mathematical resources at one's disposal—had always been considered important. The National Council for Teachers of Mathematics (NCTM) endorsed 'problem solving' and we had a theoretical understanding of how to decompose and teach heuristic strategies, although the process had not been done and curricula supporting problem-solving instruction had not been built. The roles of metacognition and belief systems were understood, as were the causes of counterproductive beliefs (Schoenfeld, 1985, 1992). The obstacles to bringing these ideas into the classroom were structural and (socio)political.

The 1990s and the math wars

If there is one phrase to describe the 1990s in mathematics education, it's "the math wars". I've written extensively about this (Schoenfeld, 2004, 2008; Schoenfeld & Pearson, 2009) so I won't repeat the details but will make some observations.

People have multiple reasons for aligning with or leading political 'movements', as has become all too clear in the intertwining of White supremacy, structural racism, fervid, and sometimes 'post factual' (e.g., QAnon) belief, and personal advantage in Trumpian politics. The same was the case, albeit not as blatantly, in the math wars. There is no question that some of the participants considered the integrity of mathematics to be at stake and felt that they were protecting it. There is also no question that partisanship gave some people, both inside and outside the mathematics community, opportunities for personal advantage such as visible prominence and political advancement. Here I want to point to some more structural issues.

The first issue is financial. I was a member of the group that wrote the 1992 California Mathematics Framework. Our meetings were public. They were sparsely attended, apart from one group—there were always representatives from major publishers at our meetings. They delivered one clear message: 'reform' is impossible because it would be too expensive. It cost $25 million to develop and produce a K-8 textbook series, they said, and no publisher was going to risk that much money on an untried concept. They were right. What did happen was that the National Science Foundation (NSF) realized that the lack of suitable textbooks was a roadblock to progress, so NSF issued a funding initiative for the production of 'standards-based' or 'reform' texts.

Reform texts catalyzed the math wars, which raged over much of the 1990s. To understand whether politics or substance matters, it is essential to note that the math wars were waged largely in the absence of hard data. The motivation for reform was clear: there was undeniable evidence of the shortcomings of 'traditional' instruction and a decade of small-scale reform-oriented studies suggested that the directions in which the NSF-funded curricula were headed were likely to be productive. The hard evidence to support this hypothesis didn't really start coming in until 2000, however. The case for reform became stronger when Sharon Senk and Denisse Thompson's (2002) summary volume indicated across-the-boards wins for reform. (The one-line summary: students using standards-based materials did roughly the same on tests of skills as students who received 'traditional' instruction; they outperformed such students on tests of problem solving and conceptual understanding (Schoenfeld, 2002)). The fact that the wars persisted for so long in the absence of hard data indicates that the forces that drove much educational policy were political rather than grounded in data and research.

Moreover, although the issues are far less direct, issues of race and its ofttimes inextricable partner, socio-economic status, were also implicated. One of the underpinnings of the Standards movement, and an explicit goal of some Standards-based curricula, was to move toward more equitable instruction.[5] To state things directly, there are those who believe that excellence and equity are in conflict—that there is a gradation of mathematical talent, and that an attempt to enfranchise all students mathematically is a disservice to those talented students who would profit from more 'rigorous' training. From that perspective, if less 'talented' students fall off the mathematical ladder, that's their problem; there is 'enough' mathematical talent to advance the nation's interests, and one should not dilute instruction to serve the masses.

The math wars were fomented in California by a group called 'Mathematically Correct', whose website still exists.[6] It is no accident that the wellsprings of Mathematically Correct were San Diego and Palo Alto. San Diego was a hotbed of right-wing conservatism, partly because of its proximity to the Mexican border and the fact that the Spanish-speaking population was increasing rapidly. Immigration backlash included the sponsoring of California's state Proposition 227, essentially an 'English only' mandate for the schools. The analysis from 'Ballotpedia', an independent analysis group, summarized Prop 227 as follows:

> Proposition 227 changed the way that 'Limited English Proficient' (LEP) students are taught in California. Specifically, it:
>
> Required California public schools to teach LEP students in special classes that are taught nearly all in English. This provision had the effect of eliminating 'bilingual' classes in most cases.
>
> Shortened the time that most LEP students stayed in special classes.
>
> Eliminated most programs in the state that provided multi-year special classes to LEP students by requiring that (1) LEP students move from special classes to regular classes when they had acquired a good

---

5 NCTM's stance on equity and diversity has been problematized. See, e.g., Martin (2009); for an update, see Martin (2019). However partial or inadequate the NCTM position may have been, it was a flash point for controversy, as discussed above.

6 See http://www.mathematicallycorrect.com

> working knowledge of English and (2) these special classes not normally last longer than one year. (1998 California Proposition 227, 2022)

This was one of many policies implemented by the right wing, tapping into the xenophobia that ultimately metastasized into Donald Trump's immigration policies. In San Diego in the 1990s, the White voting majority felt threatened by a growing Latino minority. Claiming that high quality learning was being threatened by untested equity-driven mathematics programs, with the tacit implication that the new programs were tailored to minority students, was a perfect wedge issue to mobilize White voters.

Palo Alto represented similar issues in a different way. The area had a mixed but separated demographic: Palo Alto itself was upper-upper middle class, and East Palo Alto had a largely minoritized population consisting in large measure of the people who cooked for, cleaned for, and maintained the homes of those (literally!) on the other side of the tracks. Here the issue wasn't fear of disenfranchisement, but loss of privilege. The 'good schools' in Palo Alto reliably sent their students to the best schools and universities. Why tinker with success, for abstract reasons of equity and diversity? If it ain't broke (for your children, that is), don't fix it. (N.B. My rhetoric is mild, but the rhetoric of the political battles in Palo Alto was anything but.)

The tensions remain. The same right-wing players who brought us the math wars are now manufacturing a battle over the anti-tracking stance in the 2021 draft California Mathematics Framework.

The 2000s, part 1: No Child Left Behind

On a purely personal note, I want to bring up one of my signal failures. I was a lead author of the successor volume to the 1989 NCTM *Standards*. The process by which *Principles and Standards* was created and vetted was beautifully managed and the endorsement of the process by all of the major mathematics societies quieted the math wars.

All too aware of the impact of testing and the import of WYTIWYG, I argued that our assigned task, writing standards and providing examples of interesting classroom activities, was good but not good enough: the wording of the *Standards* (on the order of 'students will understand X') was somewhat vague and aspirational, and could be misconstrued. At the time the standards movement nationwide was morphing into a test-based accountability movement, so the nature of assessments was vitally important. It would be easy, I argued, to craft assessments that bastardized our intentions. I proposed to the writing team that we incorporate sample assessments into our document. It was put to a vote, and I lost 25 to 1.

The result was a disaster. Had NCTM produced sample assessments, it would have taken the lead in saying 'This is what we want students to be able to do with the mathematics they learn'; there is a good chance that such tests would have shaped statewide assessments, and thus shaped instruction. When NCTM failed to do so, the vacuum was filled by the No Child Left Behind Act (commonly known as NCLB). Under NCLB, states built their own assessments. Most of those assessments, in line with traditional assessments, focused on low-level skills. This effectively undermined the goals of *Principles and Standards.*

Some good intentions motivated the creation of NCLB. Each state defined its own standards, assessments, and performance targets. The idea was to ratchet up performance standards gradually and to provide support and rewards for reaching those standards. Disaggregation mechanisms—*every* demographic group had to meet the standards—assured attention to the performance of all students. And, there were carrots and sticks. The carrots were that schools that lagged behind would be given significant resources to improve. The sticks were that if they failed to improve for 'too long', penalties would be imposed. Individual students would be left back; teachers would be dismissed; if a school failed to meet progress goals for a number of years in a row it would be dismantled, and whole districts would be put in receivership.

That approach might have been workable (although highly punitive) if the carrots were in place, so that districts that faced challenges in making adequate progress were provided with resources to address the challenges they faced. But guess what? In the congressional sausage-making process, the penalties for failing to make progress were carved in stone but the resources to support failing districts were never authorized.

Problems abounded. There was huge variability in the sets of standards and assessments built by the states. Most of the assessments were of low quality. Some states gamed the system, demanding minimal progress until right before the 2013–2014 year; somewhat paradoxically, setting low standards allowed them to avoid penalties. But the main issue was structural inequality. Rich districts had the resources to do well enough, at least at the beginning. That's not to say that testing didn't bend schools out of shape. When I went to observe some classes in late February, a teacher I knew from my having worked in her school told me not to bother watching instruction—"all we're doing is prepping for the test, there's no real teaching going on". But that was a district that could afford 'business as usual'. The challenges faced by poor and minoritized districts were far worse.

Because the enacted version of NCLB failed to provide fiscal support for 'failing' districts but did penalize them, under-resourced and minoritized districts quickly found themselves being penalized. A local district I was working in was the first to be put in receivership. The results were devastating, adding injury to injury. The district, with a 90% minoritized population and hardly any resources, was forced into continuous test-prep mode. The result is that the students who were in the greatest need of meaningful instruction were systematically denied it. This was but one example of the structural racism within the system. It is consistent, of course, with many equally blatant examples (e.g., Kozol, 1992; Rothstein, 2017).

The bottom line: seemingly reasonable policy decisions can have significant negative impact on people's lives. This, again, is the issue of 'learning loss'.

The 2000s, part 2: The What Works Clearinghouse

Some background on testing is necessary before I proceed here. Testing is *not* a neutral measure of proficiency. Any test assesses what is declared to be important to some degree, depending on how artfully the test is constructed. But there's great variation. Take literacy as an example. If you define "literacy" as having a specific vocabulary, you give vocabulary tests and teachers wind up drilling their students on vocabulary. If you define "literacy" as the ability to analyze text, you develop a very different kind of test; kids read and think. The nature of the test is consequential, because students are declared to be "literate" (or not) based on their test scores.

It's the same in math. One definition of mathematical proficiency is skills- and knowledge-based. The basic idea is that students should be able to execute the skills taught in the curriculum. A more inclusive definition calls for having students demonstrate proficiency in skills, conceptual understanding, and problem solving. Depending on which approach you assess, you get very different results.

Table 14.1 shows the differences in the two approaches. The SAT-9 was a skills-based assessment used across California in the 1990s. The MARS test was a test of skills, concepts, and problem solving. Ridgway and colleagues (2000) administered both tests to more than 5000 students at each of grades 3, 5, and 7. The patterns are the same across grades.

Table 14.1 Student proficiency as reflected by the MARS and SAT-9 tests (Ridgway, Crust, Burkhardt, Wilcox, Fisher, & Foster, 2000).

	SAT-9	
MARS	**Not proficient**	**Proficient**
	Grade 3 (N = 6136)	
Not proficient	27 %	21 %
Proficient	6 %	46 %
	Grade 5 (N = 5247)	
Not proficient	28 %	18 %
Proficient	5 %	49 %
	Grade 7 (N = 5037)	
Not proficient	32 %	28 %
Proficient	2 %	38 %

Let's take grade 3 as an example. If a student was declared proficient on the MARS test, there's a 46/52 = 88% chance that the student would be declared proficient on the SAT-9. That looks like pretty good alignment. But if a student was declared proficient on the SAT-9, there was a 46/67 = 69% chance that the student will be declared proficient on the MARS test. To put this more directly, 31% of the students declared to be "proficient" by California's official test turn out to be "not proficient" when conceptual understanding and problem solving were assessed in addition to skills.

This result is a big deal if you care about conceptual understanding and problem solving. It's also a big deal if you care about curricula. If you use the SAT-9 to assess grade 3 performance, it looks like 67% of the students are proficient. That's not wonderful, but it seems within bounds; there appears to be a base for improvement. If you use the MARS test to assess grade 3 performance, however, you get a very different story. Only 52% of the students test proficient; you'd better make some radical changes. In short, *what you test matters.* It's shocking that I have to say this—but read on.

The background just provided establishes the context for some general comments and then a description of my specific experience with the "proficiency-based" testing.

No Child Left Behind was only one of the educational policy initiatives put in place during George W. Bush's presidency. There was also the misguided attempt on the part of the U.S. Department of Education to define randomized controlled trials (RCTs) as the 'gold standard' of educational research (see, e.g., U.S. Department of Education, 2003). The issue is not that randomized trials aren't an excellent way of conducting research under certain circumstances. The issue is that in pragmatic terms, the Department of Education discounted almost all other forms of research—it didn't consider evidence produced by alternative methods to be adequate evidence of effectiveness. For a broader discussion of alternative methods and their validity, see *Scientific Research in Education* (National Research Council, 2002), which was produced in rebuttal to the Department of Education's agenda (although, of course, it didn't say so); see also Schoenfeld (2007), which lays out criteria for rigorous and meaningful research and problematizes the use of randomized controlled trials in educational research.

If this discussion were merely 'academic', that would be one thing. But the relevant issues turn out to be Political, with a capital P. As part of its agenda to certify high quality instructional materials, the Department of Education's Institute of Educational Sciences created the What Works Clearinghouse. WWC's mandate was to certify when instructional interventions had been validated by rigorous means. If a curriculum or other instructional treatment had been evaluated by means of some formal assessment, WWC staff would evaluate the quality of the evaluation. A carefully conducted randomized controlled trial would get you the highest marks—the equivalent of the 'Good Housekeeping seal of approval' for the curriculum. WWC was established to conduct a large literature review, identifying interventions that 'worked'.

In 2003 I was appointed the WWC's Senior Content Advisor for Mathematics. Basically, my responsibility was to ensure the intellectual integrity of the enterprise. (Staff did the work.)

An early article I produced for WWC was intended to serve as part of a large technical document that framed WWC's approach to certifying instructional materials. The article explained the history of mathematics curriculum development and assessment. It predicted slim pickings for the WWC mathematics literature review, because very few instructional treatments had been subjected to the kinds of randomized controlled trials that the WWC used as its evaluation standard. The Institute for Educational Sciences (IES), which funds WWC, instructed WWC to remove my article from the document. When I complained, WWC said that I would have the opportunity to revise the document for publication when some instructional treatments had been reviewed and WWC was further along in the process.

I waited. After the WWC staff produced its first series of evaluations, I was informed that the Clearinghouse planned to work with a journal to create a special journal issue characterizing WWC's work. I was told to update my article and submit it to the journal. When I went through the new data, the predictions I had made in my earlier piece were confirmed: very few studies met WWC's criteria. More importantly, as I worked through the data I discovered a fundamental flaw in WWC methodology. WWC had not analyzed what the assessment measures used in the studies actually assessed. Thus WWC's certifications of quality had little meaning. When an instructional treatment was judged to meet WWC criteria, it was impossible to know what exactly the treatment did well. Did students learn skills, or problem solving, or conceptual understanding, or something else? There was no way of knowing without conducting a content analysis of the assessment. Because there was only a handful of certified programs, I urged WWC to conduct the relevant content analyses. They refused.

I revised my article. The revision, like its antecedent, contained the history of mathematics curricula and my prediction of slim pickings. It documented the accuracy of the predictions and contained a discussion of why WWC's refusal to perform content analyses was deeply problematic. I submitted my revision to the journal and waited for reviews. After a very long delay I was informed that, after conducting a "prepublication review," IES had instructed WWC to remove from the journal every single paper that had been written by WWC staff. Of course, it made no sense to publish my paper as a stand-alone; the journal issue was cancelled.

The only way I can see to interpret the sequence of events that I have just described is that IES killed the special issue in order to prevent the publication of my piece. (This is not the first case of a federal agency blocking publication of 'inconvenient truths.' There is a history of such actions with regard to climate change and other areas.) I resigned from my role as a WWC advisor and published the details of the story in *Educational Researcher* together with a rejoinder from WWC and my response (Schoenfeld, 2006a; Harman et al., 2006; Schoenfeld, 2006b).

The 2010s

The previous section described the negative impact of deliberate policy choices. What follows in this section features the laws of unintended consequences—the epitaph for which is a Robert Burns' quote, "the best laid schemes o' mice an' men / Gang aft a-gley". I include these stories because they help frame my final discussion of the goals of mathematics instruction.

In math-ed terms, the 2010s can be considered the decade of the Common Core and the assessment systems that enforce it. Despite the best of intentions and some positive outcomes (e.g., greater consistency in nation-wide goals for instruction) there are deeply problematic aspects to both. I appreciate the challenges faced by the authors of the Common Core, who had little time to compile their work and were doing their best to avoid rekindling the math wars. The result, in contrast to the NCTM *Standards* volumes, is a rather slender volume. NCTM's (2000) *Principles and Standards* weighed in at more than four hundred densely packed pages that described and exemplified content and processes, with equal space given to both – that is, the fundamental *processes* of problem solving, reasoning, communicating, making connections, and using mathematical representations received as much attention as the *content* that was described (number and operations, algebra, geometry, measurement, and data analysis and probability).

The Common Core contains a three-page list of 'standards for mathematical practices' and a seventy-four-page list of 'standards for mathematical content'. Here is a sample from the beginning of the grade 6 content description:

> Use ratio and rate reasoning to solve real-world and mathematical problems, e.g., by reasoning about tables of equivalent ratios, tape diagrams, double number line diagrams, or equations.
>
> a. Make tables of equivalent ratios relating quantities with whole-number measurements, find missing values in the tables, and plot the pairs of values on the coordinate plane. Use tables to compare ratios.
>
> b. Solve unit rate problems including those involving unit pricing and constant speed. *For example, if it took 7 hours to mow 4 lawns, then at that rate, how many lawns could be mowed in 35 hours? At what rate were lawns being mowed?*
>
> c. Find a percent of a quantity as a rate per 100 (e.g., 30% of a quantity means 30/100 times the quantity); solve problems involving finding the whole, given a part and the percent.
>
> d. Use ratio reasoning to convert measurement units; manipulate and transform units appropriately when multiplying or dividing quantities. (Common Core State Standards Initiative, 2010, p. 42)

In my experience, such neat clean descriptions get turned by school districts into curricular checklists—"Have we worked problems on tables of equivalent ratios? Yes, check. Have we practiced on tape diagrams? Yes, check". And so on. That is, it's easy to go from bullet points to scope and sequence. You get content 'coverage' in the narrowest sense.

And what about important practices and processes? The Common Core discusses eight key practices on pages 6–8, and the list of those practices is reprinted at the beginning of each content chapter. We've known for decades that problem-solving success hinges on: students' knowledge base; their access to problem solving strategies; effective metacognition, specifically monitoring and self-regulation; and productive belief systems, about mathematics and about oneself vis-à-vis mathematics—in today's language, productive mathematical identities (Schoenfeld, 1985). To speak bluntly, the Common Core offers no meaningful support for anything but content. Functionally, there is no support for problem solving. There is no support for the development of metacognitive skills, and beliefs and mathematical identities are not addressed.

When discussing 'real-world' implementations of curricula, Ann Brown and Joseph Campione (1996) observed that all curricula and frameworks undergo mutations when they move into classrooms. The challenge, she said, was to avoid lethal mutations. Unfortunately, (a) the Common Core's list of bullet points is easily converted into a curricular scope-and-sequence, (b) the Common Core offered no meaningful support for mathematical processes and practices, and (c) it provided no exemplification of rich and interesting mathematical problems and discussions. In consequence, in addition to not providing direct support for ambitious curricula, the Common Core left the door wide open for lethal mutations.

There are two ways out of this dilemma. The first is large-scale curriculum development, a process that takes many years and large investments. For the most part, that just didn't happen. It is the case that some good *Standards*-based curricula were retrofitted to the Common Core, and some ongoing projects are providing good materials. The problem is that high stakes assessments were going to be implemented soon after the Common Core was adopted. School districts needed Common-Core-consistent curricula as soon as possible. The results were mostly cut-and-paste disasters. This is a systemic failure.

The second way out of the lethal mutations dilemma could be the use of well-constructed assessments. Given WYTIWYG (What You Test is What You Get), a set of robust assessments that interpreted the Common Core in the right ways could have driven instruction in the right directions. Hugh Burkhardt and I were asked to head the team that drafted the specifications for the Smarter Balanced Assessment Consortium (SBAC) (Schoenfeld & Burkhardt, 2012), which contracted with about half the states in the US to implement Common-Core-consistent assessments. We were excited about the possibilities because they promised two fundamental changes. First, we constructed a system that was able to provide meaningful and reliable sub-scores regarding students' knowledge of: (1) concepts and procedures, (2) problem solving, (3) communicating reasoning, and (4) modelling and data analysis. The point of such sub-scores is that they can highlight particular students' or schools' strengths and weaknesses rather than providing single numerical grades. Second, we exemplified the assessment with a collection of mathematical tasks that embodied the mathematical richness we wanted students to engage with. The tasks were taken largely from task banks constructed by the Mathematics Assessment Project.[7] They have been used in secure testing situations for many years.

The 2012 SBAC specs are no longer on the SBAC website. In fact, the specs were never implemented in ways consistent with Burkhardt's and my expectations. The problem is that the SBAC Governing Board always planned to move toward computer-graded exams, which are cheaper, more 'reliable', and more 'secure' than person-graded exams.[8] SBAC built what it could and implemented it. In my opinion, the transition to computer-based exams de-natured the mathematics in the exams to the point where the exams fail to represent the mathematical richness that we had built into the exam specifications. (To be fair to SBAC, I am measuring them against high standards. The assessments produced by the other national assessment consortium, Partnership for Assessment of Readiness for College and Careers, are far worse.)

The point is that there have been opportunities to orient the system toward richer and more engaging—even if somewhat traditional—mathematical content. For largely systemic and political reasons, that hasn't been done.

The 2020s

The first few years of the 2020s have already given us more than a decade's worth of challenges. As I wrote in Schoenfeld (2022):

> The murders of George Floyd, Breonna Taylor, Trayvon Martin, Sandra Bland, Ahmaud Arbery, and numerous other Blacks at the hands of police and white supremacists laid bare for all except those who refuse to acknowledge it the structural racism that underpins American society (Center for American Progress 2019, Urban Institute 2020, Wilkerson

7 See https://www.map.mathshell.org

8 There are mechanisms for hand-grading exams that are comparatively inexpensive and secure – see the arguments in Burkhardt & Schoenfeld 2019. AI-graded exams could have been phased in gradually.

> 2020). It's not that such issues were unknown; it's that the murders and the protests they engendered made them much more difficult to ignore. The reality that many minoritized people live in a world apart from White America, with different and much more devastating expectations for quality of life (including education) has been rendered day after day in high resolution.

If anything, the situation has gotten worse in the years since I penned those words. The completely manufactured 'controversy' over teaching critical race theory in schools represents a full-fledged attempt to ban the teaching of the history of oppression described in the previous paragraph. Such actions and their consequences reach into every mathematics classroom.

In much of my previous work I theorized about what took place inside ostensibly closed systems—people solving problems in isolation, teachers making decisions, actions in the classroom. My problem-solving research asked: "What are the aspects of thinking and understanding that need to be examined in order to determine the success or failure of any individual's attempt to solve a problem?" The (theoretically complete) answer was: "You need to know about the individual's knowledge base, problem solving strategies, metacognitive behavior, and belief systems" (Schoenfeld, 1985, 1992). Similarly, my research on teachers' decision-making asked: "What do you need to know in order to model the in-the-moment choices a teacher makes during instruction?" The (theoretically complete) answer was: "If you know the teacher's resources, orientations, and goals in very fine detail, then you can produce a detailed model of the teacher's choices by using a specific probability-based decision mechanism" (Schoenfeld, 2010). My ongoing classroom research asks the question: "Which dimensions of classroom interactions are necessary and sufficient to ensure that students will emerge from instruction as knowledgeable, resourceful, and agentive thinkers and problem solvers?" The (theoretically complete) answer is: "It suffices to examine the five dimensions of the Teaching for Robust Understanding (TRU) Framework: the quality of the mathematics; opportunities for productive struggle; equitable access to meaningful engagement with core content; opportunities for the development of agency and positive mathematical identities;

and formative assessment" (Schoenfeld 2013, 2014; Schoenfeld & The Teaching for Robust Understanding Project, 2016).

The challenge is that these closed systems, while allowing for theoretically complete solutions, also wind up finessing key questions of causality. In problem solving, where do knowledge and belief systems come from? In a society where students are stereotyped, tracked, and provided very different opportunities to learn, such issues matter: it's not just what individuals bring to a problem situation, it's what shaped their knowledge and belief systems before they sat down to work on the problem. It's the same with teaching: Where do teachers' orientations come from? For example, we have to think about what led a teacher to say, when I asked him whether he'd ever consider giving his class a problem and let them grapple with it, "not these students, it would just confuse them. I do that with my honors students". And, when we think about the construction of powerful learning environments along the lines of the TRU Framework, we have to think about the distributions of opportunity to do so. These are massive societal issues. The solutions to the closed system problems point to what needs to be done, but larger systemic issues need to be taken into account when we consider what caused things to be as they are and how we might address them.

Nowhere is the set of larger social issues clearer than when we consider the national impact of COVID-19. Essential workers—disproportionately people of color—were forced to work but were not given prioritization for vaccination. The consequences are all too predictable, as indicated by a piece in the *New England Journal of Medicine* entitled "Structural Racism, Social Risk Factors, and Covid-19—A Dangerous Convergence for Black Americans" (Egede & Walker, 2020).

Similarly, children of poverty and children of color suffer the academic impacts of COVID disproportionately: see "Addressing Inequities in Education: Considerations for Black Children and Youth in the Era of COVID-19" (Gaylord-Harden, Adams-Bass, Bogan, Francis, Scott, Seaton, & Williams 2020). That article has the following section heads:

> Systemic Racism is the Pre-Existing Condition Affording COVID-19 the Opportunity to Disproportionately Impact the Black American Community
>
> Black Families are Facing More Severe Economic Consequences

Black Children Face Disadvantages in Remote Learning Settings

Schools That Serve Black Children are Less Able to Provide Remote Learning Experiences

Black Children are Experiencing Elevated Levels of Stress

There are similar findings for a wide range of minoritized populations.[9]

Despite consistent evidence along these lines, the predominant concerns in the media are focused on 'learning losses'. That's what makes headlines. Earlier this year I wrote an editorial to that effect, titled "It's Time for an Academic Reset". It made the following arguments:

What really matters? First and foremost, students' mental and emotional well-being. COVID's impact has fallen disproportionately on communities of color and people who are economically disadvantaged. Privileged students often have good technology, good Wi-Fi, and nice places to study. One of my former students, who teaches in a low-income, highly diverse district, had to find her students to give them electronic tablets they could work on; then some of those students had to park themselves outside of schools to get a Wi-Fi signal. The current crisis magnifies longstanding inequities. Making believe we can make 'normal' progress under these circumstances without doing serious damage to the most disadvantaged students is just plain crazy. We need to find modes of schooling that support students socially and academically.[10]

The editorial was rejected by the *New York Times*, the *Washington Post*, the *Sacramento Bee*, *Education Week*, and more. Well, OK, maybe they weren't interested in the arguments put forth by a lone academic. So, I worked with the Laureate Chapter of the education honor society Kappa Delta Pi—a Who's Who of scholars and equity advocates—to craft an updated version of the editorial. No luck. The challenge is that 'learning loss' sells in policy terms, while thoughtful examinations of underlying issues are a tough sell. And inequities persist.

9 See the collection of SRCD policy papers at https://www.srcd.org/research/briefs-fact-sheets/statements-evidence

10 See https://gse.berkeley.edu/news/its-time-academic-reset for the editorial. See also McKinney, de Royston, & Vossoughi, 2021.

What really matters in mathematics learning?

In this section I put forth a somewhat radical proposal based on (a) reflections on my recent experiences using mathematics in my personal life, (b) recent political events, and (c) reflections on the evolution of ways in which I have been teaching my course on mathematical thinking and problem solving.

Let me start with my roots. I'm a math person. My Ph.D. is in topology and measure theory and I truly love mathematics. I have spent my entire professional career aimed at the goals described in the opening paragraph of this chapter.

Over the years, my thinking about what matters has broadened. There are problem-solving strategies; there are issues of monitoring and self-regulation; there are belief systems. There are what I've called "productive patterns of mathematical thinking" (Schoenfeld, 2017) or, more traditionally, mathematical practices (Schoenfeld, 2020a). Then there are questions of what kinds of learning environments support students in developing such understandings, and what it takes to teach for robust understanding of mathematics (Schoenfeld 2020b; Schoenfeld and the Teaching for Robust Understanding Project, 2018.) So, my roots are firmly planted in (relatively pure) mathematical soil. But...

When I ask the question "what mathematics have I used in my non-professional life that was important and consequential?", the answer is "almost nothing I learned in school". And yet, I have made very meaningful use of straightforward mathematics. Here are two examples.

Example 1

I chair the Coronavirus Advisory Committee (CAC) for a residential program that serves adults who have developmental and other disabilities. CAC is responsible for setting policies and protocols for residents and staff that concern vaccinations, safety, masking and distancing, travel, and testing. Establishing and updating these policies takes place in the context of rapidly changing and often incomplete or contradictory information and recommendations from available sources. When you look closely, it becomes clear that some policy decisions, including those from US government agencies such as the FDA and CDC, are politically influenced. Indeed, within a period of weeks, new guidance issued by these agencies has conflicted with earlier recommendations without the evidence base having changed substantially.

This is a 'real-world' problem of some significance. How do you think about issues of COVID rationally, based on available information? How do you cut through conflicting information to make sane policy decisions? Here's a problem which I have discussed:

> It is now generally accepted that the primary mechanism of Covid-19 transmission is the inhalation of aerosol particles. Under most circumstances 6 feet of physical distancing is considered a safe distance to avoid infection. Let's take those as scientifically established for the sake of discussion. The other day as I was out for a walk (wearing a cloth mask) I was irritated by cigarette smoke produced by a smoker who was across the street, a good 30 feet away. If an aerosol irritant could bother me at a distance of 30 feet, why is 6 feet of physical distancing considered safe for COVID? (Schoenfeld, 2021, p. 397)

Have fun with this problem if you wish (or see my solution in Schoenfeld, 2021). In broad-brush terms, here's how I thought about the problem. I don't know much biology, but that's not an issue regarding this problem—if I could frame the underlying issues in the right ways a Google search would give me reasonable data. What I needed to do was figure out the right questions to ask. These questions were all I needed to address the issue. Regarding COVID transmission: how big are infectious COVID-transmitting particles and how far are they likely to travel? How dense are they in an infected person's exhalations? Similarly, regarding cigarettes: how big are cigarette smoke particles and how far are they likely to travel? How dense are they in a smoker's exhalations? Answers to those questions were easy to find and to triangulate. Once I had them, some elementary mathematics resolved the issue. (Smoke particles waft, and there are tons of them. There are way less COVID-transmitting particles, which are much larger, and sink.) This type of thinking with emerging data has helped our Coronavirus Advisory Committee establish and modify appropriate safety protocols.

Example 2

This case of mathematical thinking concerns my personal health. I was diagnosed as having type 2 (adult onset) diabetes more than twenty years ago. It's not a major concern; my blood sugar levels are easily kept within bounds by a combination of diet, pills, and exercise. When I was first diagnosed I started keeping track of what I ate and how my blood sugar levels changed.

I quickly learned that the general dietary guidance provided by nutritionists is of limited use because the dietary categories in the recommendations are too broad and there are significant differences in metabolism from individual to individual. White rice sends my sugar skyrocketing, for example, but brown rice is fine; my favorite Chinese restaurant noodle dish sent my sugar through the roof and I had to stop eating it, while my homemade pasta wasn't a problem. Simple data tracking revealed which of my pleasures I could enjoy without significant risk. It also revealed, contrary to dietitians' dogma, that a reasonable quantity of wine with dinner of wine lowered my average blood sugar rather than raising it. To settle a longstanding point of contention with my doctor, I went for three weeks without wine and compared my sugar levels with those of the previous three weeks. Wine won over abstinence!

A more serious issue arose recently when my doctor suggested substituting a new diabetes pill (medicine A) for a pill I'd been using (medicine B), because the newer medicine offers increased protection against heart disease. To my dismay but not my surprise, no information was available regarding how doses of medicine A and medicine B compare. So, my doctor and I had to proceed empirically.

Medicine A comes in doses of 10 and 25 mg. Our first empirical trial involved a roughly half-and-half switch: I added the small (10 mg) dose of medicine A to my daily regimen and cut back half on B. (To give my metabolism time to stabilize, each of the empirical trials described here took about three weeks.) The numbers from the half-and-half switch looked pretty good.

The next question was, is 10 mg of A enough by itself? To find out I stopped taking B. That didn't work well; my blood sugar rose above the levels we wanted. That led us to consider the 25 mg dose of medicine A.

Under the natural assumption that the impact of A would be proportional to dosage, we expected the extra 15 mg of A to provide a very good reduction in blood sugar levels. So, I stayed off medicine B but increased medicine A to 25 mg. The result was a surprise. There wasn't nearly as much effect as we expected—the 25 mg of A didn't reduce my blood sugar levels much more than the 10 mg dose had. That meant that we had to reconsider our basic model. My doctor had said that medicines A and B used two different mechanisms to remove sugar from people's systems. Since the move from 10 mg of A to 25 mg of A didn't help that much, it was now reasonable to assume that the mechanism by which A worked had maxed out at a little more than 10 mg. On the other hand, since medicine B worked by a different biological mechanism, its impact might well be *in addition to* that of medicine A. That's why 10 mg of A plus half of the B I'd been taking had been effective.

I won't run through the numbers here, but I will say they're compelling. What I want to focus on is the process that produced the results. My doctor and I faced a situation for which there was no medical guidance, but for which short-term experimentation was low risk (I could stop taking any combination of medicines immediately if my blood sugar numbers looked bad). We did some simple experiments assuming that the impact of the drugs would be proportional to the dosage, and then revised our assumptions when the data didn't turn out as expected. The result is a much better medical regime for me.

The kind of thinking described in examples 1 and 2 could literally be matters of life and death. In both cases I wondered if the situation at hand could be modelled using some simple proportional reasoning. And—and this is the critical part—in both cases I had the sense of agency that led me to build the models and see if they explained things. The odds are that a very small percentage of people would think in these ways or have the personal agency to do this kind of mathematically based experimentation. That's a very big problem.

I believe that problem comes in large part from the insularity of the curriculum and from the lack of agency that students develop because of the ways we teach. By insularity, I mean that students historically learn to solve only the categories of problems we explicitly prepare them to solve. Rather than thinking of the mathematics they've learned as tools that could apply in a wide range of situations, they think of that mathematics as applying to very narrow classes of problems—specifically, the kinds of problems they've been taught to solve. Just as students learned to expect that "all problems can be solved in five minutes or less" on the basis of their classroom experience (see Schoenfeld, 1985), students also learn to expect that "the math I learn in school is not applicable in meaningful ways to issues that take place outside the classroom". With such expectations, they don't think to use the math they know in situations like those in examples 1 and 2. The problem is compounded by the fact that most students have almost no experience pursuing mathematical ideas on their own. If you haven't done so in the classroom, why would you do so outside the classroom?

Mathematical agency is a fundamentally important issue. I am, once again, teaching my problem-solving course this semester. Over the years I've found myself 'covering' less and less, in that my students and I work fewer problems than before—but we work them much more deeply, exploring the mathematical issues and connections they might suggest. This semester my students and I were playing with the mathematics of 3x3 magic squares. In looking at possible extensions and generalizations a student conjectured that the sum of 9 consecutive integers would always be divisible by 3. That student ultimately argued that (a) the sum of 3 consecutive integers could be shown to be divisible by 3; (b) 9 consecutive integers could be divided into 3 triples, each of which is divisible by 3; (c) since 3 was a factor of each triple, 3 was thus a factor of the sum.

The student's observation and our reflections on it led to other questions. What about the sum of 5 consecutive integers? What about the sum of n consecutive integers, if n is odd? What if n is even? Things got complicated as we played with examples. Some numbers could be obtained as sums of consecutive integers, but some (4 and 8, for example) couldn't. That led to this question: which integers *can* be expressed as the sum of consecutive integers? The class was off and running, in directions I hadn't expected. They worked through the class break, wrote about the problem passionately in our class logs, and ultimately followed their ideas until they produced a complete solution to the problem. Now, in this particular instance my students produced a solution to a known problem, but that doesn't matter.[11] They were doing mathematics, and it was exhilarating. More important than the fact that they solved a particular problem was the fact that they saw themselves as honest-to-goodness mathematical sense-makers. When you have that sense of yourself, you're empowered to tackle new problems—and if doing so becomes enough of a habit, you might feel empowered enough to take on the kinds of COVID and health-related problems I discussed at the beginning of this section. That is: students who have engaged in that kind of generative mathematical thinking throughout their academic careers are much more likely to be mathematically agentive.

Let me try to pull the various themes of this discussion together. First, structural inequities in schooling have worsened during COVID. Society at large has done its best to ignore the issue, focusing on meaningless 'learning loss' instead. Second, we know that myriad students are disaffected from mathematics. There are multiple reasons for this, including its perceived irrelevance and perceived inaccessibility. Third, if people can't use elementary mathematics to reason about what are literally life-and-death issues, mathematics as taught is a dismal failure. Fourth, if people have no mathematical agency, they won't use what they 'know', so their school knowledge is irrelevant.

If you take these issues seriously, radical reform is in order. For mathematics to be personally meaningful to students, it must be more exploratory; a sense of agency simply can't come from being trained to apply methods and ideas you've been taught. And, for mathematics to be meaningful, it must be more personally relevant. Here I don't mean the superficial relevance of topics drawn from 'real life', for example, discussions of sharing pizza equitably when students are learning fractions.

Many meaningful examples can be drawn from real life, and they can be mathematized. That, in part, is the general issue of "mathematical literacy" (see Burkhardt & Schoenfeld, in preparation). Issues of social justice can and must be mathematized as well. There is a small body

11 Historical note: my students have, at times, derived new mathematics when pursuing ideas they found interesting, and the results have been published. That didn't happen this semester, but that's not the point. What matters is that these students saw themselves as capable of creating new mathematics and took great pleasure in it.

of research and resources along these lines (see, e.g., Gutstein 2006, Gutstein & Peterson 2015), and there needs to be much more. But there's more to be considered than mathematizing real world and social justice contexts in classrooms. The challenge is to design ways for students to do that mathematizing in ways that result in their empowerment—the feelings of agency and identity that make it *natural* to see oneself as someone who can approach meaningful problems and make sense of them. What if we thought about organizing curricula with these goals in mind? I think there are possibilities, if only hypothetical for now. In what follows I briefly outline the pie-in-the-sky version, and then suggest that it isn't impossible.

Imagine a massive research and development project centered around the creation of multiple-days-to-weeks-long units that feature:

a. potentially meaningful issues to be addressed or resolved;
b. a student-centered pedagogy supporting exploration in ways consistent with the development of student agency; and
c. scaffolding for teachers that helps them engage with issues (a) and (b) in increasingly powerful ways over time.

Imagine, further, that the units address a broad range of issues, including

- interesting and important mathematical concepts and practices;
- meaningful challenges from the 'real world' that can profit from mathematizing; *and*,
- issues of social justice.

And, as long as we're imagining things, imagine building the kinds of professional networks that support teachers in leveraging what they've learned from working with such instructional units.

This vision isn't impossible. Evidence shows that carefully designed instructional materials can result not only in student learning, but in teacher change—at scale. The Formative Assessment Lessons (FALs) developed by the Mathematics Assessment Project are two- to three-day units that present students with one or more challenges to address, in exploratory fashion.[12] The teacher support provided in the FALs consists of twenty-page lesson plans that structure the explorations and help teachers support the students in those explorations. The lesson plans include descriptions of students' likely misconceptions and ways to address them, while maintaining an ambience of inquiry. Studies of FAL implementation indicate significant student learning gains (Herman et al., 2014) and teacher learning (Research for Action, 2015). The fact that there are twenty FALs per grade (in grades 6 through 10) means that it is possible to build fifty to sixty days of instruction per grade in this mode. That's a third of an academic year. If you can do that, it's possible to build a full year's worth of instruction in similar fashion.

The FALs were constructed to be aligned with the Common Core. What if we were to treat some meaningful real-world problems the same way? What if we were to treat some social justice issues the same way? What if we were to craft an entire curriculum with a mix of centrally important mathematics, social justice, and applied units? On the one hand, I think that such materials could make a significant difference—and that a funding agency with a sense of vision could help to make some of this happen. On the other hand, I can imagine the prospect of the first complete social justice unit being caricatured on *Fox News* and catalyzing the next round of the culture wars. I could say more, but this isn't the place to go into such ideas in depth. My intention here is to plant some seeds for thought. Perhaps some of them can be helped to grow.

Acknowledgement

The author gratefully acknowledges comments on earlier versions of this manuscript from Abraham Arcavi, Brantina Chirinda, Heather Fink, Brian Greer, Markku Hannula, Siqi Huang, Xinyu Wei, and Sandra Zuñiga-Ruiz.

12 See https://www.map.mathshell.org/lessons.php

References

Brown, A. L., & Campione, J. C. (1996). Psychological theory and the design of innovative learning environments: On procedures, principles, and systems. In L. Schauble & R. Glaser (Eds.), *Innovations in learning: New environments for education* (pp. 289–325). Lawrence Erlbaum.

1998 California Proposition 227. (2022, August 1). *Wikipedia*. https://en.wikipedia.org/wiki/1998_California_Proposition_227

Burkhardt, H., & Schoenfeld, A. H. (2019). Formative assessment in mathematics. In R. Bennett, H. Andrade, & G. Cizek (Eds.), *Handbook of formative assessment in the disciplines* (pp. 35–67). Routledge. https://doi.org/10.4324/9781315166933-3

Burkhardt, H., & Schoenfeld, A. H. (in preparation). Assessment and mathematical literacy: A brief introduction. *International Encyclopedia of Education, 4th Edition.*

Center for American Progress. (2019, August 7). Systemic inequality: Displacement, exclusion, and segregation. *American Progress*. https://www.americanprogress.org/issues/race/reports/2019/08/07/472617/systemic-inequality-displacement-exclusion-segregation/

Common Core State Standards Initiative. (2010). *Common Core State Standards for Mathematics.* http://www.corestandards.org

Egede, L., & Walker, R. (2020). Structural racism, social risk factors, and Covid-19: A dangerous convergence for Black Americans. *New England Journal of Medicine, 383*(12). https://doi.org/10.1056/NEJMp2023616

Gamow, G. (1947). *One Two Three… Infinity.* Viking.

Gaylord-Harden, N., Adams-Bass, V., Bogan, E., Francis, L., Scott, J., Seaton, E., & Williams, J. (2020, September 9). Addressing Inequities in Education: Considerations for Black children and youth in the era of COVID-19. *SRCD.* https://www.srcd.org/research/addressing-inequities-education-considerations-black-children-and-youth-era-covid-19

Gutiérrez, R. (2013). The sociopolitical turn in mathematics education. *Journal for Research in Mathematics Education, 44*(1), 37–68. https://doi.org/10.5951/jresematheduc.44.1.0037

Gutstein, E. (2006). *Reading and writing the world with mathematics: Toward a pedagogy for social justice.* Taylor & Francis.

Gutstein, E., & Peterson, B. (Eds.). (2015). *Rethinking mathematics: Teaching social justice by the numbers.* Rethinking Schools.

Halmos, P. (1980). The heart of mathematics. *American Mathematical Monthly, 87*(7), 519–524. https://doi.org/10.1080/00029890.1980.11995081

Herman, J., Epstein, S., Leon, S., La Torre Matrundola, D., Reber, S., & Choi, K. (2014). *Implementation and effects of LDC and MDC in Kentucky districts.* University of California, National Center for Research on Evaluation, Standards, and Student Testing.

Herman, R., Boruch, R., Powell, R., Flesichman, S, & Maynard, R. (2006). Overcoming the challenges: A response to Alan H. Schoenfeld's 'What doesn't work'. *Educational Researcher, 35*(2), 22–23. https://doi.org/10.3102/0013189X035002022

Kozol, J. (1992). *Savage inequalities.* Harper Perennial.

Lappan, G., & Phillips, E. (2009). A designer speaks. *Educational Designer, 1*(3). http://www.educationaldesigner.org/ed/volume1/issue3/article11

Martin, D. B. (2009). Researching race in mathematics education. *Teachers College Record, 111*(2), 295–338. https://doi.org/10.1177/016146810911100208

Martin, D. B. (2019). Equity, inclusion, and antiblackness in mathematics education. *Race, Ethnicity and Education, 22*(4), 459–478, https://doi.org/10.1080/13613324.2019.1592833

Mason, J., Burton, L., & Stacey, K. (1982). *Thinking mathematically.* Addison-Wesley.

McKinney de Royston, M., & Vossoughi, S. (2021, January 18). Fixating on pandemic "learning loss" undermines the need to transform education. *Truthout*, https://truthout.org/articles/fixating-on-pandemic-learning-loss-undermines-the-need-to-transform-education/

National Council of Teachers of Mathematics. (1989). *Curriculum and evaluation standards for school mathematics.* NCTM.

National Council of Teachers of Mathematics. (2000). *Principles and standards for school mathematics.* NCTM.

National Research Council. (2002). *Scientific research in education.* National Academy Press.

Pólya, G. (1957). *How to solve it.* Princeton University Press. (Original work published 1945)

Pólya, G. (1954). *Mathematics and plausible reasoning.* Princeton University Press.

Pólya, G. (1981). *Mathematical discovery.* Wiley. (Original work published 1962-1965)

Research for Action. (2015). *MDC's influence on teaching and Learning.* Research for Action. https://www.researchforaction.org/publications/mdcs-influence-on-teaching-and-learning

Ridgway, J., Crust, R., Burkhardt, H., Wilcox, S., Fisher, L., & Foster, D. (2000). *MARS report on the 2000 tests.* Mathematics Assessment Collaborative.

Rothstein, R. (2017). *The color of law*. Liveright.

Schoenfeld, A. H. (1985). *Mathematical problem solving*. Academic Press.

Schoenfeld, A. H. (1988) When good teaching leads to bad results: The disasters of well taught mathematics classes. *Educational Psychologist*, *23*(2), 145–166. https://doi.org/10.1207/s15326985ep2302_5

Schoenfeld, A. H. (1989). Explorations of students' mathematical beliefs and behavior. *Journal for Research in Mathematics Education*, *20*(4), 338–355. https://doi.org/10.5951/jresematheduc.20.4.0338

Schoenfeld, A. H. (1992). Learning to think mathematically: Problem solving, metacognition, and sense-making in mathematics. In D. Grouws (Ed.), *Handbook for research on mathematics teaching and learning* (pp. 334–370). MacMillan.

Schoenfeld, A. H. (2004). The math wars. *Educational Policy*, *18*(1), 253–286. https://doi.org/10.1177/0895904803260042

Schoenfeld, A. H. (2006a). What doesn't work: The challenge and failure of the What Works Clearinghouse to conduct meaningful reviews of studies of mathematics curricula. *Educational Researcher*, *35*(2), 13–21. https://doi.org/10.3102/0013189X035002013

Schoenfeld, A. H. (2006b). Reply to comments from the What Works Clearinghouse on What Doesn't Work. *Educational Researcher*, *35*(2), 23. https://doi.org/10.3102/0013189X035002023

Schoenfeld, A. H. (2008). Problem solving in The United States, 1970–2008: Research and theory, practice and politics. *ZDM Mathematics Education*, *39*(5-6), 537–551. https://doi.org/10.1007/s11858-007-0038-z

Schoenfeld, A. H. (2017). Teaching for robust understanding of essential mathematics. In T. McDougal (Ed.), *Essential mathematics for the next generation: What and how students should learn* (pp. 104–129). Tokyo Gagukei University.

Schoenfeld, A. H. (2020a). Mathematical practices, in theory and practice. *ZDM Mathematics Education*, *52*(6), 1163–1175. https://doi.org/10.1007/s11858-020-01162-w

Schoenfeld, A. H. (2020b). Reframing teacher knowledge: A research and development agenda. *ZDM*, *52*(2), 359–376. https://doi.org/10.1007/s11858-019-01057-5

Schoenfeld, A. H. (2021). Reflections on 50 years of research & development in science education: What have we learned? And where might we be going? In A. Hofstein, A. Arcavi, B.-S. Eylon, & A. Yarden (Eds.), (2021). *Long-term research and development in science education: What have we learned?* (pp. 387–412). Brill. https://doi.org/10.1163/9789004503625_017

Schoenfeld, A. H. (2022). Why are learning and teaching mathematics so difficult? In M. Danesi (Ed.), *Handbook of cognitive mathematics*. Springer. https://doi.org/10.1007/978-3-030-44982-7_10-1

Schoenfeld, A. H., & Burkhardt, H. (March 20, 2012). Content specifications the Summative Assessment of the Common Core State Standards for Mathematics. https://portal.smarterbalanced.org/library/en/mathematics-content-specifications.pdf

Schoenfeld, A. H., & Pearson, P. D. (2009) The reading and math wars. In G. Sykes, B. Schneider, & D. Plank (Eds.), *Handbook of education policy research* (pp. 560–580). Routledge. https://doi.org/10.4324/9780203880968-51

Schoenfeld, A. H., & The Teaching for Robust Understanding Project. (2018). *An Introduction to the Teaching for Robust Understanding (TRU) framework.* Graduate School of Education. Retrieved from https://truframework.org.

Senk, S. L., & Thompson, D. R. (Eds.). (2002). *Standards-based school mathematics curricula: What are they? What do students learn?* Erlbaum.

Urban Institute. (2020). Structural racism. https://www.urban.org/features/structural-racism-america

US Department of Education. (2003). Identifying and implementing educational practices supported by rigorous evidence: A user friendly guide. https://ies.ed.gov/ncee/pubs/evidence_based/evidence_based.asp

Wilkerson, I. (2020). *Caste.* Random House.

15. Networks, controversies, and the political in mathematics education research

Lisa Björklund Boistrup and Paola Valero

The stories about what constitutes the field of mathematics education research are threaded in a network of institutions, people, and materialities that both produce and sustain them. In such distributed network, controversies concerning these stories are constantly negotiated. Drawing on Latourian concepts and analytical strategies, such stories, network and controversies are explored in an attempt of understanding the political in mathematics education as a 'matter of concern'. An analysis is deployed of the contemporary controversy on the justification for school mathematics in the school curriculum as it is played out in research that engages with the Organisation for Economic Co-operation and Development (OECD)'s Program for International Student Assessment (PISA) as an event shaping the political reasoning about mathematics education. Using the format of a play, the results show the positions entangled in the controversy surrounding mathematics education in current societies. Casting light to these controversies helps trace the multiple entanglements between mathematics education and the cultural politics and economy of our times.

Stories in/on mathematics education

What to say about mathematics education as a domain of scientific research depends on the perspective from which one decides to look at the field. Already in 1998, Jeremy Kilpatrick and Anna Sierpinska

©2024 L. B. Boistrup et al., CC BY-NC 4.0 https://doi.org/10.11647/OBP.0407.15

(1998) published an ICMI-Study[1] volume on mathematics education as a research domain wherein a number of recognised researchers at that time reflected on the core of the research field. Since then, a number of overview publications in handbooks and in special collections in books and journals have produced new insights and discussions about the advances and limitations of the field of research (e.g., Inglis & Foster, 2018; Niss, 2019). In these meta-reflections, that show the field's reflexivity regarding its practices and results, one can find stories about its history, origins, development, and evolution, as well as its hopes and aspirations for its future. To call these accounts 'stories' does not imply any kind of diminishing of their veracity, accuracy, or foundation. It only signals that there may not be an absolute and objective description of mathematics education research and its results, but that there will always exist localised attempts by storytellers to articulate an account of the people, materialities, and practices that, in specific time-space configurations, are considered key elements in shaping mathematics education as a field of investigation. That stories are diverse and are told from different vantage points does not, however, mean that those stories do not have a resonance. Indeed, they do, as they become part of what the many people involved in the activities of the field come to express when referring to mathematics education research.

Traces of those stories are to be found in the very same way that the people involved think and talk about the field: a relatively new area of research with a place in universities, of interdisciplinary nature but identified as a social science, where mathematics in a broad sense plays a role. It is a field of academic inquiry in search of an identity, with local and regional roots, but also highly international, with the overall aim of understanding teaching and learning practices and contributing to their improvement. It is necessary as a foundation and support for bettering teacher education and actual teaching and learning in schools. It contributes to achieving higher results in large-scale measurements, a part of the national strategies to increase interest in the Science, Technology, Engineering, and Mathematics (STEM) fields and promoting and sustaining individual progress, social development, national economic competitiveness, and so on. Out of all these traces,

1 International Commission on Mathematical Instruction (ICMI).

and plagued by a mathematical desire to find order or at least a secure point to hold onto, one could succumb to the temptation of producing a definition that most people – if not all – could endorse:

> Mathematics education is [...] concerned with the technologies of learning and teaching in institutionalized pedagogic settings. It [also] includes researching mathematics education in sites beyond the classroom (e.g., local communities and families, workplaces, policy making, the media, textbook production) and research activities that describe and theorize these practices, including research that is directed towards studying the social, economic and political conditions and consequences of those practices. (Jablonka et al., 2013, p. 43)

This definition, broader than the one proposed by Kilpatrick and Sierpinska in 1998, features a widely encompassing variety of elements, resembling the sense of a network of practices of mathematics education (Valero, 2010). From another perspective, inspired by the Anthropological Theory of Didactics (e.g., Artigue & Winsløw, 2010), there are the multiple, embedded levels of praxeologies that organise mathematics education practices and also its study. Yet other resonant accounts of the field are to be found in the onto-semiotic approach to the didactics of mathematics (e.g., Godino et al., 2019) or the socioepistemology of mathematics teaching and learning practices (e.g., Cantoral, 2020), to mention a few. All these accounts are frequently seen as theoretical frameworks that articulate notions about what constitutes mathematics education, identifying the people, processes, materialities, and institutions involved in its making. These stories about mathematics education and, concomitantly, mathematics education research unfold particular sensibilities towards the focus of attention. They inevitably foreground some elements and shade others. What is common to all these accounts, however, is that each one of them actualises ways of conceiving of the elements that constitute mathematics education; and through such actualisation, the stories, in fact, actively shape what counts – and what does not count – in research. In other words, these stories do something; they have agency; they effect power.

When we come to the discussion of how to determine what counts – or not – and what is possible to think and do in mathematics education, we usher in the political question of how the stories of mathematics education research constantly contest one another and how they perform the very same objects and relationships that they intend to portray. Thus, what is said in a field of study about itself is subject to discussion. It is a controversial issue, more than a matter of fact. It is indeed here that thinking about mathematics education research with Bruno Latour (e.g., 1999, 2005, 2018) may help map new territories. In particular, this approach may help us when considering the power effects of the accounts. What could be possible? What may become new potential imaginations for mathematics education research?

In this chapter we embrace some notions and analytical strategies in the work of Latour to think about the stories of mathematics education as a field of research. While his insights on the functioning of the natural sciences as a terrain of practice have informed studies on science and technology in society and, to some extent, have also illuminated directions in science education (e.g., Elam et al., 2019; Kwak & Park, 2021), the use of Latourian ideas in mathematics education has been limited (De Freitas, 2016; Valero, 2019). As Latour's recent work on the intermeshing of the multiple crises facing humanity poses unavoidable questions about the political orientation of the world (Latour, 2018), we find that troubling the political stories of mathematics education as a field, bringing it in conversation with Latourian concepts and analytical strategies, is a fruitful and compelling step to take.

The chapter starts with an account of some of Latour's ideas – such as actor-network theory, controversies and globalisation in a time of climate change – that we adopt when discussing mathematics education research. Then we concretise these ideas as analytical moves to explore a central controversy in mathematics education nowadays, namely the justification for school mathematics in the school curriculum. With an interest in mathematics education research around the globe, we made the decision to pay specific attention to a global phenomenon, which has had significant influence on mathematics education, namely the Organisation for Economic Co-operation and Development (OECD) Program for International Student Assessment (PISA) (Jablonka, 2016). Assessment in various forms clearly shapes what mathematics and mathematics education may be about (Boistrup, 2017) and in these international comparisons the political aspects of assessment also on a societal level are highlighted. Working with Latourian tools, we performed a limited empirical investigation of how mathematics education research texts from 2004 to 2020 establish relationships to PISA and which controversies are noticeable in the research. We conclude with some remarks about how Latourian tools offer us a different set of concepts and strategies to understand mathematics education research as an actant in sustaining particular possibilities and stories about practices in the field.

A new political look at mathematics education research

How many hours, if one were able to count them one by one or do a rough estimation, do the children of the world sit in a mathematics lesson? (Far too many?) And of that enormous number of hours, in how many did children follow, listen, and actually grasp a mathematical idea? (Far too few?) If far too much time is being spent with little result, such time for children and all those involved could be seen as a poor investment... any savvy capitalist mind would say. Wouldn't it be more productive or generate a better outcome to do something else instead? Doing exercise to improve health, mastering a practice, or serving the community could give a more profitable return. And still, all around the globe there seems to be a sustained political clamour to increase the allocated hours of mathematics in compulsory school curricula. The expected effects have the attention of a wide range of people, all hoping wishfully to score the jackpot of a mathematically talented child.

Few school subjects cause as much of a stir as mathematics, and few have so many contradictions. It suffices to look at local newspapers around the world every time there are new mathematics test results (e.g., Barwell & Abtahi, 2019; Lange, 2019). The question that emerges is: What sustains the ways of doing that generate this situation? 'The social' would be one answer. If we follow Latour's provocative challenge (Latour, 2005) to the social sciences, the use of the adjective 'social' to refer to the fuzzy 'something among people that makes things happen' is inaccurate, even not productive. Instead of assuming the existence of the 'social', he proposes to identify and trace the relationships that take place when things happen, and understand how those relations among people, artefacts, and other types of materialities, institutions, etc. are instances of performance and enactment and, at the same time, the moments that, repeated over time, sustain how we collectively do things and think about them. Furthermore, such relationships are not simple one-to-one exchanges or points of contact, but rather extended, changeable networks with more or less strong connections among a wide range of actants (Latour, 2005). 'Actants' is also a term that signals the attempt to not only focus on human actors, but also on the wide diversity of things that can mobilise agency. Networks are unstable, fragile arrays that depend on the multiple materialities that allow connections to become established within a fully local universality (Latour, 2011). And rather than an existing entity, the network can better

> designate a mode of inquiry that learns to list, at the occasion of a trial, the unexpected beings necessary for any entity to exist. A network, in this second meaning of the word, is more like what you record through a Geiger counter that clicks every time a new element, invisible before, has been made visible to the inquirer. (Latour, 2011, p. 799)

In other words, networks make visible the arrays in which things – human and non-human – emerge as significant and powerful. In this sense, Latour means that notions that were coined centuries ago to designate some kind of 'phantom' forces that steer or regulate people – such as 'nature, society, or power, notions that before were able to expand mysteriously everywhere at no cost' (Latour, 2011, p. 802) – can finally be pinned down to the localised configuration of relationships that constitute collective existence in all its manifestations.

With these ideas in mind, we can now refine our question into more specific inquiries. What relationships among actants (human and non-human) sustain the ways of doing in mathematics education, with its successes and failures? In particular, what relationships sustain the heightened focus on school mathematics, and the desire to politically steer it towards an expected benefit of individuals, communities, and nations? To explore the questions above, an empirical investigation that follows the actants and their relationships would be appropriate. There could also be many points of entry into the exploration. Latour (2005) suggests that the identification of a dispute, a controversy on what seems to be central for the whole arrangement of practice, can be a productive point of departure.

Controversies constitute important jolts from which the functioning and doing of science can be entered (Latour, 2005). While some views of scientific knowledge and practices would emphasise the production of facts and truth as the main result of the scientific endeavour, a relational form of inquiry makes it possible to reveal that results stand, not just because of their intrinsic veracity, but rather because there is a network of people, institutions, and materialities that sustain their production and their legitimacy as reliable discoveries or significant factual revelations. This, in no way, means that the results of science are simply 'social constructions' or fabrications of discourse without a real material existence. On the contrary, the point is rather that the formulation of scientific statements is the instantiation of different concrete scientific work, doings and crossings of the elements, human and non-human, that produce them. Their being is not in the fact, but in the network of relationships that supports the fact. Such production is full of discussions, ranging from the methods and artifacts used in the investigation to the support for findings and their dissemination in society. Suffice to say that the discussions that scale to serious controversies have to do with the fact that scientific results do not just stay as debates among academics – in scientific journals or conferences – but are part of larger connections that mobilise resources, influence, and even the belief in their rightness and adequacy. As knowledge and scientific practices are entangled in the broad network of actions and decisions in society in issues that are at stake for different actants, science is no longer a matter of finding true facts. As Latour argues, the doing of science and its result has become a matter of concern. And, as such, science – of any type – is not an external observer, nor a privileged vantage point to tell the world – but one of the many forces is in the midst of politics and of the effecting of power.

The controversies of science come close to all people, even in instances that do not seem so evidently clear. Scientific controversies of different types are at the core of democracy in times where governing is deeply enmeshed with, and steered through, expert knowledge. This is the characteristic that Michel Foucault had already pointed to concerning the entanglement of knowledge and power in modernity (Foucault & Faubion, 2000). Recent times have made this clear: Is it safe and preferable to be vaccinated for the COVID-19 virus? Which of the vaccinations is best and for whom? These have been quite large controversies of global reach during a massive scientific mobilisation following the outburst of a pandemic in the years 2020–2021.

Still, other minor controversies could be: Which type of assessment of students' mathematical learning is more desirable? Which one is fairer, or which can be more inclusive? This controversy has been discussed in terms of the configuration of an assessment dispositif, encompassing multiple associated discourses and practices (e.g., Boistrup, 2017). Apparently, some controversies are more 'important' than others, some more 'scientific' than others. However, in different times and scales and for different people, these issues come closer and can have different effects. In the terrain of mathematics education, controversies that have to do with knowledge and science constantly emerge and are negotiated. This is why one can consider the network of mathematics education as a field of cultural politics (Diaz, 2017; Valero, 2018) where constant issues of concern are under dispute to be defined. One of these issues is why school mathematics is important to keep as a central subject in the school curriculum – despite its many sustained failures. The controversy on the justifications for school mathematics does not only occupy researchers in mathematics education (e.g., Niss, 1996), but also concerns politicians, economists, educators, local authorities, and of course the very many children who enjoy/suffer it and ask: Why do we have to learn mathematics?

This controversy is pivotal for mathematics education research and the many stories about its purpose, objects, and methods discussed at the beginning of this chapter. The controversy lies at the heart of how the reasons for school mathematics are articulated – implicitly or explicitly – through the relationships among the multiple actants involved in directing mathematics education. With that also come the types of knowledge and research that are deemed valuable and useful to operate in the network. Different stories about mathematics education as a field of research link in particular ways to the clamour for more mathematically competent populations to secure a large enough workforce qualified in STEM. For example, a recurring assumption in the field is that the mathematical knowledge learned in the classroom can easily be transferred to other fields, such as technology and even everyday use. This assumption has theoretically and empirically been contested, addressing how transfer is a simplistic idea connected to particular views of knowledge and learning (e.g., Lave, 1988; Lobato, 2006). What takes place is rather the transformation (recontextualisation or transposition) of mathematics into other fields for it to be relevant, for instance into vocational contexts (FitzSimons & Boistrup, 2017) or, even in university, interdisciplinary contexts (Valero & Ravn, 2017). The point is that mathematics, when entering other fields and connecting to other knowledge-tools and practices, does not remain 'the same' as before, but is actively re-assembled with elements and overarching ideas of the new context (Boistrup & Hällback, 2022).

The narrative of the power of mathematics residing in its transferability – and direct usability – to almost all fields of knowledge, in turn, has been supported by an array of actants such as governments and changes in educational policy and school curricula. Also, by professional associations and economic interest groups demanding the production of a qualified workforce, and by international organisations, such as the United Nations Educational, Scientific, and Cultural Organization (UNESCO) and the OECD, providing quite concrete tools for action to make STEM education a clear element of modernisation (e.g., Zheng, 2019). In particular, the earlier Organisation for European Co-operation (OEEC), which in 1961 turned into the OECD, as part of the support of education for technological development and building of human capital (Rizvi & Lingard, 2009), has systematically sponsored both national school reforms in mathematics and the establishment of collaborative sites of meeting between practitioners in schools and people who started studying and developing mathematical pedagogy and curricula at teacher education or universities. The role of OECD in the boosting of mathematics education has been discussed in the case of various European countries (e.g., De Bock & Vanpaemel, 2019; Gispert, 2014).

The clearest example of this support was the realisation of the Royaumont Seminar in 1959 (OEEC, 1961), which is recognised as an important event, a point of controversy regarding the purpose of mathematics and mathematics education in the context of educational modernisation for economic reconstruction. Research on the history of mathematics education, particularly at the time of the New Math movement (Prytz, 2020) has documented and studied its impact in mathematics education. Historians of education have also contextualised the event as a point in the creation of the scientific modernisation of education, central to the medicalisation of educational research (e.g., Tröhler, 2015). While for mathematics education researchers the Royaumont Seminar initiated a controversy on the ideal views of mathematics that should inform a New Math curriculum, for a historian of education this is an important point of configuration of a 'technocratic culture characterized by confidence in experts rather than in practicing professionals' (Tröhler, 2014, p. 749). The network of connections around what should count for (mathematics) education allowed a 'particular organistic understanding of the social reality [to be] taken for granted and research [to be] conducted under the mostly undiscussed premises of this particular understanding' (p. 749). Within this configuration, mathematical competence has come to be perceived as a key factor in individual and national development (e.g., Tsamadias, 2013). Mathematics education is an area of the curriculum that can be used to monitor and govern differences among people and populations. Research in mathematics education is expected to produce the expert knowledge to improve schooling and to support the growing desire to make populations mathematically literate and competent, a central asset in the production of human capital (Valero, 2017).

At this point, the question emerges of the significance of these connections for recent mathematics education research and for the stories produced about the field. Tracing the networks of mathematics education is an investigation strategy in which we (e.g., Boistrup & FitzSimons, in press; Valero, 2017) and others (e.g., Andrade-Molina, 2021; Ziols & Kirchgasler, 2021) have previously engaged. With this strategy, we explore the controversy surrounding the justifications for mathematics education that emerge in research related to OECD's PISA, given its current salience in locating mathematics education at the centre of educational governing (Popkewitz, 2022).

Researching connections and controversies in mathematics education research

As we will illuminate, there have been a variety of positions over the years in the field of mathematics education research, as revealed when authors connect to international comparisons by OECD. Some have justified the relevance of mathematics through the existence of such international comparisons, while others have been more critical towards their presence and effects. Even when taking a critical perspective towards PISA (or the connected PIAAC, the Program for the International Assessment of Adult Competencies, a form of PISA adapted for adults aged sixteen to sixty-five), there are different positions adopted. For instance, it is possible to take a critical stance as to how PISA/PIAAC limits what mathematics might be conceived as. In Boistrup and FitzSimons (in press), this kind of critique is expanded, with inspiration from Latour's (2018) way of conceptualising globalisation. Boistrup and FitzSimons take Latour's two versions of globalisation, minus and plus, as a starting point for discussing globalisation in relation to matters concerning vocational mathematics education. The authors illuminate how PIAAC, particularly in the construction of survey questions for text takers, is a clear example of globalisation minus, in line with the following quote:

> The term is used to mean that a single vision, entirely provincial, proposed by a few individuals, representing a very small number of interests, limited to a few measuring instruments, to a few standards and protocols, has been imposed on everyone and spread everywhere. It is hardly surprising that we don't know whether to embrace globalization or, on the contrary, struggle against it. (Latour, 2018, pp. 12–13)

In the chapter, the authors discuss how the limitations of PISA/PIAAC have affected local contexts of the world, such as how mathematics vocational education in Australia has undergone a shift towards a restricted and limited view, far from acknowledging the complexities of mathematics in workplace contexts. Their conclusion is that even if the data from OECD's international assessments may be used to gain some interesting insights – which has been challenged by, for example, Anna Tsatsaroni and Jeff Evans (2014) and more recently by Chiara Giberti and Andrea Maffia (2020) – the negative political effects still outweigh any benefits of such international comparisons.

Alexandre Pais and Paola Valero (2014) in their commentary on a special issue on social theory and research in mathematics education also address PISA/PIAAC critically, when pointing out that it is not enough for a critical (or social, as they put it) approach to mathematics education to criticise the misuses to which both teachers and different policies put this school subject. They argue that mathematics itself 'has to be problematized by means of understanding its importance, not in itself – problem solving, utility, beauty, cultural possibilities, and so on

– but in terms of the place this subject occupies within a given societal arrangement' (p. 5).

Giberti and Maffia (2019) have, similarly to us, an interest in how OECD's PISA is used in mathematics education research, and they present a comprehensive literature review. They address the relevance of critical research into the effects of PISA: 'As a conclusion, we suggest that critical research into the effect of PISA can be developed further, especially in those countries that have joined the OECD survey in recent years' (p. 266). When reading further into the article, it becomes clear that what these authors mainly focus on is not the field in general vis-à-vis PISA, but how, de facto, the test and data from PISA are being used in research. Among their findings they present a list of topics, in order of occurrences in the articles analysed, where the PISA test and data were utilised for the purpose of analysis:

- comparative studies at national level;
- teacher education;
- comparative studies on tests;
- curriculum development;
- gender;
- affect and motivation;
- modelling;
- technology;
- equity;
- language;
- textbooks;
- lifelong education;
- other.

On the one hand, the authors address critical discussions of PISA, mainly when they refer to Clive Kanes, Candia Morgan and Anna Tsatsaroni (2014), and how technologies produced by the OECD can be understood as constituting the 'PISA mathematics regime'. On the other hand, Giberti and Maffia (2019) did not include the article by Kanes et al. when composing their identified list of topics, since these authors did not statistically analyse data from PISA. We infer that the kind of critical research into PISA that Giberti and Maffia call for would concern comparisons of, for example, gender differences in PISA outcomes. This means that our interest in this section of the chapter is quite different from the interest of the comprehensive study by Giberti and Maffia. We rather want to understand the field of mathematics education research by tracing how PISA has been discussed over the years in published mathematics education research, and what claims regarding mathematics and mathematics education have resulted from these discussions, in terms of connections and controversies between different human and non-human actants.

Tracing research controversies

The data used in this study are derived from a selection of published research where PISA has a central role. The selection started with a full text search of the word 'PISA' in the journal *Educational Studies in Mathematics* (*ESM*) up to, and including, the year 2020. *ESM* was chosen as one of the broadest international journals in the field of mathematics education. The result was 101 articles. Editorials and commentary texts were excluded. We then selected the articles with four or more mentions of PISA in the text, excluding the references. The criterion of four entries was chosen after examining a number of articles, which revealed that three or fewer entries appeared in articles where PISA was not addressed in a significant way. The result then was twelve full articles.

In a second stage, for each of the twelve articles, we selected the paragraphs where PISA is mentioned. We also added related paragraphs that explain the reasoning connected to paragraphs mentioning PISA. When pasting these paragraphs into one document, including titles and abstracts, the total data set consisted of almost 30000 words.

For each article we analysed the selected paragraphs, addressing the following analytical questions:

- How is PISA mentioned in the text? For instance, PISA results may be used to justify the relevance of a study.
- What claims in relation to PISA are possible to read from the text? For instance, PISA results can be regarded as telling the

truth about the quality of the teaching practices in mathematics in a country.

- What connections do the authors make between different actants involved? For example, the PISA tasks, governments, media, researchers.
- What controversies do the authors address between actants etc.? For instance, are there key tensions or disagreements, such as PISA being seen as embodying a good type of education, and traditional school mathematics as representing a bad type.

For all articles we identified connections and controversies among the twelve analysed articles, and between actants addressed in the articles. Following Latour, we have aimed to stay close to the data to avoid creating a presumed 'social'. Instead, we focus on how the authors address PISA, while tracing associations among statements.

Connections and controversies around PISA in ESM

We start by briefly presenting the twelve articles following a timeline of publication. This has the purpose of giving voice to each of the publications, while simultaneously providing the reader with some overview of the content of the analysed articles. We then articulate a network of connections and controversies identified in the data set, also addressing the roles of different actants, as construed in our analysis. We have chosen to do this in the form of a play, where the actants, human and non-human, are the characters in a play, displaying glimpses of connections and controversies.

Articles addressing PISA in ESM

The first article addressing PISA in *ESM* is by Uwe Gellert (2004), who critically reflected on the use of didactic material in mathematics classes. A PISA task is here presented as an example of non-inclusive tasks, which are best solved putting everyday knowledge aside (like how to share a pizza). A year later, Anna Sfard (2005) drew heavily on PISA when presenting the results of a Survey Team study at the International

Congress of Mathematics Education (ICME-10), on the relations between mathematics education research and practice. A significant claim is that the mathematics education research field has much to gain from adding quantitative analysis drawing on PISA to the more frequent qualitative studies. Four years later, César Sáenz (2009) presented an analysis of the difficulties Spanish student teachers have in solving the PISA 2003 released items. Sáenz adopted the PISA methodology, through the use of tasks, but also the conceptualisation of mathematical competence developed in the PISA framework, and its testing procedures.

Oduor Olande (2014) also drew on the PISA methodology when examining Nordic students' school performance on items containing graphical artefacts. This article by Olande is the first of several during 2014, in which PISA was addressed. The article by Paul Andrews, Andreas Ryve, Kirsti Hemmi, and Judy Sayers (2014) has its main focus on PISA in a critical analysis of the successful Finnish PISA results compared to an analysis of the authors' own interview and classroom data. The authors also base their argumentation on the fact that the Finnish results on another international comparison Trends in International Mathematics and Science Study (TIMSS) were rather mediocre. Throughout the article, the authors problematise the taken-for-granted view of the Finnish PISA results as a sign of the quality of the teaching.

In a special issue on social theory and research in mathematics education, two articles focus on OECD's international comparisons. Kanes et al. (2014) adopted theoretical tools from Basil Bernstein and Foucault to analyse the 'PISA regime', comprising both the knowledge structures produced by the regime but also the ways in which students, teachers and other agents may be produced as subjects. They propose critical research on how to better understand the forms and the mechanisms of PISA in different local contexts, rather than using the PISA shock in society and media for justification of research on how to enhance practice. Tsatsaroni and Evans (2014) also adopted a framework based on Bernstein and Foucault to study PIAAC, while also addressing PISA in their writing at some points. On the one hand, they argue that the version of mathematical competence in PIAAC is far from the complexities of mathematics in adult life, and that PIAAC/PISA require serious consideration and debate in mathematics education research with a focus on power relations. On the other hand, they advocate for the use of PISA/PIAAC data in further studies, for example related to demographic data. Ariyadi Wijaya, Marja van den Heuvel-Panhuizen, and Michiel Doorman (2015) adopt the PISA methodology when seeking the explanation in the national context (as reflected in textbooks) for the low Indonesian PISA result on context-based mathematics tasks.

The above studies are followed by three studies which all address gender differences vis-à-vis mathematics in PISA. A study by Zvia Markovits and Helen Forgasz (2017) draws on PISA results on gender differences in performances as part of the background of the study. The study is then carried out on different data. Yan Zhu, Gabriele Kaiser, and Jinfa Cai (2018) make use of PISA data to carry out a secondary analysis on the Chinese PISA 2015 data to examine gender equity in Chinese students' mathematical achievement. They focus on societal aspects at the individual level (e.g., students' socio-economic status) and systemic aspects. Trine Foyn, Yvette Solomon, and Hans Jørgen Braathe (2018) describe in the introduction how the Norwegian PISA results display no gender differences in mathematics performances, as opposed to results in other contexts. PISA then paves its way into the data set, in a focus group interview, where girls discuss a newspaper article about the PISA survey with the headline 'Norwegian Girls Have Maths Anxiety'. The girls in the study described the boys in the high ability group as more self-confident in mathematics.

The final article in our data is Merrilyn Goos and Sila Kaya (2020). They presented a comparative review of research on understanding and promoting students' mathematical thinking. They analyse papers from *ESM* during two periods: 1994–1998 and 2014–2018. Their review is guided by an analysis of conceptualisations of 'mathematical thinking' proposed in the research, wherein the PISA 2021 assessment framework is one part.

Actants and actors in play

When going through the twelve articles from *ESM* which address PISA, we noticed that the authors bring in different actants (Latour, 2005). Some of these lean more towards what Latour labels non-human, for example the PISA tasks, PISA framework, education systems, and media, while others are human, in the sense of different kinds of actors, such as researchers, politicians, teachers, leaders, and students.

We organise the account around the main groups of actants in the data for which there are connections and controversies vis-à-vis other actants. We now also start to introduce the play; for each actant, we also provide a description in the form of its role as a character in a theatrical play. The play is subsequently presented, reflected through a selection of scenes.

Actants – Actors	**Characters in the play**
PISA methodology refers to the overall framework of PISA, where mathematical competence is described. We also refer to test items that are made public, and hence possible for researchers to use in research. In the PISA methodology we include the data from PISA, which is possible, when permission is given, to use for reanalysis. In different ways, the PISA methodology is present in almost all of the twelve articles.	*PISA methodology*, with test items and assessment procedures, sometimes acts as a ghost affecting others while not showing itself. Sometimes it plays openly, declaring its concerns about essential mathematical content for students' adult life, or the desire for testing etc. It aims at becoming bigger and stronger.
PISA results refer to the results of the tests and questionnaires which are made public by the PISA administration. Most of the articles address PISA results, but in different ways.	*PISA result* is a character with a strong voice, almost yelling its important message around the world.
(*School*) *Mathematics* is present in about half of the articles when addressing how PISA connects to claims about what mathematics (e.g., school mathematics) is or should be.	*Mathematics* has different appearances, mainly as part of the stage design, with different versions of mathematics outlined in writing on screens. In certain scenes it is in the spotlight, other times it is backgrounded. This main character takes many shapes, similar but also very different.

Education systems refer to the diversity of state organised education systems in the PISA participating countries. Those systems are presented as being often quite stable over time, which is addressed in connection to PISA in around a third of the articles.	*Education system* is a powerful character, who moves and changes very slowly. It has much impact on many of the other characters. It fears PISA results.
Media (e.g., newspapers) is a significant actant in relation to PISA in about a third of the articles, often in the background, where the societal effects of PISA are described.	*The media* shouts out messages as a speaker for a variety of characters when it presents PISA results. Simultaneously this actor has a will of its own as it chooses what to shout out to the world, and what to keep quiet about. This character is driven by a wish to be seen and heard, even at the expense of creating a real stir in society.
Governments with politicians are addressed in a couple of articles as those steering education through policy in relation to PISA.	*Government with politicians* is a powerful character, representing a broad range of governments of the world. It is in charge of some of the other characters, such as the education system, leaders and teachers. Simultaneously, the government is afraid of PISA results. It is similar to education systems and belongs to the same family.
The practices of teaching and learning mathematics with leaders, teachers, and students. The articles address and connect to this group of actants in different ways. For example, the practice of teaching mathematics in a country may be connected to its PISA results.	*Leaders, teachers, and students* who practice and learn mathematics is a group of characters that communicate mainly among themselves. However, many other groups like to talk about this group.

Researchers and research fields. Researchers individually and as part of a field of research are actants. Through their writing they connect to each other and to other actants in different ways, representing a variety of research fields.

Powerful researcher speaks from a privileged powerful position, both acting and speaking on behalf of a significant research organisation, but also for arguing for what is necessary in a broad research field. This character speaks for a group of researchers that trusts PISA and mainly sees the benefits of PISA for mathematics and mathematics education.

Critical researcher represents a group of researchers who rather sees the PISA characters as threats to many of the characters in the play, including themselves.

Trusting researcher represents a group of researchers who wants to be friends with the PISA methodology and the PISA results, or at least tries to avoid acting in opposition to these characters. This character is a follower of the powerful researcher.

The chorus comments on the events on the stage, both foreseeing what will come and commenting on what has happened.

'Tell us the truth, oh PISA, and we will follow'

Scene 0

The chorus (chanting as a remote incessant whisper, far away in the distance, not visible on stage):

Where is the truth to follow?

Where is the truth to follow?

Tell us the truth and we will follow.

Scene 1: The everyday and mathematics

Critical researcher: Look everybody! (pointing at different local contexts). Look at our world. Is it not quite remarkable, in all its complexities?

Government: Yes, maybe. But what has that to do with my people?

Education system: What does it have to do with my mathematics?

PISA methodology: Well, I think that this is really relevant. Look at our PISA tests. We have, finally, managed to grasp and measure young people's competence to do something with knowledge. That is what PISA testing is about, to test mathematics in use.

Mathematics: I have been for so long lost in the world of ideas, so hidden in the mind. But now (looking thankfully at PISA methodology) I get a true body: me in this complex world of everyday life. Me in use.

Trusting researcher: Wow! So great, so beautiful. PISA tests are created for the good of mathematics in everyday life, by powerful people. Hurray!

Critical researcher: But, hey, look at the test items, they are not about everyday life. If we really look at what they are about, they do not reflect everyday lives of real people.

Trusting researcher (not looking at the critical researcher, but in awe at PISA methodology): In my research I can rely on PISA. I will take the PISA items as authentic. Then I can trust that my research gets to have good quality.

The chorus (chanting, getting a little bit closer):

Tell us the truth and we will follow.

Tell us the truth about mathematics in everyday life.

Tell us the truth and we will follow.

[Scenes 2–4 omitted]

Scene 5: The quality of mathematics education research

(On a high pedestal, *PISA methodology* and *PISA result* stand holding one hand and lifting the other victoriously.)

Powerful researcher (Enters the scene. Walks slowly, with a straight back, to an elevated position on the stage. Turns to all present researchers): Hear what I have to say. The research in mathematics education needs to do better, and for that, we should learn from PISA. We cannot continue with all these small qualitative studies. Instead, we need more data, solid data, consistent data, laaaarge DATA.

The chorus (chanting, same distance as previously):

Tell us the truth and we will follow.

Tell us the truth about the need for more PISA data.

Tell us the truth and we will follow.

Critical researcher: But, hey, wait. First, we need to establish if the data is of good quality. What claims can we make if the PISA test does not bring out relevant data? Please let us not be hasty here. Actually, my colleagues...

Trusting researcher (interrupting the critical researcher): We should trust PISA! This is what I mean (looks at the powerful researcher)! I can use the data that PISA produces (looks at PISA methodology in awe), and by that do quantitative analyses, and then produce research that is counted as solid, good quality, secure research.

Government (turning to other governments, PISA methodology, and also teachers): Look, look, look. We... some of us are doing good. *We* (with emphasis in the voice and turning its back to some governments) are improving mathematics education and researchers are making good use of the PISA data.

PISA methodology (looking at all others): This is what we told you. This is what we wanted. Now more studies based on PISA can spread around the world, advocating for the one and only version of mathematical competence, the one that really matters for all around the world. Now we can really count!

The chorus (chanting, a little bit closer):

Tell us the truth and we will follow.

Tell us the truth about how PISA will save us.

Tell us the truth and we will follow.

[More scenes omitted]

Scene *n*: Effects of the debate around PISA

(*PISA result* whispers to *The media.*)

The media (runs frenetically around the stage, holding *PISA result* by the hand, shouting): I have news, great news: Now we know who the winners and losers are, in the competition game of mathematics! Breaking news, listen to me!

Governments together: Tell us! Tell us! Was my country successful? Are my people good? Did we win? Are we better in mathematics now?

The media (pointing at different governments): You are a winner; you are a loser; you are better than last time; you are worse than last time. You are just OK; you should try harder. And you... you have no hope.

Government A (looking at *Teacher A*): Look at the results from PISA! You need to be better, so our country will get a better result next time the PISA competition runs, sorry, I mean comparison, not competition.

Government B: And I mean that it is important that the students in our country learn better mathematics. Anyway, it is your responsibility (pointing at a teacher)! And yours (pointing at school leader), and especially yours (pointing at *Student B*). Anyway, you all go and FIX IT!

Mathematics (placed behind *Teacher A* with both hands on her shoulders): How difficult can I be? Me in use, me for competition, me for a better world. Just fix it!

Teacher A (looking down): I am doing my best... (puts an arm around *Student A*'s shoulders)

(*Student A* looks up at *Teacher A* and sighs.)

Trusting researcher (to other researchers): Have you thought about all the fuss that PISA creates in media. Should we not address this?

Critical researcher: Yes, I agree. PISA actually restricts how we view mathematics and the PISA shock that spreads around the world is not really relevant. Instead, we should problematise the effects of PI...

Powerful researcher (interrupting critical researcher): Well, well, well my little friend... We acknowledge the sometimes-non-beneficial attention PISA gets in media. But we should rather celebrate the large attention and interest in mathematical matters. It is not only we, researchers in mathematics education, who care about this important subject area of mathematics, it is everybody. We are thriving well! (Takes PISA methodology and mathematics by the hand, smiles and looks around.)

Trusting researcher: Yes, you are right. And look at all the data that is there free to use. I will tell my colleagues that this is the way forward to a good career and good research, to make all our dreams come true.

The chorus (entering the stage):

Tell us the truth that we can follow.

PISA tells us the truth and we will follow.

PISA tells us the truth and we will trustfully (4/5 of the choir sings) follow.

PISA tells us the truth and we will critically (1/5 of the choir sings) follow.

Stories of power of/in a research field of controversy

What constitutes mathematics education, as a domain of research, is a question to which one can respond in many ways, through different stories about the people, the practices, the materialities, and institutions that support ideas and the concrete everyday activities of the many actants involved. Even though it is not so controversial anymore to state that such stories are political – in the sense that they carry with them particular directions about the whole series of elements and connections that form part of the field, and also in the sense that such stories agentively effect and bring to life the very same relationships, objects, and phenomena that they study – there is still a discussion, almost a kind of controversy, in mathematics education research about how to think and how far to go when conceiving of the network of mathematics education as political. Such controversy revolves around a very core issue, namely what counts as mathematics for the education of people and what justifies its prominence in contemporary, state-governed school curricula.

Many of the stories about the field that we mentioned at the beginning of this paper have more or less explicit positions about how political the field is, and why. Concomitantly, each of these stories articulates a position on the question above. The answers to this question have broadened since the publication of the ICMI volume *Mathematics Education as a Research Domain: A Search for Identity* (Sierpinska & Kilpatrick, 1998). Eva Jablonka et al. (2013) included more sites and practices to count as part of mathematics education research, including the study of 'the social, economic, and political conditions and consequences of those practices' (p. 43). As a result of an overview of the growth around the turn of the twentieth century of theories to study the social, cultural, and political dimensions of mathematics education, the authors concluded that:

> From our interrogation we see signs of a shift away from cognitive psychology and evidence of critical questioning, of the creation of new ideas, and new ways of doing things, as well as a tolerance for multiplicity. All of these observations will contribute to the development of a body of professional knowledge in our discipline, informed by theory rather than driven by policy. We believe the international research community holds the reins of exciting potential for further development of leading edge knowledge in mathematics education. (p. 62)

The impulse that Jablonka et al. identify towards multiplicity of identities – or different competing stories, sometimes in controversy – can be supported nowadays. Indeed, our attempt to tease these stories of the field from a Latour-inspired perspective brought us to focus on the distributed network of relationships within which mathematics education unfolds. The question of what counts as mathematics for education and how it is justified can be then addressed tracing the connections between a variety of actants and figuring out what comes to be stated as stories and what becomes disputed, in other words, which controversies emerge in such a network.

While many of the stories of mathematics education as a field of research tend to delimit the network of actants that define it narrowly around the people and materialities more directly linked with teaching and learning in classrooms, a Latourian move brings us to open up the network in search of other significant actants in the network. In previous research, we have argued that the striving to understand the conditions and consequences of mathematics education practices brings us outside of the comfortable space of didactical and pedagogical relationships to locate research in the field of cultural politics, including the governing dispositives of our time (e.g., Boistrup, 2017; Valero, 2018). The entanglement of mathematics education and the striving for economic growth is to be traced in the close ties between economic agendas, the increasing governing of school mathematics education, curricular reforms across and within countries, and the very same research stories that the field produces.

In our empirical investigation we set out to explore the controversies present in published research as we examined how researchers have related to OECD's PISA program. Since its launch in 2000, PISA and its series of materials, functioning, and institutions has become an authoritative voice in the governing of mathematics education. Thus, its connections to research in mathematics education would show important aspects of the stories about the field at this moment. When going through the twelve articles from *ESM* published between 2004 and 2020, which address PISA, we could distinguish an assemblage in which the authors bring in different actants. Some of these lean more towards what Latour (2005) label non-human – the PISA tasks, PISA framework, education systems, and media – while others are human actors such as researchers, politicians, teachers, leaders, and students. As it was shown above, these actants, non-human and human, have different kind of roles in the assemblage and in the controversies.

In our tracing of a network of mathematics education, we focused on controversies within the twelve articles, which allowed us to create a play that intends to make explicit some of the actants and relationships present in the examined research text. In relation to the question of what counts as mathematics for education and what justifies its prominence in the curricula of education systems, the play grasps different controversies in mathematics education, where some researchers take on a critical perspective vis-à-vis international comparisons such as PISA, while other researchers – a larger number – adopt PISA as a truth teller, embracing the important role of mathematics that PISA advocates for. One controversy is about PISA items in relation to the real world. While critical researchers (henceforth, CR) address how the test items neither reflect any contextual reality in a relevant sense, nor any cultural, local aspects, more trusting researchers (henceforth, TR) take the PISA framework for granted and accept PISA's claims of testing mathematics in everyday life.

Another controversy in the analysed texts revolves around the opportunities of quantitative studies adopting PISA methodology with respect to the diversified trend of smaller, qualitative studies. Following TR, PISA as a source for research should be embraced, since mathematics education practices are in need of improvement. CR, on the other hand, question the 'PISA regime' and its effect on the research field. Yet another controversy concerns the relationships between PISA, teachers, and the media, and whether teachers are to be blamed or not for PISA results. CR call for a mathematics education field which 'feels for' the teacher, while TR use the PISA results to evaluate teachers' work. Included in this controversy is whether to trust and build on (or not) connections between teachers' subject knowledge and teaching effectiveness as measured by PISA, or between PISA results and the state of teaching practices in a country. The controversy around mathematics is also present as PISA-defined mathematical knowledge appears as 'good' and desirable, and traditional school mathematics as 'bad' and in need of change. Following TR, PISA gets the role of telling the truth about what school mathematics should be. CR questions this and advocates for PISA as a threat to a view of mathematics as a plurality of mathematical practices incorporating cultural and contextual aspects.

Finally, a larger controversy that becomes evident in the analysis and as a result of the whole paper is the ties between mathematics education and capitalism. The push for mathematical qualifications to be central for the economic growth and development of the individual, communities, and nations has become a central point of controversy around what counts as mathematics education and why it is important, and OECD's PISA has become a clear actant here. Several recent studies have paid attention to this issue and have drawn pointed to the benefits and dangers of the link between mathematics education and growing capitalism—and its associated, brutal in(ex)clusions. For example, Mark Wolfmeyer, in Chapter 16 of this volume, examines the 'assessment spread' connected to TIMSS and its related technologies and institutions and shows the connections that sustain a global view for generating a consumerist trained human capital through mathematics. The clear emphasis on a critique of mathematics education (research) to serve particular economic organisations is a way of exploring how power is effected within the networks that constitute mathematics education practices. Our Latourian exploration binds us in connection to the larger network that constitutes the narratives of our field, which we cannot ignore.

References

Andrade-Molina, M. (2021). Narratives of success: Enabling all students to excel in the global world. *Research in Mathematics Education, 23*(3), 293–305. https://doi.org/10.1080/14794802.2021.1994453

Andrews, P., Ryve, A., Hemmi, K., & Sayers, J. (2014). PISA, TIMSS and Finnish mathematics teaching: An enigma in search of an explanation. *Educational Studies in Mathematics, 87*(1), 7–26. https://doi.org/10.1007/s10649-014-9545-3

Artigue, M., & Winsløw, C. (2010). International comparative studies on mathematics education: A viewpoint from the Anthropological Theory of Didactics. *Récherches en Didactiques des Mathématiques, 30*(1), 48–82. https://revue-rdm.com/2010/international-comparative-studies

Barwell, R., & Abtahi, Y. (2019). Mathematics education in the news: Introduction. *Canadian Journal of Science, Mathematics and Technology Education, 19*(1), 1–7. https://doi.org/10.1007/s42330-019-00043-z

Boistrup, L. B. (2017). Assessment in mathematics education: A gatekeeping dispositive. In H. Straehler-Pohl, N. Bohlmann, & A. Pais (Eds.), *The disorder of mathematics education: Challenging the socio-political dimensions of research* (pp. 209–230). Springer. https://doi.org/10.1007/978-3-319-34006-7_13

Boistrup, L. B., & FitzSimons, G. E. (in press). Vocational mathematics and competence: Effects of and resistance to globalization. In A. Chronaki & A. Yolcu (Eds.), *Troubling notions of global citizenship and diversity in mathematics education*. Routledge.

Boistrup, L. B., & Hällback, M. (2022). Designing and researching vocational mathematics education. In L. B. Boistrup & S. Selander (Eds.), *Designs for research, teaching and learning: A framework for future education* (pp. 61–81). Routledge. https://doi.org/10.4324/9781003096498

Cantoral, R. (2020). Socioepistemology in mathematics education. In S. Lerman (Ed.), *Encyclopedia of Mathematics Education* (pp. 790–797). Springer. https://doi.org/10.1007/978-3-319-77487-9_126-4

De Bock, D., & Vanpaemel, G. (2019). *Rods, sets and arrows: The rise and fall of Modern Mathematics in Belgium*. Springer. https://doi.org/10.1007/978-3-030-20599-7

De Freitas, E. (2016). Bruno Latour. In E. de Freitas & M. Walshaw (Eds.), *Alternative theoretical frameworks for mathematics education research: Theory meets data* (pp. 121–148). Springer. https://doi.org/10.1007/978-3-319-33961-0_1

Diaz, J. D. (2017). *A cultural history of reforming math for all: The paradox of making in/equality*. Routledge.

Elam, M., Solli, A., & Mäkitalo, Å. (2019). Socioscientific issues via controversy mapping: bringing actor-network theory into the science classroom with digital technology. *Discourse: Studies in the Cultural Politics of Education, 40*(1), 61–77. https://doi.org/10.1080/01596306.2018.1549704

Foucault, M., & Faubion, J. D. (2000). *Power*. New Press.

Foyn, T., Solomon, Y., & Braathe, H. J. (2018). Clever girls' stories: The girl they call a nerd. *Educational Studies in Mathematics, 98*(1), 77–93. https://doi.org/10.1007/s10649-017-9801-4

Gellert, U. (2004). Didactic material confronted with the concept of mathematical literacy. *Educational Studies in Mathematics, 55*(1), 163–179. https://doi.org/10.1023/B:EDUC.0000017693.32454.01

Giberti, C., & Maffia, A. (2020). Mathematics educators are speaking about PISA, aren't they? *Teaching Mathematics and its Applications, 39*(4), 266–280. https://doi.org/10.1093/teamat/hrz018

Gispert, H. (2014). Applications: Les mathématiques comme discipline de service dans les années 1950–1960. In D. Coray, F. Furinghetti, H. Gispert,

B. Hodgson, & G. Schubring (Eds.), *One hundred years of L'Enseignement Mathématique: Moments of mathematics education in the twentieth century* (pp. 251–270). L'Enseignement Mathématique.

Godino, J. D., Batanero, M. d. C., & Font, V. (2019). The onto-semiotic approach: Implications for the prescriptive character of didactics. *For the Learning of Mathematics, 39*(1), 38–43.

Goos, M., & Kaya, S. (2020). Understanding and promoting students' mathematical thinking: A review of research published in *ESM*. *Educational Studies in Mathematics, 103*(1), 7–25. https://doi.org/10.1007/s10649-019-09921-7

Inglis, M., & Foster, C. (2018). Five decades of mathematics education research. *Journal for Research in Mathematics Education, 49*(4), 462–500. https://doi.org/10.5951/jresematheduc.49.4.0462

Jablonka, E. (2016). Mathematics education as a matter of achievement. In M. A. Peters (Ed.), *Encyclopedia of Educational Philosophy and Theory*. Springer. https://doi.org/10.1007/978-981-287-532-7_521-1

Jablonka, E., Wagner, D., & Walshaw, M. (2013). Theories for studying social, political and cultural dimensions of mathematics education. In M. A. Clements, A. J. Bishop, C. Keitel, J. Kilpatrick, & F. K. S. Leung (Eds.), *Third International Handbook of Mathematics Education* (pp. 41–67). Springer. https://doi.org/10.1007/978-1-4614-4684-2_2

Kanes, C., Morgan, C., & Tsatsaroni, A. (2014). The PISA mathematics regime: Knowledge structures and practices of the self. *Educational Studies in Mathematics, 87*(2), 145–165. https://doi.org/10.1007/s10649-014-9542-6

Kwak, D.-J., & Park, E. J. (2021). Mediating process for human agency in science education: For man's new relation to nature in Latour's ontology of politics. *Educational Philosophy and Theory, 53*(4), 407–418. https://doi.org/10.1080/00131857.2020.1838273

Lange, T. (2019). Unpacking the emperor's new policies: How more mathematics in early childhood will save Norway. *Canadian Journal of Science, Mathematics and Technology Education, 19*(1), 8–20. https://doi.org/10.1007/s42330-019-00041-1

Latour, B. (1999). *Pandora's hope: Essays on the reality of science studies*. Harvard University Press.

Latour, B. (2005). *Reassembling the social: An introduction to actor-network-theory*. Oxford University Press.

Latour, B. (2011). Networks, societies, spheres: Reflections of an actor-network theorist. *International Journal of Communication, 5*, 796–810.

Latour, B. (2018). *Down to earth: Politics in the new climatic regime*. Polity.

Lave, J. (1988). *Cognition in practice: Mind, mathematics, and culture in everyday life*. Cambridge University Press.

Lobato, J. (2006). Alternative perspectives on the transfer of learning: History, issues, and challenges for future research. *Journal of the Learning Sciences, 15*(4), 431–449. https://doi.org/10.1207/s15327809jls1504_1

Markovits, Z., & Forgasz, H. (2017). 'Mathematics is like a lion': Elementary students' beliefs about mathematics. *Educational Studies in Mathematics, 96*(1), 49–64. https://doi.org/10.1007/s10649-017-9759-2

Niss, M. (1996). Goals of mathematics teaching. In A. J. Bishop, K. Clements, C. Keitel, J. Kilpatrick, & C. Laborde (Eds.), *International handbook of mathematics education* (pp. 11–47). Kluwer. https://doi.org/10.1007/978-94-009-1465-0

Niss, M. (2019). The very multi-faceted nature of mathematics education research. *For the Learning of Mathematics, 39*(2), 2–7.

OEEC. (1961). *New thinking in school mathematics*. OEEC.

Olande, O. (2014). Graphical artefacts: Taxonomy of students' response to test items. *Educational Studies in Mathematics, 85*(1), 53–74. https://doi.org/10.1007/s10649-013-9493-3

Pais, A., & Valero, P. (2014). Whither social theory? *Educational Studies in Mathematics, 87*(2), 241–248. https://doi.org/10.1007/s10649-014-9573-z

Popkewitz, T. S. (2022). International assessments as the comparative desires and the distributions of differences: Infrastructures and coloniality. *Discourse: Studies in the Cultural Politics of Education*, 1–23. https://doi.org/10.1080/01596306.2021.2023259

Prytz, J. (2020). The OECD as a booster of national school governance. The case of New Math in Sweden, 1950–1975. *Foro de Educación, 18*(2), 109–126. https://doi.org/10.14516/fde.824

Rizvi, F., & Lingard, B. (2009). The OECD and global shifts in education policy. In R. Cowen & A. M. Kazamias (Eds.), *International handbook of comparative education* (pp. 437–453). Springer. https://doi.org/10.1007/978-1-4020-6403-6_28

Sfard, A. (2005). What could be more practical than good research? *Educational Studies in Mathematics, 58*(3), 393–413. https://doi.org/10.1007/s10649-005-4818-5

Sierpinska, A., & Kilpatrick, J. (1998). *Mathematics education as a research domain: A search for identity*. Kluwer.

Sáenz, C. (2009). The role of contextual, conceptual and procedural knowledge in activating mathematical competencies (PISA). *Educational Studies in Mathematics, 71*(2), 123–143. https://doi.org/10.1007/s10649-008-9167-8

Tröhler, D. (2015). The medicalization of current educational research and its effects on education policy and school reforms. *Discourse: Studies in the Cultural Politics of Education, 36*(5), 749–764. https://doi.org/10.1080/01596306.2014.942957

Tsamadias, C. (2013). The mathematical capital and its economic value. In E. G. Carayannis & G. M. Korres (Eds.), *European socio-economic integration* (pp. 43–50). Springer. https://doi.org/10.1007/978-1-4614-5254-6_3

Tsatsaroni, A., & Evans, J. (2014). Adult numeracy and the totally pedagogised society: PIAAC and other international surveys in the context of global educational policy on lifelong learning. *Educational Studies in Mathematics, 87*(2), 167–186. https://doi.org/10.1007/s10649-013-9470-x

Valero, P. (2017). Mathematics for all, economic growth, and the making of the citizen-worker. In T. S. Popkewitz, J. Diaz, & C. Kirchgasler (Eds.), *A political sociology of educational knowledge: Studies of exclusions and difference* (pp. 117–132). Routledge. https://doi.org/10.4324/9781315528533-8

Valero, P. (2018). Human capitals: School mathematics and the making of the homus oeconomicus. *Journal of Urban Mathematics Education, 11*(1&2), 103–117. https://doi.org/10.21423/jume-v11i1-2a363

Valero, P. (2019). Mathematics education in the 'New Climatic Regime'. In L. Harbison & A. Twohill (Eds.), *Proceedings of the Seventh Conference on Research in Mathematics Education in Ireland* (pp. 7–13). Dublin City University.

Valero, P., & Ravn, O. (2017). Recontextualizaciones y ensamblajes: ABP y matemáticas universitarias [Recontextualizations and assemblages: PBL and university mathematics]. *Didacticae, 1*(1), 4–25. https://doi.org/10.1344/did.2017.1.4-25

Wijaya, A., van den Heuvel-Panhuizen, M., & Doorman, M. (2015). Opportunity-to-learn context-based tasks provided by mathematics textbooks. *Educational Studies in Mathematics, 89*(1), 41–65. https://doi.org/10.1007/s10649-015-9595-1

Zheng, L. (2019). A performative history of STEM crisis discourse: The co-constitution of crisis sensibility and systems analysis around 1970. *Discourse: Studies in the Cultural Politics of Education, 42*(3), 337–352. https://doi.org/10.1080/01596306.2019.1637332

Zhu, Y., Kaiser, G., & Cai, J. (2018). Gender equity in mathematical achievement: The case of China. *Educational Studies in Mathematics, 99*(3), 245–260. https://doi.org/10.1007/s10649-018-9846-z

Ziols, R., & Kirchgasler, K. L. (2021). Health and pathology: A brief history of the biopolitics of US mathematics education. *Educational Studies in Mathematics, 108*(1), 123–142. https://doi.org/10.1007/s10649-021-10110-8

16. Globalisation of mathematics education and the world's first monoculture: Assessment spread's association with consumerism and human capital

Mark Wolfmeyer

The global spread of mass schooling supports ideologies of human capital and consumerism that we can consider as the world's first monoculture. Educational organisations with global reaches, such as the International Association for the Evaluation of Educational Achievement and the Organisation for Economic Co-operation and Development, spread particular mathematics education goals that present opportunities for analysis and critique by mathematics educators who seek to advance causes beyond or in opposition to the consumer-industrial complex. In this chapter I utilise Joel Spring's review of perspectives on globalisation and education to motivate extended analysis of one example of 'assessment spread' within mathematics education, namely the Trends in International Mathematics and Science Study. Complementary methods of analysis (historical/contextual and content-based) reveal a strong association between mathematics education assessment spread and the rise of a world culture emphasising human capital and lifelong consumerism.

Introduction

In this chapter I take up themes on the purpose of mathematics education that resonate with other chapters in this volume, in which

©2024 Mark Wolfmeyer, CC BY-NC 4.0 https://doi.org/10.11647/OBP.0407.16

mathematics education is viewed as relating to global capitalism. I suggest the possibility that mathematics education practices are not just responsive to the context of global capitalism; rather, they are perceived by the world's power elite as a means to achieve their objectives. In my contribution, I collocate perspectives on globalisation and globalisation of education with phenomena that we, as mathematical enthusiasts and mathematics educators, know all too well: international mathematics educational assessments like the Trends in International Mathematics and Science Study (TIMSS). I argue that such phenomena, what I refer to as 'assessment spread' to highlight the historical context of increasing participation, serve as one of many vehicles through which a global culture can permeate the world, creating a first monoculture. Specific tenets of this monoculture include commitments to laissez-faire capitalism, markets dominated by consumerism, and education systems oriented to those ends. In other words, with the global economy as its foundation, the first monoculture's twin features are increasing the number of people in the world that 1) enter the wage labour market and 2) become lifelong consumers of goods and services in a market-based system. Mathematics education is seen by the power elite as an opportunity to extend this vision for the world.

I begin with the perspectives on globalisation and globalisation of education as reviewed by Joel Spring (2014). Among the options he presents, which include 'world system' and 'culturist', what he terms 'the world culture theory' on globalisation provides the most suitable means for analysing the assessment spread phenomena at the heart of this project. In utilising this framework, I turn the theory on its head by describing the exact nature of this world culture that is spreading: a first monoculture dominated by laissez-faire capitalism and the supports required by it, such as education for human capital. Existing literature from globalisation and mathematics education studies contributes important connections to the theories on globalisation as well as laying the ground for my present inquiry into assessment spread.

In my focus on TIMSS, I begin with a narrative to document this ongoing activity as an example of assessment spread. This includes tracing the growing list of participating countries, with attention paid to their relative engagement with the global economy. Along with this historical narrative, I provide a contextual analysis (similar in goal but different in method from my other projects, e.g., Wolfmeyer, 2014) to describe the motives represented in this project, mostly by analysing the implications presented by the World Bank's involvement in TIMSS. A primary finding of interest emerges from this historical and contextual analysis: several participating TIMSS countries increased their engagement in the global economy during their periods of participation. Specifically, the vast majority of participating countries maintained *high-level income* or *significantly increased* their per-capita income levels during the years of participation in TIMSS. As I will argue, this data presents a perfect instantiation of the World Bank's vision for a global society with everyone participating in the wage labour market and consumer society. To be sure, I dare not make any claim of causality, that TIMSS participation actually caused countries' citizens to better engage in the global economy, but suggest that the association displayed by the data carries weight for my argument that TIMSS and assessment spread relate to an emerging world culture of human capital and consumerism.

In the final section, I complement the historical and contextual analysis with content analysis; I engage with Houman Harouni's (2015) notion of the political economy of mathematics to analyse released mathematics assessment items from a recent implementation of TIMSS. This content analysis again confirms the relationship of TIMSS to the spread of the global economy. The vast majority of content items within TIMSS are associated with mathematical preparation for human capital, with a very limited scattering of alternatives, thereby signifying to users of this assessment that mathematics could not be applied to anything but behaviours relating to wage labour and being a consumer. Thus, taking content and context analyses together, I suggest that the mathematics education assessment spread over time is associated heavily with a global economy of laissez-faire capitalism and human capital existing primarily for a consumerist, profit-driven world culture.

Perspectives on globalisation and education

In this section I review Spring (2014) for a variety of perspectives on globalisation, with specific attention to their relationship to the globalisation of education. This review presents several options for consideration when analysing the assessment spread phenomena in mathematics education and ultimately, what Spring terms the 'world culture' perspective seems the most fitting to the spread's aims and origins. In addition, I also provide some analyses in existing literature regarding globalisation of education as it pertains to the world culture theory as well as a few contributions from among the mathematics education literature that engage the globalisation perspectives as reviewed here.

Spring (2014) reviews existing bodies of scholarship on globalisation of education to produce three broad categories: world culture theorists, world system (postcolonial/critical), and culturist. These reflect his understanding of the ways that scholars engage in the activities of spreading education practices across the globe. World culture theory resonates most closely to the assessment spread of a global mathematics education because it insists 'all cultures are slowly integrating into a single global culture' (p. 7). However, review of each is important for understanding the opportunities at play when analysing global phenomena in education.

Clear examples of how educational systems are unifying globally are abundant. For instance, Spring point out that most school systems comprise a sequential ladder from the elementary to middle level to secondary years of schooling, with groupings of students by age and with achievement as indicating progress through the ladder. However, he notes how John Meyer, David Kamens, and Aaron Benavot (1992) suggest that the spread of the nation state required education of the citizenry to ensure political stability and economic growth. The authors argue that 'the gradual rationalisation of the Western polity, the modern curricular structure became a take-for-granted "model" by the turn of the twentieth century' (p. 72). As the Western concept of the nation state spread, 'the standard model of the curriculum has also diffused throughout the world, creating a worldwide homogeneity in the over-all categorical [curriculum categories] system' (Spring, 2014, p. 8).

The spread of educational practices across the globe corresponded, thus, to the spread of governing people through the structure of the nation-state. A main feature of the globalisation of education is what Spring highlights, from the work of Francisco Ramirez (2003), as the 'credentialed society'. By tying educational achievements to the opportunity to obtain employment, global education spread commits not only to the needs of a nation's citizenry but to the capitalist economy and markets that rose alongside it. A key feature of such developments is increasing the number of people engaging in the wage labour structure. The naturally occurring relationship between the nation state and capitalism is well documented in literature. For example, Arun Ghosh (1997) states: 'At any rate, historically, capitalism and nation states evolved and prospered together' (p. 683). Furthermore:

> Ramirez locates the origins of world cultural theorists in the work of John Boli, Frank Lechner, George Thomas, and Immanuel Wallerstein. These theorists argue that a world culture began with the spread of Western Christian ideas in the late nineteenth century and escalated after the end of World War II. (Spring, 2014, p. 8)

World cultural theorists primarily focus on the relationship between the spread of nation state governing structures and the similar schooling systems set up across these nation states. What appears explicitly absent in a world culture's emphasis on nation state is the acknowledgment that a modern nation state requires a capitalist economic structure as well. This could be sometimes suggested by those utterances that confuse 'the development of capitalism with the development of "liberal" thought' (Ghosh, 1997, p. 683). The suggestion of credentialing as a key feature within world culture's schooling systems implies this entanglement between capitalism and the modern nation state. My contribution here recognises the entanglement of developing nation states and free-market capitalism, not always going hand in hand, but very often co-developing, as illustrated by analysis of TIMSS and its association with the World Bank.

Spring also describes critiques of globalisation via postcolonial theory, also referring to related scholars as 'world systems theorists' who 'argue that the richest nations legitimise their power by imposing their educational values on other nations. These educational values include schooling for economic growth and developing workers for a free market economy' (Spring, 2014, p. 10). Specific discussions here include the ways that the World Bank enacts educational programs in its broader mission to enact laissez-faire capitalism across the globe (Spring, 2004). The World Bank and their networked actors put forth the suggestion of a knowledge economy as the key for emerging economies in so-called 'developing nations'. This policy features in multiple World Bank publications, such as *Building Knowledge Economies: Advanced Strategies for Development* (World Bank, 2007). Similarly, the global economic elite, as represented by the Organisation for Economic Co-operation and Development (OECD), declares an education for human capital among the wealthy nations as the primary educational imperative (as in Keeley, 2007).

Viewing world culture theory alongside world systems and postcolonialist critiques of globalisation immediately reveals important insights for understanding a globalised mathematics education, as observed through assessment spread phenomena. I will argue that the assessment spread and globalised mathematics education practices are *generated from* a world culture theory framing: they situate in a movement towards a unified monoculture emphasising the nation state and its concomitant laissez-faire capitalism and consumerism. By asserting this, then, I align with world systems theorists' portrayal of the state of affairs we have witnessed over the last hundred years. In other words, both theoretical framings engage with my present work and, as a rhetorical strategy, my chapter title and framing assert the dominance of world cultural theory as a perspective for signifying just what is at stake with such assessment spread phenomena, namely the dominance of a unifying world culture emphasising human capital and consumerism.

The remaining theorists on globalisation and education of interest to Spring are worth considering, though they are less significant to this project. The culturist perspective provides a viewpoint suggesting agency across parties in the globalisation of education. Spring points out their rejection of movement towards a unified practice in education and declaration in favour of a concept of 'educational borrowing and lending' (Spring, 2014, p. 11). Culturists represent a variety of interpretations; they differ, however, in how helpful they are in understanding assessment spread in mathematics education. One culturist example, as indicated by Spring, occurred with the late 1980s educational fad in the United States that looked to the schooling practices in Japan. Implicit in this discussion, however, is the framing of a quest to perfect a system across the globe in service of nation-state and capitalist economies; the practice was less about borrowing and lending than it was about deepening commitments to capitalism and the nation state.

Other culturist examples provide optimism and balance to the significant and weighty suggestions by world culture theory and the world systems critique. As Spring suggests, culturists argue that there is not just one world education model:

> Anderson-Levitt argues that there are two competing world models for education. World culture theorists, she argues, consider the goal of the world education model to be preparing students to be workers in the global economy. She identifies two competing world education models, which I will label 'Economic Education World Model' and 'Progressive Education World Model'. I propose the existence of two other competing global models, which I label 'Religious Education World Models' and 'Indigenous Education World Models'. These last two global models openly reject world models of schooling based on Western education. (Spring, 2014, p. 13)

These competing global models for education are worth considering in mathematics education, as I note below, although will not factor into my discussion of assessment spread. Mathematics education as a global practice, via assessment spread and closely associated curricular convergence, appears to me as a dominant 'Economic Education World Model' that seeks to define and claim a particular set of human behaviours that are identified as mathematical.

Concepts relating to this review of globalisation and education have been applied to studies in the globalisation of mathematics education prior to my contributions here. For example, Spring (2014) frames his discussion on globalisation and mathematics education with the dominating discourse of the knowledge economy (in part via the World Bank, as I note above). This appears as a critical world systems approach, as reviewed above, especially when he suggests an 'export' of Eurocentric models of education and higher education across the globe and, specific to my focus, a suggested 'appropriation effect' via the assessment spread, as in:

> [Projects] can also be focused on applying some predetermined framework as in international assessment and comparative studies (e.g., SIMS, TIMSS, PISA) typical of scientific paradigm research. In both of these types of research there is what I shall term the *appropriation effect*. In this, locally gathered knowledge from 'developing' countries is appropriated for academic and other uses in 'developed' countries. (Ernest, 2016, p. 40)

As another example, Bill Atweh and Phillip Clarkson (2010) frame inquiries into internationalising mathematics education with an open perspective on globalisation, as somewhat of a commitment to Spring's culturist perspective, but with clear intentions to be critical of the negative effects, as in the world systems and postcolonial approach:

> [Some] internationalization and globalization processes may be good, whereas others may be less desirable and should be contested. Further, what is considered good aspect of internationalization and globalization for a particular group of people may very well be at the expense of other groups. (p. 80)

And in provoking discussions about differing models of global mathematics education systems, perhaps resonating with a culturist approach, I would promote the work of Peter Appelbaum and Susan Gerofsky (2013), with their vision of an alterglobalisation in mathematics education. They suggest 'another world is possible' by reviewing the alterglobalisation 'movements of people who are extremely concerned about the neoliberal agenda of a self-regulating free-market, and the linkage between global cultures and a dominant Western culture that often runs counter to many of the values and principles by which people live their lives' (p. 27). Asserting a stance that is not quite anti-globalisation, they emphasise a need for global 'renewal of political citizenship and activism' and, specific to mathematics education, suggest a world model of mathematics education grounded in the arts and participatory democratic practice.

Finally, recall that the mathematics education community initiated an early critique of the internationally comparative mathematics tests! Hans Freudenthal (1975) provided pointed arguments against the early efforts that ultimately led to the TIMSS, emphasising many points such as that mathematicians and mathematics educators were not involved in the design, and dubious conclusions are often made from the data.

And what could be blunter than the summary by Christine Keitel and Jeremy Kilpatrick (2012) that:

> The studies rest on the shakiest of foundations – they assume that the mantel of science can cover all weaknesses in design, incongruous data and errors of interpretation. They not only compare the incomparable, they rationalize the irrational. (p. 254)

In the present project, I aim to take up these conversations critiquing TIMSS yet again, and specifically by associating the project with its broader policy objectives. Providing insights and critique of significant projects like TIMSS (and other examples of mathematics education spread) aims to complement the literature for mathematics educators and enthusiasts who recognise the relationships between mathematics educational practice and social life, such as scholars working among the international Mathematics Education and Society group.

Reviewing perspectives on globalisation and education has uncovered a promising framework for investigating assessment spread in mathematics education across the globe. Situating the phenomena as grounded by a world-culture perspective allows us to consider how dominating and powerful agencies envision a global, unified culture and practice in service of particular elite goals. In this manner I put forth a critical, some might say postcolonial, analysis of the power embedded in assessment spread of mathematics. With this critique, I aim to motivate rejections of the seemingly benign or neutral assessment spread that we witness in mathematics education. My contribution complements the culturist contributions from mathematics education, reviewed above, that provide much-needed optimism when we have such a formidable force in mathematics education with which to contend. I next turn to tracing the history and social context of the TIMSS practice before offering a content analysis of released test items from that assessment's recent iteration.

From 10 to 60+: TIMSS, the World Bank, and a global economy

In this section I survey the history and context of TIMSS to provide additional material for analysing the global spread of assessment in mathematics education. In its origins, the practice began with only a handful of similar countries, but over time, the entanglement of the World Bank with the main organisation behind TIMSS, IEA (International Association for the Evaluation of Educational Achievement) suggests a new context in line with the world-culture perspective that frames my inquiry. In this section, I will trace this history and context as well as present the data on participating countries and their participation rates in the global economy, as measured by per-capita income levels.

The IEA practices began with their pilot study of twelve countries in 1959, which included subjects beyond mathematics and science. The countries were Belgium, England, Finland, France, (Federal Republic of) Germany, Israel, Poland, Scotland, Sweden, Switzerland, United States, Yugoslavia, and the language that IEA uses to describe the original project suggests a culturist framing:

> The founders of the IEA viewed the world as a natural educational laboratory, where different school systems experiment in different ways to obtain optimal results from educating their youth. They assumed that if research could obtain evidence from across a wide range of systems, the variability would be sufficient to reveal important relationships that would otherwise escape detection within a single education system. (IEA, n.d., n.p.)

In other words, although mostly comprising a list of nations powerful at the time, the originators of this practice suggested that there was 'no one way' to teach but that there was one way to assess. This would allow for best practices to be shared and analysed for efficacy; the contradiction arises from the assumption that one assessment implies consensus on educational goals. This assumption in the pilot study, albeit framed through a culturist lens, suggests a global objective to unify both the goals and the practices of education. At this point in the history, these goals were less explicit and more open than at later stages, as the assessment spread continued and especially once the World Bank became involved.

The following tables indicate the countries that have been brought into IEA's assessment practice in mathematics over time. After the pilot study in 1959, we have the FIMS (First International Mathematics Study) of 1964, the SIMS (second) of 1980–1982, the TIMSS (third and including science) of 1995, and finally, the renaming of the practice to TIMSS (Trends in International Mathematics and Science Study) and its iterations via a four-year cycle with the last completed in 2019. Table 16.1 below shows the countries that participated prior to the renaming. Table 16.2 includes the countries participating in the four-year iterative assessments since 1995, with an additional variable presented, namely the World Bank's analytical classification for each country based on the Gross National Income (GNI) per capita. I will use these classifications to make arguments about the significance of the assessment spread. In Table 16.2, the categories of economic classification are high, upper middle, lower middle, and low income. Thresholds for these categories vary slightly from year to year and increase over time due to inflation of the US dollar. The thresholds for 2019's analytical classifications were: Low – GNI per capita less than $1035 US, lower middle – between $1036 and $4045, upper middle – between $4046 and $12535, high – above $12535. Sources to create these tables include the IEA (n.d.) website with lists of participating countries and World Bank Group (n.d.), a data set with historical classifications according to the economic groups listed above.

To help illustrate the geographic spread of a globalised mathematics education, I accompany each table with a dymaxion world map indicating participating countries. I chose the dymaxion two-dimensional projection of the world sphere to destabilise typical projections' assumptions of north-south superiority/inferiority as well as to more accurately represent land mass proportionally. The projection I used is an adaptation of Buckminster Fuller's Airocean projection (1954), adapted by Visioncarto (2018) and free for all uses. A participating country is indicated by a circle on the dymaxion map. The darker the circle, the earlier the country began participation. In Figure 16.1, to represent the first three iterations, the participating countries have darker circles; accordingly, in Figure 16.2 (to accompany Table 16.2), the additional countries added in the iterations of the assessment occurring from 1995 to 2019 are indicated with lighter gray circles.

Table 16.1 Participating countries in IEA's early mathematics assessments.

1959 (Pilot)	1964 (FIMS)	1982 (SIMS)
Belgium	Australia	Belgium
England	Belgium	Canada
France	England	England/Wales
W. Germany	Finland	Finland
Israel	France	France
Poland	W. Germany	Hong Kong

Scotland	Israel	Hungary
Sweden	Japan	Israel
Switzerland	Netherlands	Japan
US	Scotland	Luxembourg
	Sweden	Netherlands
	US	New Zealand
		Nigeria
		Scotland
		Swaziland
		Sweden
		Thailand
		USA

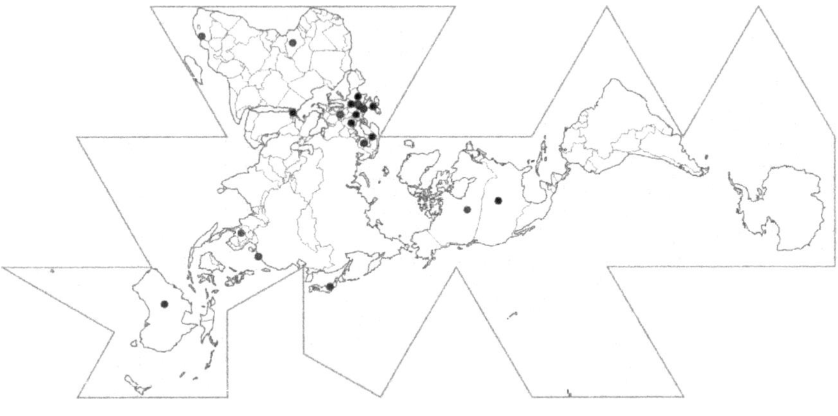

Fig. 16.1 Participating countries through 1982. Adaptation of Buckminster Fuller's Airocean projection (1954), adapted by Visioncarto (2018).

The next table shows countries of participation from (19)95 to (20)19, with economic classification by the World Bank. (All country names are as stated in the TIMSS reports, and are not necessarily the same as the name used by either the World Bank, the nation state itself, or its people at the time.) The first batch of countries are those with the highest income level per capita. Their years of participation are indicated by 4s in the appropriate column which also signifies their income level.

The next batch of countries are those that only participated in one year. Their year of participation is marked in the column of the year and with the income level 1 (low), 2 (lower middle), or 3 (upper middle). Since these countries participated for only one year, we cannot make any conclusions about the relationship between participation and the country's income level. The next batch of countries in Table 16.2 are those that participated and did not increase income level. It should be noted that for these countries (fifteen plus Palestine, for which income data were not available) they remained at their income level during their years of participation. The last batch of countries in Table 16.2 are those that increased their income levels during TIMSS participation. Twenty-three countries in total are included in this final batch, signifying that the majority of countries with lower income levels who participated in TIMSS did increase their income levels over time.

Table 16.2 Participating countries in TIMSS, years 1995 to 2019, with income level.

	95	99	03	07	11	15	19
Participating countries with high per-capita income levels							
Australia	4	4	4	4	4	4	4
Austria	4			4	4		
Bahrain				4	4	4	4
Belgium	4	4	4		4	4	4
Canada	4	4	4	4	4	4	4
Chinese Taipei		4	4	4	4	4	4
Cyprus	4	4	4	4		4	4
Denmark	4			4	4	4	4
England	4	4	4	4	4	4	4
Finland		4		4	4	4	4
France	4					4	4
Germany	4			4	4	4	4
Hong Kong SAR	4	4	4	4	4	4	4
Iceland	4						
Ireland	4				4	4	4
Israel	4	4	4	4	4	4	4
Italy	4	4	4	4	4	4	4

Japan	4	4	4	4	4	4	4
Korea	4	4	4	4	4	4	4
Kuwait	4			4	4	4	4
Malta				4	4	4	4
Netherlands	4	4	4	4	4	4	4
New Zealand	4	4	4	4	4	4	4
Northern Ireland					4	4	4
Norway	4		4	4	4	4	4
Oman				4	4	4	4
Poland					4	4	4
Portugal	4				4	4	4
Qatar				4	4	4	4
Scotland	4			4			
Singapore	4	4	4	4	4	4	4
Spain	4		4	4	4	4	4
Sweden	4		4	4	4	4	4
Switzerland	4						
USA	4	4	4	4	4	4	4
Participating countries – only one year of participation							
Algeria				2			
Azerbaijan					3		
Bosnia and Herzegovina				2			
El Salvador				2			
Estonia			3				
Greece	3						
Honduras					2		
Mexico	3						
Mongolia				2			
Participating countries – No increase in income level							
Argentina	3		3			3	3
Botswana			3	3	3	3	3
Colombia	2			2			
Croatia					4	4	4
Egypt			2	2		2	2
Kazakhstan				3	3	3	3

Lebanon			3	3	3	3	3
Malaysia		3	3	3	3	3	3
Moldova		1	1				
Palestine			P	P	P	P	P
Philippines	2	2	2				
Serbia			P	3	3	3	3
South Africa	3	3			3	3	3
Syria			2	2	2		
Ukraine				2	2		
United Arab Emirates				4	4		
Participating countries – Increase in income level							
Armenia			2	2	2	2	3
Bulgaria	2	2	2	3		3	3
Chile		3	3		3	4	4
Czech Republic	3	3		4	4	4	4
Georgia				2	2	3	3
Ghana			1	1	2		
Hungary	3	3	3	4	4	4	4
Indonesia	2	1	2	2	2	2	3
Iran	2	2	2	2	3	3	3
Jordan		3	2	2	3	3	3
Latvia	2	2	3	3			
Lithuania	2	2	3	3	3	3	3
Macedonia		2	2		3		
Morocco		2	1	2		2	2
Romania	2	2	2	3	3		
Russian Federation	2	2	2	3	3	3	3
Saudi Arabia			3	4	4	4	4
Slovak Republic	2	3	3	4	4	4	4
Slovenia	3	4	4	4	4	4	4
Thailand	2	2		2	3	3	3
Tunisia		2	2	2	3		
Turkey		2		3	3	3	3
Yemen			1	1	2		

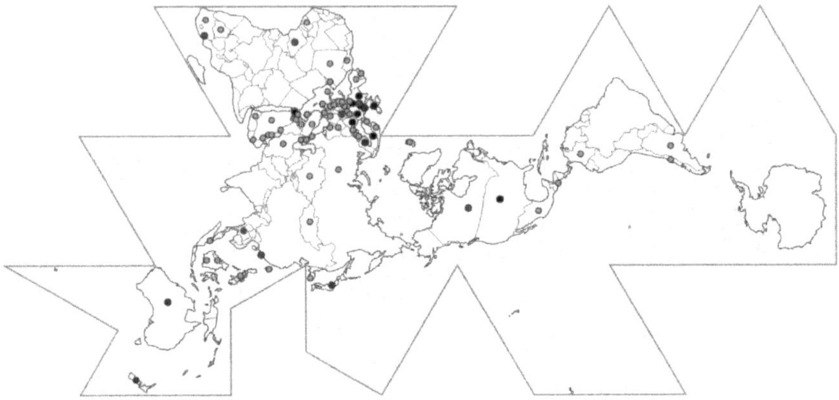

Fig. 16.2 Participating countries in TIMSS through 2019. Adaptation of Buckminster Fuller's Airocean projection (1954), adapted by Visioncarto (2018).

The tables reveal several interesting phenomena after careful consideration of the years and involvement over time. The simple and clear statement that this assessment has spread across the globe throughout the years cannot be understated and is best viewed when looking back and forth between Figures 16.1 and 16.2. What began as an initial pilot study to compare a handful of countries is now a consistent practice with about sixty-one countries participating on average in the last four iterations. Looking geographically, what began as eight European countries, the United States and Israel, now also routinely includes several countries from North, Central and South America, the Middle East, Asia, Oceania, and Africa. And economically, what began as a group of predominantly wealthy nations has expanded to include a diverse group of lower, lower middle, upper middle, and higher-income countries. The phenomenon of mathematics education assessment spread is laid bare by these tables; providing them in their complete form is a clear method to capture the spread.

However, the significance of the spread moves beyond the increasing geographic diversity and simple glimpse at economic diversity. To the latter, I first offer further consideration of the World Bank's economic classifications for each country by relating more details and analysis regarding the World Bank's involvement with TIMSS and education in general. The World Bank is a regular funding source for IEA and is regularly listed as a funding partner for each cycle of TIMSS. The first year that this was displayed on their reports and/or websites was in the 1999 TIMSS cycle. The significance of World Bank's involvement in TIMSS is best understood by considering the World Bank's involvement in global education generally.

Earlier, I pointed to the World Bank's push for a global economy of human capital. Spring (2004) offers poignant analysis to the framing in which World Bank takes educational action:

> [Their] educational ideology contains a particular vision about how society should be organized. For many people, this vision is just assumed to be a necessary part of the advancement of world societies. It is an image of the good society that is often unquestioned because of its promise of economic abundance for all [...] As envisioned by the World Bank, a good society is one based on the mass production of consumer goods within a global economy. Each region or nation contributes to mass production through factory and agricultural goods. The production of agricultural goods is done on large corporate farms or plantations. Small family agricultural units are replaced by large units with factory-like organization. Workers in these larger units are trained for specialized roles and work in corporate teams. Those who previously worked on family farms either work on corporate agricultural units or move to urban centers. [...] From the viewpoint of the World Bank, the problem is that many countries have not reached a high enough level of economic development to participate in the mass consumer society. The role of education is to help them make this leap. (pp. 40–41)

As Spring continues, he provides examples of the types of educational activities that the World Bank has enacted over time. These include a variety of curricula, funding and loans, and assessment programs. In this chapter, I am pointing to specifically one of these, namely the World Bank's funding of and entanglement with TIMSS. It's clear that TIMSS has been enacting the World Bank's educational ideology since at least 1999 when they acknowledged the World Bank as a significant funding source. With deeper analysis, however, by returning to the economic data from Tables 16.1 and 16.2, the participating countries and their economic classifications are a perfect instantiation of the World Bank's vision for the globe. The data speak volumes especially for their connections to the vision for a monoculture of human capital and consumerism.

Looking at the diversity of GNI per capita among the countries in the 1995 TIMSS, we see a majority of participating countries with high levels of income but also seven that are upper middle and eleven that are lower middle income. In the next few iterations of TIMSS, the numbers of lower middle income increase (to thirteen participating countries in years 1999 and 2003) and the participating countries include those with lower income now as well (two countries in 1999 and four in 2003). The pattern of increasing economic diversity holds steady as the cycles move on, but interestingly toward the later years we begin to see fewer numbers in the lower and lower middle income categories. In the last few years, the expansion of participating countries has not kept pace with the rapid additions we observed earlier. However, the data reveals a striking fact: taking together the full list of participating countries in TIMSS cycles, there are eighty-four in total, and while many of these countries have high-income classifications, some do not. Exactly twenty-three countries from among all participating increased their GNI per-capita income category as classified by the World Bank during the years of participation in TIMSS, many holding steady at their higher income levels in the last few cycles of TIMSS. As another way of looking at this data, of the eighty-four participating countries five began participation classified as 'low income', of which four increased their income category during the years of TIMSS participation. To be sure, participating countries increased their economic productivity within the context of increasing global capitalism. The point is not that these countries are distinguished by participation in TIMSS but that their economic growth occurred at the same time as they participated. It would be much more interesting to find that the majority of TIMSS participating countries' economics remained stagnant or 'lowered' as they increased participation in TIMSS.

The countries with increased income levels over the years were Armenia, Bulgaria, Chile, Czech Republic, Georgia, Ghana, Hungary, Indonesia, Iran, Jordan, Latvia, Lithuania, Macedonia, Morocco, Russia, Romania, Saudi Arabia, Slovak Republic, Slovenia, Thailand, Tunisia, Turkey, and Yemen.

The increase in income level corresponds exactly to Spring's description of the World Bank's global vision. A country that increases its per-capita income levels over time means that more people in the country are entering the wage labour market and have the ability to consume goods and services in market economies. When the World Bank looks at data for countries like these, it sees its vision realised: more people entering the global marketplace as both the human capital that markets need and as consumers in an ever-increasing demand for products.

Looking specifically at the TIMSS participating countries, we see an increase in GNI per capita for these twenty-three countries equating to *about one quarter of all participating countries in TIMSS*. Furthermore, all twelve high-income countries hold steady at the high-income level. These are remarkable data; the World Bank is surely pleased to see that countries participating in its activities are increasing the numbers of people entering the wage labour market, or at the very least increasing their salaries by engaging in global capitalism, and increasing demands for global consumption. The World Bank and IEA might even make a leap in suggesting that the TIMSS mathematics assessment, in close association with converging curricula, causes new mathematics education practices that in turn enables a country's citizens to enter the global economy more readily. However, such an assertion would be poor mathematics. In other words, I am not pointing to any cause-and-effect relationship between TIMSS participation and the increase in income levels of participating countries; I am pointing to the association between the two. A nation state's choice to participate is likely indicative of several other actions they are taking, many of them additionally with support of the World Bank, to increase their engagement with the global economy. At the very least, participating in TIMSS represents a country's willingness to engage in the ideology that Spring describes and it should come as no surprise, then, that the majority of countries who could increase their income levels actually did. Although there can be no suggestion of cause and effect between TIMSS and increasing global economic activity, it will serve useful to indicate further how the two are associated. This appears in the next section when I lay out the content of TIMSS and how it clearly accords with the world's first monoculture.

The commercial-administrative mathematics of TIMSS

Given TIMSS's history within the context of the World Bank, which is clearly linked to the goals of increasing human capital and consumerism, as argued above, I next offer content analysis of TIMSS material to further explicate the association between mathematics assessment spread and the global economy. Although I am careful not to suggest a cause-and-effect relationship here, at the very least my content analysis below displays that TIMSS practice is very much in line with global capitalism. The discussion here will reveal that the majority of mathematical activities referred to in TIMSS material displays a narrow-minded view of the opportunities within the array of mathematical behaviour. To do so, I first review Harouni's (2015) categories of mathematics, since I use these as a tool to code the contents of test items released by TIMSS.

Harouni (2015) traces the history of mathematics/mathematics education in terms of its direct correspondence to engagement in the political economy. Thus, the categorisation of mathematical behaviours he provides are highly relevant to the nature of my present inquiry. His four categories of mathematical behaviour are commercial-administrative, artisanal, philosophical, and social-analytical.

Commercial-administrative mathematics centres on the mathematical activity of counting and always implies the market economy's use of currency, even when not made explicit. He writes: 'Ultimately, in modern curricula, when a textbook question talks about apples and oranges, it does not mean apples and oranges. It means money' (p. 62). Furthermore, explicit examples of this type abound in mathematics education; very often, mathematics students practice problems to do with interest rates, buying and selling, breaking even, and the like.

Artisanal mathematics, as Harouni suggests, centres on the mathematics of measurement and uses numbers and geometry to engage in the three-dimensional world as creators and doers.

Philosophical mathematics, centring on patterns, is 'the one corresponding to the math practiced in universities. This type of math stands neither inside nor beside productive labour. Its product is neither an object nor an interaction in the world but an order in the mind' (p. 64). For example, mathematics students sometimes engage in creative discovery of mathematical concepts, like determining a pattern about the three sides of any right triangle.

Finally, *social-analytical mathematics* takes the materials and tools from among the three others and engages the social world for understanding patterns and making group decisions. Harouni clarifies that in some ways, this category can be used to extend the objectives in the commercial-administrative category, such as in a corporation making decisions based on big data. However, Harouni also points to the traditions in critical mathematics education as examples of the category that move in a different social-analytical direction (e.g., Frankenstein, 1983; Gutstein, 2006).

Harouni's categories emerge from his discussion of the political economy's relationship with mathematics education. In associating mathematical practice to the ways that the political economy emerged over the course of history, these categories provide an opportunity to classify mathematical practices that exist in the current historical moment. For this reason, I located TIMSS-released items from their grade 8 assessment in 2011 (IEA, 2013) and used these categories to analyse the types of problems that appear in the document; in all there were seventy-seven mathematical tasks to code. The document organised the mathematical tasks according to content strands, including Number, Algebra, Geometry, and Data and Chance. As I refer to specific problems as examples, I will use the code for each problem as it appears in the document, typically the letter M followed by a six-digit number.

I first split the tasks into two lists, those mathematical tasks that were situated *in context* versus those that had *no context*. The total number of tasks with no context, that is, no relationship to a real-world situation or scenario, was fifty-four. For example, one mathematical task that is without a context asks the test-taker to find a fraction equivalent to a given decimal (item M042059). As another example, test-takers are asked to determine the largest value that can be obtained when the product of any pair of two-digit numbers is found using the digits 3, 5, 7, and 9 exactly once (item M042002). There were twenty-three mathematical tasks that had an explicit context to relate to the mathematical action to be performed. Examples of these kinds of tasks included calculating the area of a garden (item M052173), using a histogram to interpret data about soda sales (item M032701), and calculating the cost of a taxi trip (item M032477).

I chose to start with context because of the implications of mathematical purpose that context can reveal. The scenarios that test-takers read for uses of mathematics suggest the types of mathematical behaviour that TIMSS promotes. I will return to the no-context problems later but will continue for now with the twenty-three items coded as in-context. For these, I next used Harouni's insights (2015) to specify the explicit connections to the real world and purposes for mathematics that are displayed by the task. For the twenty-three context-driven mathematical tasks, I coded nine as commercial-administrative, eight as artisanal, and six as social-analytic. However, in working with the data it became immediately obvious that a second round of coding was needed for several items because of implications towards commercial-administrative, as suggested by Harouni. For example, the social-analytic can be used for commercial-administrative means. It will be helpful to look at an example item to make transparent these coding decisions that I made.

Figure 16.3 reproduces item M032695, a task denoted as 'Make a pie chart with labels' and with the topic paraphrased as 'Where people go after secondary school'. I coded the task as social-analytic because the mathematics requires the test-taker to look at counts of people in relation to the whole and make a visual display to present details, with the potential that a viewer could make inferences about some social data. To be clear, the task does not require this kind of critical thinking (making inferences from data) and amounts to not much more than a test-taker's recall of the procedures in making a pie chart. However, as seen in the example, notice the categories of people to be analysed. Students in a school are grouped by their next choices in life with a small percentage attending university, a greater percentage going immediately into the workforce, and over half attending either business or trade school. There is not any other option or any 'other' category. The options for after school are all in keeping with a human capital and consumerist vision for every individual. With the aims of the World Bank and their motives in entangling with TIMSS as laid out in the previous section, it is almost as if the World Bank itself wrote this mathematical task.

TIMSS 2011 8th-Grade Mathematics Concepts and Mathematics Items

Content Domain	Main Topic	Cognitive Domain
DATA AND CHANCE	Data Organization and Representation	Applying

Make a pie chart with labels

Of the 400 students in a school, 50 plan to go to university, 100 to a polytechnic school, 150 to a business college, and the remainder plan to enter workforce.

Use the circle below to make a pie chart showing the proportions of students planning to do each of these. Put labels on your chart.

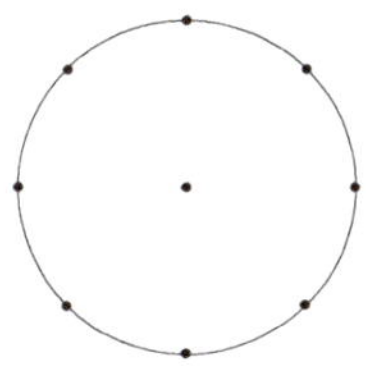

Item Number: M032695

SCORING

Correct Response

- Pie chart correctly divided and labeled

(1 section – university; 2 sections – polytechnic; 2 sections – workforce; 3 sections – business college)

Partially Correct Response

- Four sections with at least two, but not all, of correct size and correctly labeled
- Four sections of correct size but no labels, or labels 50, 100, 150, 100

Incorrect Response

- Four sections with one or none of correct size
- Other incorrect (including crossed out, erased, stray marks, illegible, or off task)

Overall Percent Correct

Education system	Percent correct
Singapore	84 ▲
Chinese Taipei-CHN	80 ▲
Japan	77 ▲
Korea, Rep. of	77 ▲
Hong Kong-CHN	74 ▲
Finland	70 ▲
Russian Federation	67 ▲
Australia	65 ▲
Slovenia	64 ▲
Hungary	63 ▲
England-GBR	61 ▲
New Zealand	60 ▲
Lithuania	58 ▲
Norway	57 ▲
Israel	55 ▲
Sweden	54 ▲
Italy	53 ▲
United States	**53** ▲
International average	45
Malaysia	45
Ukraine	44
Turkey	43
Thailand	43
Chile	43
United Arab Emirates	35 ▼
Romania	35 ▼
Kazakhstan	34 ▼
Iran, Islamic Rep. of	33 ▼
Macedonia, Rep. of	30 ▼
Jordan	29 ▼
Bahrain	29 ▼
Tunisia	28 ▼
Oman	27 ▼
Indonesia	26 ▼
Palestinian Nat'l Auth.	25 ▼
Qatar	25 ▼
Armenia	25 ▼
Georgia	23 ▼
Saudi Arabia	23 ▼
Lebanon	20 ▼
Syrian Arab Republic	19 ▼
Morocco	13 ▼
Ghana	10 ▼

Benchmarking education system	
Massachusetts-USA	70 ▲
Quebec-CAN	69 ▲
Ontario-CAN	63 ▲
Minnesota-USA	61 ▲
North Carolina-USA	61 ▲
Connecticut-USA	60 ▲
Colorado-USA	60 ▲
Alberta-CAN	58 ▲
Indiana-USA	58 ▲
California-USA	47
Florida-USA	43
Dubai-UAE	39 ▼
Alabama-USA	39 ▼
Abu Dhabi-UAE	36 ▼

▲ Percent higher than International average
▼ Percent lower than International average

Copyright © 2013 International Association for the Evaluation of Educational Achievement (IEA) 100

Fig. 16.3 Item M032695, 'Make a pie chart with labels'. SOURCE: TIMSS 2011 Assessment. Copyright © 2013 International Association for the Evaluation of Educational Achievement (IEA). Publisher: TIMSS & PIRLS International Study Center, Lynch School of Education, Boston College, Chestnut Hill, MA and International Association for the Evaluation of Educational Achievement (IEA), IEA Secretariat, Amsterdam, the Netherlands.

Similarly, all of the items classified as social-analytic received a second coding of commercial-administrative because they did not correspond to any sense of social analytics beyond individual competition and/or profit-related contexts. The same was true for a bulk of mathematical tasks coded as artisanal. Many of these problems that present a real-world scenario have explicit or immediate implications in relation to the consumer economy or wage-labour market. For example, item M042041 requires the test-taker to calculate measurements of a pipe, presumably for fitting to a plumbing scenario or other such need for a pipe. The exact text in the question signifies the central character who needs to know this measurement as the 'workman'. A different word choice might position this person in relation to their craft and thereby be coded as artisanal, but IEA's choice here reflects a position as a wage labourer who cuts pipes for some profit-driven situation in which he serves as a cog in the wheel (let alone the poor wording that reinforces gender norms). Other examples of artisanal tasks that I aligned also to commercial-administrative include items M052061 ('Packing eggs in boxes') and M052206 ('Number of books to fill the box'). With the second round of coding, it turns out that only three items do not contain commercial-administrative as the primary or secondary code. The table below summarises the codes for the items that had context, with the vast majority having a primary or secondary connection to commercial-administrative.

Table 16.3 TIMSS Released items with context, coded by type.

Commercial-administrative	9
Artisanal (2nd: Commercial-administrative)	5
Social analytic (2nd: Commercial-administrative)	6
Artisanal	3

Therefore, TIMSS-released items indicate that, where overt displays of mathematical behaviour are connected to the real world, these are most likely to be related to commercial-administrative goals.

However, what can be said about the remaining tasks, those that do not display a connection to the real world? For these, the items that I coded with 'no context', I relied on Harouni's notion of philosophical mathematics, what he describes as the mathematics of universities and researchers in the field. For many of the fifty-four tasks that had no context, however, this immediately proved to be controversial. Coding a mathematical task that requires the test-taker to convert a decimal to a fraction (as in item M042032) seemed hardly to be what Harouni imagines for philosophical mathematics. For help in distinguishing what I could code as truly philosophical, I turned to literature in mathematics education that distinguishes between procedural mathematics and 'doing mathematics' like that of research mathematicians. Mary Kay Stein and Margaret Schwan Smith (1998) offer clarity in comparing procedural mathematics with non-routine tasks that require higher order thinking by students, making connections, noticing patterns, etc.

Thus, to further code the problems with no context in the TIMSS-released items, I clarified which problems were 1) procedural problems that (when considering mainstream middle school textbooks) are often taught to be routines that students memorise and without connections to mathematical concepts and 2) problems that were not routine and required activating either conceptual connections and/or deductive approaches. As examples, the two items I mentioned earlier as no-context tasks provide an example of the two types. Item M042032 (decimal to fraction) is a typical routine that students practice *ad nauseum* in the classroom. On the other hand, M042002 requires test-takers to use their understanding of place-value and problem-solve by carefully considering some cases to determine which product will be greatest. While the solution can be found by a routine (trying out each case), this is not a typical problem like converting from a decimal.

With this coding scheme in hand, I could then distinguish within the fifty-four TIMSS-released items that were devoid of context. Exactly forty-seven of them were coded as procedural problems with no conceptual connections and only seven remained as nonroutine problems requiring problem solving, deductive reasoning, and/or conceptual connections. I argue that these tasks overwhelmingly represent routine mathematics requiring speed and memorisation rather than analytical, higher cognitive demand mathematics. I argue these forty-seven routine, no-context tasks resonate strongly with a mathematics education for human capital and consumerism. Recall Harouni's argument that counting oranges and apples is a stand-in for counting money. In the same way, the emphasis on procedural mathematics, devoid of context, devoid of meaning, implies a context that is yet again *commercial-administrative.* By emphasising the automatic processes in mathematics, the lack of thought and connection, these tasks commit to mathematical practices that wage labourers need to perform quickly and without critical thought.

Taking all the codes and implications thereof together, TIMSS mathematics items from 2011's grade 8 test contain sixty-seven items with overt and/or implied connections to the *commercial-administrative* purpose for mathematics with only ten that fit outside this norm. Three items are purely *artisanal* and seven items are purely *philosophical*. The exhaustive coding process I have detailed thus indicates a content analysis to complement the context analysis earlier in this chapter. The content of assessment spread is commercial-administrative, the increased use of these materials across the globe also increases the suggested associations between mathematical behaviour and corporate-profit, competition, and consumerism. Although it is far too big a leap to suggest that the content in TIMSS caused the increased income levels of participating countries, revealing this content analysis does suggest how such participation, one among likely many other actions taken by participating countries, at the very least resonates heavily with the goal that said country engages more fully with the world's emerging monoculture of the global economy, of human capital and consumerism.

Assessment spread in mathematics education is not unique to TIMSS. A similar analysis could focus on the Program for International Student Assessment (PISA) created by the Organisation for Economic Co-operation and Development (OECD). As an assessment spread that initially focused its reach on elite countries and now has extended its spread to include 'developing countries' as well, future research areas can consider how contextual analysis and content analysis, as above, might reveal the extent to which assessment spread via PISA provides a complementary understanding of assessment spread.

As another, related topic, assessment spread reveals several open questions regarding the ways that the mathematics education and educational policy worlds interpret and discuss the findings. For example, future research might suggest that assessment practices and their ensuing competitions in reporting results unify both a global mathematics education and, thereby, implications for a monoculture emphasising laissez-faire capitalism and consumerism as noted throughout this analysis. The repeated outperformance of particular nation states in these global competitions results in leadership and guidance of particular nation states, with some interesting potential patterns for exploration. Not only do these monocultural features (laissez-faire capitalism, human capital, exploitation, and consumerism) appear highly present in the 'best' performing countries, but many countries who are seen as leaders in the global competition of mathematics education also ascribe to the most severe governmental structure associated with capitalism, namely, authoritarian capitalism. I began to initiate these arguments in Wolfmeyer (2014, Chapter 3), however a fresh discussion is warranted given the emergence of new geopolitical spaces and discussions about authoritarian capitalist governments across the globe in recent years.

In this chapter, I have articulated via two research methods that TIMSS as a phenomenon of assessment spread is concomitant with the spread of the global economy. I conclude that a dominant global mathematics education associates with the twin goals of increasing the number of people who enter the wage labour market and become lifelong consumers. On the one hand, critical mathematics educators must develop clear arguments, as presented here, to assert their claims about these entanglements rather than passively accepting the reports from TIMSS, PISA, and the like (as dramatically conveyed in Chapter 15 of this volume). On the other hand, we can use these analyses to motivate culturist perspectives in mathematics education, such as an alter-globalisation mathematics education (Appelbaum & Gerofsky, 2013), that would resist a global and dominating mathematics education solely dedicated to support human capital and consumerism.

References

Appelbaum, P., & Gerofsky, S. (2013). Performing alterglobalization in mathematics education: Plenary in the form of a jazz standard. *Quaderni di Ricerca in Didattica (Mathematics)*, 23, 23–48.

Atweh, B., & Clarkson, P. (2010). Internationalization and globalization of mathematics education: Toward an agenda for reseach/action. In B. Atweh, H. Forgasz, & B. Nebres (Eds.), *Sociocultural research on mathematics education: An international perspective* (pp. 77–94). Routledge. https://doi.org/10.4324/9781410600042-12

Ernest, P. (2016). Mathematics education ideologies and globalization. In P. Ernest, B. Sriraman, & N. Ernest (Eds.), *Critical mathematics education: Theory, praxis, and reality* (pp. 35–79). Information Age.

Frankenstein, M. (1983). Critical mathematics education: An application of Paulo Freire's epistemology. *Journal of Education*, *165*(4), 315–339. https://doi.org/10.1177/002205748316500403

Freudenthal, H. (1975). Pupils' achievements internationally compared: The IEA. *Educational Studies in Mathematics*, *6*(2), 127–186. https://doi.org/10.1007/BF00302542

Ghosh, A. (1997). Capitalism, nation state, and development in a globalised world. *Economic and Political Weekly*, *32*(14), 683–686.

Gutstein, E. (2006). *Reading and writing the world with mathematics: Toward a pedagogy for social justice*. Taylor & Francis.

Harouni, H. (2015). Toward a political economy of mathematics education. Harvard Educational Review, *85*(1), 50–74. https://doi.org/10.17763/haer.85.1.2q580625188983p6

International Association for the Evaluation of Educational Achievement (IEA). (2013). *TIMSS 2011 Assessment.* TIMSS & PIRLS International Study Center.

International Association for the Evaluation of Educational Achievement (IEA). (n.d.). History. *IEA*. https://www.iea.nl/about/org/history

Keeley, B. (2007). *Human capital: How what you know shapes your life*. OECD publishing.

Keitel, C., & Kilpatrick, J. (2012). The rationality and irrationality of international comparative studies. In I. Huntley, G. Kaiser, & E. Luna (Eds.), *International comparisons in mathematics education* (pp. 241–255). Falmer. https://doi.org/10.4324/9780203012086-19

Meyer, J., Kamens, D., & Benavot, A. (1992). *School knowledge for the masses: World models and national primary curricular categories in the twentieth century*. Falmer. https://doi.org/10.4324/9781315225173

Ramirez, F. (2003). Toward a cultural anthropology of the world? In K. Anderson-Levitt (Ed.), *Local meanings, global schooling: Anthropology and world culture theory* (pp. 239–254). Palgrave Macmillan. https://doi.org/10.1057/9781403980359_12

Spring, J. (2004). *How educational ideologies are shaping global society: Intergovernmental organizations, NGOs, and the decline of the nation-state*. Routledge.

Spring, J. (2014). *Globalization of education: An introduction* (2nd ed.). Routledge.

Stein, M. K., & Smith, M. S. (1998). Mathematical tasks as a framework for reflection: From research to practice. *Mathematics teaching in the middle school*, *3*(4), 268–275.

Visioncarto. (2018). Buckminster Fuller's Airocean projection (1954) [Map]. *Visionscarto*. https://visionscarto.net/base-maps#&gid=1&pid=1

Wolfmeyer, M. (2014). *Math education for America? Policy networks, big business, and pedagogy wars*. Routledge.

World Bank. (2007). *Building knowledge economies: Advanced strategies for development*. World Bank.

World Bank Group. (n.d.). Historical classification by income in XLS format [Data set]. *Databank*. http://databank.worldbank.org/data/download/site-content/OGHIST.xls

17. Bringing ethnomathematical perspectives into classrooms

Swapna Mukhopadhyay and Brian Greer

In this chapter, we offer some suggestions, informed by our personal histories and experiences, as to how ethnomathematical perspectives might enrich school mathematics classrooms. We regard this as inherently political work, in terms of combatting the intellectual White supremacy that pervades the Eurocentric narrative of the history of academic mathematics and that is explicitly or subliminally everpresent in so many mathematics classrooms. Likewise, we argue that the ongoing worldwide homogenisation of school mathematics is unhealthy. Above all, we argue that school mathematics is culpably deficient in terms of its relations with other forms of mathematical activities and insofar as it does characterise such relationships, often harmfully misleading.

Introduction

> Ethnomathematics is the mathematics practiced by cultural groups, such as urban and rural communities, groups of workers, professional classes, children in a given age group, indigenous societies, and so many other groups that are identified by the objectives and traditions common to these groups. (D'Ambrosio, 2002, p. 1)

As Ole Skovsmose (2022) establishes, the concerns of Ethnomathematics are intimately related to those of critical mathematics education in general.

We begin by drawing out of our backgrounds the insights that inform our current perspectives. As presently described, Swapna's engagement goes back to the 1980s, her doctoral research being conducted in a village in India, studying children learning in a way very different from being

©2024 S. Mukhopadhyay et al., CC BY-NC 4.0 https://doi.org/10.11647/OBP.0407.17

schooled, and at a time when it was becoming clear to anthropologists and cultural psychologists that ascribing universality to local European descriptions and theories of humanity is absurd. During the 1980s and 1990s Brian was moving from studying mathematical cognition, to mathematics education, and thence to critical mathematics education. He heard Ubiratan D'Ambrosio speak for the first time in 1995 and soon after that the authors of this chapter began to work together, as sketched below.

We summarise our engagements with Ethnomathematics across our careers and relate these to what we see as important themes. Against this background, we pose the central question: In what ways, given the realpolitik of educational regimes across the world, could school mathematics classrooms be enriched by ethnomathematical perspectives appropriate to the particular socio-political contexts? We mainly address classrooms in the United States (but of course, with relevance, *mutatis mutandis*, to many other educational systems). Accordingly, this chapter contrasts with Aldo Parra's (Chapter 10, this volume) which is about directly collaborative political work with communities in South America; however, in the current state of educational regimes almost anywhere, any attempt to infuse school classrooms or initial preparation of teachers with an ethnomathematical perspective is a political act. Such work presents both challenges and opportunities (Greer, 2021; Vithal & Skovsmose, 1997).

The experiences that we relate and reflect upon are illuminating for several broad themes that we highlight. A central thrust in the inherently political nature of Ethnomathematics has always been the construction of a counternarrative to the Eurocentric myth about the development of academic mathematics, a myth that can be seen as a long-established, deeply-entrenched, and ongoing manifestation of White (intellectual) supremacy. We also address the air of intellectual superiority of the mathematical academy towards what we have termed 'the mathematics of people who make things that work' (Mukhopadhyay & Greer, in press).

Much more attention needs to be given to establishing links between school mathematics and the day-to-day lived experiences of students, their families, and communities. And the paraphernalia of the technological age, with all their associated opportunies and dangers, form a pervasive part of their lives.

Ethnomathematical studies

Studying the children of weavers, potters, and farmers in rural India

During the 1980s, when Swapna was working on her doctorate at Syracuse University, it was becoming increasingly clear, through the work of scholars such as Jerome Bruner, Michael Cole, and many precursors, that cognition is never culture-free and that, for example, Piagetian tasks should not be viewed as universal windows into children's cognition.

The fieldwork for Swapna's doctoral dissertation was carried out while living in a small village in India near the Bangladesh border. As part of the preparation for this work, she took classes in weaving and pottery, the latter being something she still does. In the village, she totally embedded herself in the lives of the community, who looked after her with great care. Coming from a middle-class big-city background, the experience was formative in so many ways, not least in leading her to respect the knowledge, skills, and adaptability of the people and to see a very different way of growing up, in which the children, from the earliest years, learn by involvement in the family's work – what Andrew Dayton and Barbara Rogoff (2016) characterise as 'learning by observation, participation, and invention'.

Specifically, she wanted to explore how this early experience might be reflected in children's cognitive, in particular spatio-mathematical, functioning. The children were aged eight to twelve, and the expectation that their contributions to the family would take precedence over going to school, meant that their exposure to schooling was minimal. The form of weaving, carried out by looking down on a two-dimensional surface contrasts with the three-dimensional pottery, made on a wheel. It was confirmed that the weavers' children performed relatively better on two-dimensional tasks and the potters' children on three-dimensional, with both groups outperforming farmer's children on spatio-mathematical tasks in general. All groups performed computational tasks embedded in story problems at comparable levels.

As one example, when she posed conservation of liquid volume tasks to girls, they accurately predicted the level that the liquid would rise to in the second container, reflecting their experience and responsibility in buying cooking oil poured into a container at the market and ensuring the price was fair.

Thus, in many respects, the work illuminated the great differences between a schooled childhood (in which learning is separated from life-related consequences) and learning while doing, though graduated apprenticeship, and with consequences for actions.

The boat-builders of Frasergunj

During a trip to the Bay of Bengal some fifteen years ago, we visited a small village called Frasergunj, where we came across a group of men building a large wooden fishing-boat. Throughout the period since, Swapna has formed a close and ongoing friendship with these craftsmen, while conducting extensive observations and interviews to try to understand how they make such complex and well-constructed boats (Mukhopadhyay, 2013).

A team of eight to ten men, of varying age and experience build a boat in for to six months during the dry season. They are carpenters, mostly unschooled and illiterate, from villages in Bangladesh, who have adapted to the specialist skills of boat-building. They work almost always without blueprints – when asked about this, they told Swapna that they can work to plans if required but 'it slows us down'. While working, there is very little speaking. Less experienced members of the team are mentored, often without speaking, and learn by doing.

When asked about how they know that some part of the process has been done correctly, for example, curving a plank by heating and pulling over a fulcrum so that it can be added in the progressive construction of the hull, their responses hardly go beyond saying something like 'we can see it'. And when asked how they judge if a boat is a good boat, they answer along the lines that it keeps the fishermen safe and lasts a long time – in other words, it works.

We do not suggest that boat-building as a topic could naturally be introduced into mathematics classrooms unless the context is appropriate – for example, Jerry Lipka and his team (Lipka, Wong, Andrew-Ihrke, & Yanez, 2012) have included canoe construction in curricular materials they developed and used with children of the Yup'ik people of Alaska. The implications are at a more fundamental level. Is there a place for teaching mathematics less in the typical style wherein the motivation for learning something is that it will be useful for learning more mathematics (and that justification can be repeated indefinitely) or for some vague work-related purpose? At the most general level, as discussed further below, could school mathematics be more relatable to the students' lives? At a theoretical level, what are the implications of the distinction made by Edwin Hutchins (2000) between 'cognition in captivity' and 'cognition in the wild'.

Tlingit culture: 'Sharing our knowledge'

A more recent formative experience for Swapna has been working with the Tlingit people of the Pacific Northwest coast of the United States, mainly in South-West Alaska. The Tlingit share the history of oppression of all Indigenous Americans, North and South, through cultural genocide, linguicide, exploitation, and intergenerational trauma. As well as being subjected to oppression by Europeans, they suffered at the hands of the Russians, from whom the United States 'purchased' Alaska in 1867.

With a population of about 600000, and spread across the border into Canada, present-day Tlingits, with the neighboring Haida and Tsimshian, strive to reclaim their cultural identity and pride. Swapna has worked closely with the Sealaska Heritage Institute in Juneau.[1] In particular, she has collaborated with Tlingit weavers and basket-makers in Juneau, Sitka, Hoonah, and with school educators, to explore how the ethnomathematical perspective might enrich the school experience of Tlingit children. For example, she has interacted intensively with the legendary Haida master-weaver, Dolores Churchill, with whom she co-taught professional development classes for teachers. Of interest is that, while Swapna's attention is drawn to the complex geometry of the finished designs, the expert weaver focusses on the line-by-line

1 See https://www.sealaskaheritage.org

creation, with a primary emphasis on counting. Numerous aspects, such as the conception of the finished product the creator has during the act of making it, and how expert knowledge is passed on, remain to be understood, and underline how ethnomathematicians must exercise caution when projecting formalised mathematics on to what are not *primarily* mathematical activities.

Another fascinating aspect is the deep ecological consciousness embedded in the complex Ethnoscience involved in the preparation of materials for basket-making (Mukhopadhyay, 2009). Prior to European contacts, these finely woven watertight spruce root baskets were widely used for food preparation by submerging heated rocks with edibles to be cooked.

Swapna was indeed very fortunate to teach summer sessions for local teachers with eminent Tlingit and Haida local scholars and elders. One of these experiences focused on STEAM (Science, Technology, Engineering, Art, Mathematics) curricula examined the evolved design and construction of traditional halibut fish-hooks (which work).

Tlingit artefacts, whether practical or ceremonial, are always for use and complexly decorated. Bilateral symmetry, whether in canoe construction, realistic carvings of living creatures for totem polls, abstract patterns in weaving and basketry, is pervasive. Moreover the concept of symmetry, related to balance, has symbolic importance within the culture, though the Tlingit language does not have an equivalent word. The Tlingits are divided into two moieties, the Eagles and the Ravens, and whenever any discussion is taking place, there is an agentive expectation that if an Eagle or a Raven speaks, he or she must be balanced by a spokesperson for the other moiety.

We have been fortunate to attend a number of Tlingit Clan Conferences in Juneau, which combine cultural events with academic papers on important issues such as the effects of intergenerational trauma, and with a major emphasis on language revival, particularly among young people; also related gatherings in various places under the wonderfully appropriate title ‘Sharing our Knowledge’, reflecting a relational conception of the interactions between cultural knowledge systems (Parra, this volume).

Digging where we stand

Teacher preparation

Throughout her career as faculty at various universities working with future teachers, Swapna has introduced students and others to the ideas of Ethnomathematics through activities in and beyond class. In particular, she developed a strong relationship with Portland Art Museum, which displays rich examples of Native American culture artefacts, and which has an outstanding outreach to public education. By taking students to the museum on field trips (once holding classes for an entire term in a room within the museum), she focused on students learning to see and analyse the mathematics in typically unconsidered contexts. Students were encouraged to bring their children and other family members to such field trips. In particular, the native artefacts, both decorative and functional, are steeped in bilateral, translational, and rotationsl symmetry (Washburn & Crowe, 1987). By asking students to examine those symmetries, and to create new designs of their own (for example by making printing blocks for generating symmetry-rich designs) she introduced formal analysis of symmetry, a topic that forms part of the national mathematical framework.

Portland Art Museum, like many museums, is going through the process of re-examining their roles. One aspect in which they have shown leadership is through exhibitions that demonstrate how contemporary Native American painters, weavers, photographers, fashion designers, and so on are establishing relational bridges with other modern traditions, thereby negating the image of Indigenous peoples as belonging only to the past.

Culturally Responsive Mathematics Education

In 2004, under the auspices of the Centre for Learning and Teaching—West, a program funded by the National Science Foundation (NSF) spanning Portland State and four other universities, a conference was held at NSF headquarters with the above title. Participants included many of the most prominent figures in mathematics education taking, broadly speaking, an ethnomathematical stance.

We took responsibility for editing a book reflecting contributions to that conference and it duly appeared, with Arthur Powell and Sharon Nelson-Barber also co-editing (Greer, Mukhopadhyay, Powell, & Nelson-Barber, 2009). As far as we are aware, the specific term 'culturally responsive mathematics education' originated in this endeavour. We illustrate the nature of the book by reference to two of the contributions.

Geneva Gay's (2009) contribution begins with analysis of reasons why many teachers have an image of mathematics as culture-independent thereby absolving themselves of the responsibility of knowing how to teach in a culturally responsive manner. Rejecting that position, she asks a question of central importance in the United States context, marked as it is by large populational discrepancies between students and teachers: 'How can middle-class, monolingual European American math teachers work better with students who are predominantly of color, attend school in poor urban communities, and are often multilingual?' (p. 189).

Lipka worked with the Yup'ik people of Alaska for four decades (Lipka, Yanez, Andrew-Ihrke, & Adam, 2009; Lipka et al., 2012). Much of his work focused on curricular development, the creation of modules linking standard mathematical content to culturally-situated activities such as canoe construction, house construction, star navigation. The account he gives (e.g., Lipka et al., 2009) makes clear how his research program was carried out very much within the rules of the system. The bottom line was that in order to secure and retain funding for research it was necessary to produce statistically significant results from rigidly designed experiments providing evidence that the approach used yielded better outcomes *as defined by test results.*

However, in retrospect, Lipka (2020), in reviewing the intersection between Indigenous knowledge from various cultures and attempts to escape catastrophe through climate change, concluded that he needed to make a significant paradigm shift from 'ethnomathematics [in] the school context to the larger social-cultural-ecological systems' thus bringing him close to the political involvement advocated by Parra (Chapter 10, this volume) and exemplifying environmental issues as one arena where promising collaborations between formal science and Indigenous knowledge are happening.

Alternative ways of knowing (in) mathematics

At Portland State University, starting in 2006, Swapna organised a public lecture series, attended by appropriately diverse audiences, under the title 'Alternative forms of knowledge construction in mathematics'. Most of the invited speakers contributed chapters to a subsequent edited volume:

> This book is about the celebration of diversity in all its human form, specifically in relation to mathematics and mathematics education: culture, ethnicity, gender, forms of life, worldviews, cognition, language, value systems, perceptions of what mathematics education is *for*. (Mukhopadhyay & Roth, 2012, p. 1, italics in original)

In this book, the contribution by Mariana Kawall Leal Ferreira (2012) exemplifies her intense political involvement with Indigenous peoples in Brazil (see especially Ferreira, 2015). The chapter contributed by Gary Urton (2012) exemplifies the interplay between mathematics and politics in the context of colonisation, in this case the Spanish invasion of the Inkan Empire. Specifically, he describes how the highly developed statecraft of the Inkas, using khipus to record data, was forcibly replaced by double-entry bookkeeping in the European style. This exemplies a theme all too common in the history of colonialisation, the suppression of often superior cultural practices by the invaders (several examples, such as navigation, are described by Raju (2007). It is very important that, following colonisers' suppression of alternative knowledge in, for example, environmental science, botany, medicine, we are beginning to see extremely important collaborations across such knowledge systems. Without elaborating, we suggest that in the case of mathematics, the historical trajectory has been different, taking more strongly the form of assimilation.

Culturally Responsive Elementary Mathematics Education (CREME)

In 2014, the Oregon Department of Education funded a dozen or so projects under the Culturally Responsive Pedagogy and Practices Grant. Our project was the only one dealing specifically with mathematics, and it focused on elementary school students because of the importance we attach to the foundational impact of the early years. It was based on intensive interactions with a small number of dedicated teachers and their students in two Portland schools. We were fortunate to have as supportive advisers Danny Martin, Geneva Gay, and Marta Civil.

We worked mostly in two contrasting schools in North Portland. Rosa Parks Elementary, and Trillium Charter School. The students at Rosa Parks (though not the teachers) are predominantly African Americans, migrants, and refugees from many countries in Africa and elsewhere. Throughout our time working there, the school was under severe pressure on account of test scores being low. Trillium, an independent charter school (rather than the profit-oriented or religiously based kind), had relative pedagogical freedom in the elementary grades since the performance of students in the later grades was good. Thus, teachers enjoyed a considerable amount of autonomy, as illustrated below.

The project proved extremely educational and fulfilling for us. Ten teachers engaged fully, and we owe them a great deal. In the process, we learned very clearly about the power of the systemic straightjacket limiting teachers at Rosa Parks, an aspect well documented by Alan Schoenfeld (see Chapter 14, this volume)—above all, how testing imposes constraints.

The overall organisers held regular meetings at which representatives of all the projects reported on their work and we exchanged views. We became aware of a degree of tension between our aims and those of the organisers as interpreters of the funders' requirements; let's just say we were not in sympathy with the crude requirement to demonstrate rises in test scores as evidence that our approach had merit.

We also learned about the daily reality of the students. One of the teachers was asked why a student was sleeping and explained that he could not sleep at night (in a car) because the street lights were too bright. We tried hard to communicate with families, following the inspiring work of Marta Civil (e.g., Civil & Quinteros, 2012) and to channel Luis Moll's concept of 'funds of knowledge' which is 'based on a simple premise: People are competent, they have knowledge, and their life experiences have given them that knowledge' (Gonzalez, Moll, & Amanti, 2005, p. ix). In these attempts we encountered substantial obstacles, including the natural guardedness of immigrants and refugees in the current political climate.

Nevertheless, on looking back, there were many highlights in our rich reactions with students and teachers. A chapter co-authored with the teachers begins with a vignette about the power of simply asking students to find out and report to their class how to count up to twenty in languages spoken at home (Ford et al., 2018, pp. 169–170). One of the teachers (Koopman, 2017) conducted a project which began by students checking the labels on their t-shirts to see where they were made. These data were recorded on a world map. Koopman then told the students how the t-shirt was introduced as a working garment at the beginning of the twentieth century. He provided them with a mass of data relating to the economic lives of the workers and their families (most of whom also worked) in relation to the cost, at the time, of essential items. With these data, the students analysed family budgets. He also elaborated on the union movement, and concluded with some data on contemporary sweatshops in Asia and elsewhere.

A major element of CREME, linked to mathematics where possible, was to counter deficit models with support for student identity and agency. For examples, students recorded lists of what they could do ('I can bath my baby brother, I can do multiplication, I can skateboard backwards') on strips of paper rolled up and kept in containers they could hang around their necks, which we called 'talismans'. In the same vein, students (with minimal guidance from teachers) drew self-portraits and wrote poems beginning 'I am from' about themselves and their families (often in languages other than English). These were collected and self-published in three volumes 'Where we are from', 'We are from', and 'We are' with book-launches attended by parents and communities of the two schools at which children read their poems.

Themes

We next identify major themes that we can illustrate from our experiences as described in the previous section. Overarching all is the intimate relationship between forms of capitalism and the political enterprise that is (mathematics) education. Gay (2009, p. 194) expressed it forthrightly: 'mathematics becomes a proxy for academic racism, ethnic inequities in educational opportunities and a means of perpetuating a class system of "haves" and "have nots"'. Although the reader should catch many resonances in what follows and elsewhere in this book, that is too massive a topic to address adequately in this chapter.

Bringing more evidential rigour to the history of mathematics

Within Ethnomathematics, since the outset, a prominent theme has been the construction of a counternarrative to Eurocentric claims about the development of academic mathematics, more specifically what Jens Høyrup (1992) calls the 'Greek myth'. Many of the contributions in the seminal early compilation by Arthur Powell and Marilyn Frankenstein (1973) address this challenge.

What might be called the social construction of 'Pythagoras' is an appropriate starting point. According to a leading student of Greek mathematics 'Pythagoras the mathematician finally perished AD 1962' (Netz, 2003, p. 272). The date refers to a book by Walter Burkert (1962/1972), which is heavily cited in the entry on Pythagoras in the online *Stanford Encyclopedia of Philosophy* (Huffman, 2018). In surveying historical writings relating to Pythagoras, it is notable that forgeries are mentioned thirteen times. The article ends by stating that the consensus among scholars is that *Pythagoras was neither a mathematician nor a scientist.*

At the systemic level, Morris Kline provided a definitive statement of the Graecocentric myth:

> Mathematics is a living plant [that] finally secured a firm grip on life in the highly congenial soil of Greece and waxed strong for a brief period. In this period it produced one perfect flower, Euclidean geometry. The buds of other flowers opened slightly [...] but these flowers withered with the decline of Greek civilization, and the plant remained dormant for one thousand years. Such was the state of mathematics when the plant was transported to Europe proper and once more embedded in fertile soil. (Kline, 1953, p. 27)

Important counters to this absurdly extreme position have been provided by, for example, George Gheverghese Joseph (1991), Chandra Kant Raju (2007), Jim Al-Khalini (2010), and throughout the work of serious historians of mathematics such as Høyrup (see Greer, 2021, for an overview of his work and its implications for mathematics education). By way of example, there is strong evidence that important elements of calculus were developed in India before Gottfried Wilhelm Leibniz and Isaac Newton (Joseph, 1991), that the channels of communication existed to carry that knowledge to Europe, and that Eurocentric historians of mathematics have sought to suppress consideration of such evidence (Raju, 2007).

Contemporary educational politics in the United States

Without taking the space to elaborate, we assert that the position so clearly enunciated by Kline in the quotation above can be seen as an intellectual component of the White supremacy that is evident in the contemporary United States and many other parts of the world. The provocations of ethnomathematicians and other critical mathematics educators in opposing this position have contributed to the dragging of mathematics education into the 'culture wars' within the United States. Relative to the earlier 'math wars' described by Schoenfeld (Chapter 14, this volume), these developments are taking an extreme form, including personal attacks on individuals (Boaler, 2022).

A pervasive problem is illustrated by Jo Boaler with a quotation from a Nobel-prize-winning physicist:

> When I talk about education, I frequently have physicists lecture me on how I am wrong [reflecting that] nearly everyone [...] believes that they are an expert on education, just by virtue of having been to school or having a child who has attended school. (Wieman, personal e-mail message, cited by Boaler, 2022)

This rings true, above all, in relation to mathematics education. We offer the observation that a great many people, in particular politicians and educational administrators, will pontificate on the importance of children being able to, for example, solve quadratic equations, while themselves being unable to do so.

The mathematics of people who make things that work

The above phrase refers to another massive swath of humanity whose expertise is generally looked down upon by mathematicians and, at least until recently, largely ignored by historians of mathematics. In ancient Athens, it has been pointed out that:

> A consensus has emerged that Greek mathematics was heterogeneous and that the famous mathematicians are only the tip of an iceberg that must have consisted of several coexisting and partially overlapping fields of mathematical practices. (Asper, 2009, p. 107)

Ray McDermott (2012) alludes to many examples showing the falsity of the 'official story' of the role of mathematics in work. A particularly striking example is a report on ethnographic studies of architects:

> In nearly a year of a fieldwork at the architecture firm, I never saw any of the architects write down an equation and then manipulate it. I rarely saw calculations more complex than the basic arithmetic operations. The extent of geometric practice was a deep familiarity with shapes, often Euclidean ones, and an ability to visually and physically transform them in design practice. (Stevens, 2010, p. 83)

This does not sound so different from the boat-builders described above!

Two other working architects are described in vividly contrastive terms:

> 'Stupid Gerry' was terrible at the fractions examination he had to pass to become a professional draftsman, BUT he was smart and excellent at drafting, even at the parts that seemed to require a green-thumb knowledge of fractions [...] 'Dumb Ted' could be found looking incompetent wherever teachers or other students were pushing mathematics [...] but his work was excellent as long as he did not have to solve mathematics problems on a test. (McDermott 2012, p. 86)

Diversity versus homogeneity

Humanity is diverse in multiple ways, including language, ways of life, epistemologies, knowledge systems, ways of conceptualising and interacting with the environment. Arguably, awareness of this multidiversity is being eroded by globalisation (Westernisation), as warned against earlier in a study of the spatial epistemology of the Navajo people:

> Through a systematic superimposition of the world view and thought system of the West on traditional non-Western systems of thought and action all over the world, a tremendous uniformization is taking hold [...] The risks we take on a worldwide scale, and the impoverishment we witness is – evolutionarily speaking – quite frightening. (Pinxten, Van Dooren, & Harvey, 1983, pp. 174–175)

Ethnomathematics points to mathematics being embedded in multiple families of activity systems. In simplistic terms, we can point to those situated within: formal (academic) mathematics; mathematically-suffused cultural practices; mathematics of work; 'everyday mathematics'; school mathematics (see Harouni, 2015). With the exception of the first and last (to a degree), each of these is a very fuzzy category.

The diversity within formal mathematics, illustrating that it is open-ended, and that its history extends to the present and future, is illustrated, for example, in Hersh (2006). That there is enormous diversity within the next three families is obvious. By contrast, school mathematics, locally and globally, is increasingly characterised by homogeneity – converging towards a monoculture, to use the term of Mark Wolfmeyer (Chapter 16, his volume), with curricular frameworks and global testing (Chapter 15, this volume) as major instruments. It is also characterised by relational isolation from other families of mathematical activities. Indeed, we argue that, as exemplified in the discussions above about the history of mathematics, about mathematics in work contexts, and about mathematical modelling of human situations, insofar as those relations are promoted, they are often misguided and harmful.

Another aspect of diversity relates to terminological usage that may have consequences in framing images. It is understandable that a great deal of writing to date refers to 'Asians', 'Europeans' and so on, but surely it is time to move beyond such essentialising. Swapna, as an Indian, feels little in common with the Japanese, for example, while Brian is acutely aware that the Irish, as well as contributing to White supremacy, were also the victims of English oppression. In a similar vein, we regard it as dangerously superficial to refer to, say, 'African culture' given the diversity of African cultures. Pre-European peoples of North America, such as the Tlingits and the Navajo, are as diverse as the environments to which they are adapted. And it is surely possible to improve on the nonsensical term 'Western mathematics'!

Lived experiences of children, their families, and communities

Django Paris and H. Samy Alim (2017, p. 7) point out that 'contemporary linguistic, pedagogical, and cultural research has pushed against the tendency of researchers and practitioners to assume static relationships between race, ethnicity, language, and cultural ways of being'. This protest is echoed by Indigenous peoples who reject the all-too-common characterisation of them in terms of what they were, rather than what they are, and will be.

It is a pervasive criticism of school mathematics that it is impoverished in terms of connections with the lives of the students, their families, and their communities. The CREME project (see above) gave us some insights into the lives of the children we were working with. When one child was asked 'Is there any mathematics at home?' the response was only in terms of worksheets brought home from school, an image of 'what is mathematics' that we tried hard to demolish. Our attempts to involve students' families in the project were very much informed by Luis Moll's conception 'funds of knowledge' (see above).

The reality of one young person's life was brought home to us when she asked how her mother could come to the launch of the students' book since she was in prison part of what Skovsmose (e.g., 2022) refers to as her 'foreground'. Moreover, as Skovsmose (2022) points out, children's (often realistic) appraisals of their educational and life possibilities may be an important factor in their engagement or lack of engagement with the mathematics presented to them in school. To put it starkly: If a young Black person is concerned about staying out of prison or even staying alive, why should he or she be interested in solving quadratic equations?

In terms of what children are exposed to in mathematics classrooms, an issue that is discussed in Chapter 13 of this volume is that of inappropriate framing of ways in which aspects of the real world can be modelled mathematically. In particular, mathematical problems as presented may have little relevance to, and indeed be incompatible with, the experience of the students. There is no reason why children could not be introduced, from an early age, to aspects of mathematical modelling, including how to discriminate between situations precisely modelled by simple arithmetical expressions, approximate cases, and purported correspondences that are absurd.

If a test item, in a context that admits of no discussion, invokes a situation to be modelled, this can have serious implications when the interpretation of the situation is culturally or class-relative (see the discussion of the 'bus fare' problem by Tate, 1995, in Chapter 13).

The technological age

If you watch the film 'Timbuktu' (directed by Abderrahamane Sissako, 2014), you will see a young girl living in a small tent in the desert in Mali climb a hill to get reception for her cell phone to talk with her brother. These devices are everywhere and consequent changes are happening so quickly we have dangerously minimal idea of what the repercussions are. In Chapter 20 (this volume), Melissa Andrade-Molina and Alex Montecino present a striking review of the images of mathematics held, by young people in particular, as revealed by comments on social media.

As Artificial Intelligence surges, there are increasing concerns about its likely effects, many of which are already strikingly apparent. What could be more chilling than the American warmonger (and winner of the Nobel Peace Prize!) Henry Kissinger declaring the possibility of 'a world relying on machines powered by data and algorithms and ungoverned by ethical or philosophical norms' (cited by Ochigame, 2021, p. 167)? Surely this represents the ultimate dehumanisation by algorithm.

Information technological developments already have had major effects on how are lives (often without any control or access to the models) are 'formatted' by 'mathematics in action' (Skovsmose's terms). Further, Ian Hacking (1999) points to the dual aspects of phenomena and the social constructs that accrete around them, including new terms (linguistic models), and mathematical models that are developed in relation to those, resulting in 'looping effects' whereby the constructs modify the phenomena, so that 'In the end, mathematics comes to constitute basic features of our life-worlds' (Skovsmose, 2022). As analysed in Chapter 13, school mathematics manifests no recognition of these aspects of contemporary life, let alone preparing students with the appropriate dispositions and agency to respond to them.

The above are minimal comments on the harmful effects on young people of technology and the implications for their mathematical education. A very different framework is what Ron Eglash and his team term 'Ethnocomputing'. As the term suggests, it relates the representational possibilities offered by computer technology and the cultural knowledge of students.[2]

So what could be done in mathematics classrooms?

As described above in relation to the CREME project and vividly related by Schoenfeld (Chapter 14, this volume), constraints on teachers who might want to introduce something of the ethnomathematical spirit into their classes are extremely severe. We suggest that the following are minimal feasible efforts, closely related to the themes listed earlier.

Being more honest about the history of academic mathematics

It is hard to imagine that any teacher would knowingly lie to the children in their class, yet it is easy to point to ways in which mathematics teachers fail to tell the truth, insofar as it is known. In terms of the specific case of the non-mathematician Pythagoras, as discussed above, many people, including mathematicians, are misinformed. At the systemic level, the same holds for the image of academic mathematics as essentially the achievement of Europeans only, as illustrated by the quotation above from Kline.

A straightforward way in which a knowledge of the history of mathematics can enhance teaching is through the use of load-bearing examples and the examination of misattributions of pieces of mathematics (the 'Theorem of Pythagoras' being a clear example). What is called 'Pascal's Triangle' (or 'Tartaglia's Triangle' in Italy) can be shown in its Chinese version dating from the thirteenth century (see the excellent resource that is Swetz, 1994, p. 328). A short test on history of mathematics might ask: 'Which of the following was a mathematician: Pythagoras, Omar Khayyam, Lewis Carroll, Florence Nightingale' to which a reasonable answer is 'All of them, except Pythagoras'.

Another pointed exercise, for example in the contexts of statements about the importance of symbolic algebra, would be to point to

---

2 Rather than attempt to summarise this very comprehensively developed theory, we refer the interested reader to https://csdt.org/publications for applications to education and https://generativejustice.org/publications for applications to economy and development.

achievements in many societies in the past which pre-dated abstract algebra and calculus.

Valorising students, allowing students to valorise themselves

In the CREME project (see above), a major emphasis was on the children's images of themselves, which, from our perspective, are liable to be greatly damaged in specific ways by school mathematics, as when performance on tests is used as a proxy indication of level of intelligence. The foundations for the current extreme situation in the United States were laid with the No Child Left Behind Act of 2002. In the words of McDermott and Kathleen Hall (2007, p. 10) that presented 'a vision for achieving progress in education through increased control and standardisation, a form of rational bureaucratic authority Max Weber [...] described as central to modernity' (see Chapter 5, this volume). Absolutely central was the use of testing, with results to be reported by ethnic categories, supposedly in the service of identifying 'failing schools' that could then be helped. Instead, as McDermott and Hall (p. 11) put it: 'Quantitative tests of aptitude and achievement have given U.S. education a way to sort children by race and social class, just like the old days, but without the words "race" and "class" front and center'. All is encapsulated in the pernicious phrase 'achievement gaps', a gloss on 'differences in test scores'. These effects are magnified in mathematics, given the political importance attached to the subject and the ease with which tests can be quantified (either you can compute 3/7 + 5/9 or you can't).

There are many other factors, including, as already discussed, the ways in which images of mathematics lend themselves to the projection of intellectual White supremacy. These reasons motivated our attempts in the CREME project (see above) to valorise the abilities and knowledge of the students. At the same time, we recognise that the children are forming understandings of the obstacles facing them in contemporary socio-political circumstances that obstruct their future possibilities in life. School mathematics education plays a major role in this regard. Accordingly, we argue that access and equity to the edifice are insufficient; it is essential to go beyond, to develop agency, in particular towards critiquing the nature of the edifice.

Relating school mathematics to the lives of students, their families and communities

Parra (Chapter 10, this volume) powerfully argues for research on Ethnomathematics to be reconceptualised as *relational*. As a parallel, a great deal of what we are attempting to articulate in this chapter can be expressed as the aim of reconceptualising school mathematics as relational, above all in relation to the children's funds of knowledge.

Critical mathematics educators argue that one aspect of making school mathematics relational is to relate it to the socio-political circumstances of the students, their families and communities. A clear instantiation of this perspective lies in the work of Rico Gutstein (e.g., 2006, 2012), influenced by Paulo Freire. He describes his work with the math for social justice class of seniors in 2008–2009 in which they collectively decided on five topics (generative themes, in Freirean terms): elections, populations displacement within their city (Chicago), HIV/AIDS, criminalisation, and sexism. All of these topics lend themselves naturally to investigation using mathematics. Gutstein (personal communication, 2010) resists characterising this work as ethnomathematical, yet it is surely related to the socio-political circumstances of the students. The choice of generative themes related to their lives makes such pedagogy relational, as does another feature of his work, namely having students make presentations to the community presenting the findings of their analyses. He summarises his work as follows:

> [...] three types of interrelated, yet distinct, knowledges relate to 'reading and writing the world with mathematics' [namely] community, classical, and critical knowledges, which all have mathematical components [...] community knowledge refers to [...] knowledge of one's own life circumstances and perspectives on reality. Classical knowledg refers to 'traditional' academic knowledge and critical knowledge means critiques and analyses of relations of power and issues of (in)justice. (Gutstein, 2012, p. 27)

In Chapter 13 of this volume, the importance of the early years in the learning of mathematics is discussed. As discussed there, children learn at an early age – through absurdly unrealistic word problems, in particular – that mathematics is not expected to cohere with their lived experience, their funds of knowledge, or even to make sense. The emphasis on early imprinting of images of mathematics also applies to issues of this chapter, including the image of mathematics as the sole creation of White (male) people and the associated beliefs about who can and cannot do mathematics.

Educate the privileged and future powerful

It is natural enough that most of the writing in the spirit of this chapter has focused on improving the education of marginalised groups. However, speaking at the 2008 ICME conference in Monterrey, Mexico, Ubiratan D'Ambrosio suggested that more attention be paid to the children enjoying privilege and likely to become powerful. This echoes themes in postcolonial writing such as Albert Memmi (1957/1965) about the deadly symbiosis between colonisers and colonised. Apart from the strategic reasons for taking this suggestion seriously in terms of attempting to speak truth to power, we may also consider the intellectual and moral harm done to privileged children (compare the discussion in Chapter 18 (this volume) on 'WhiteCrit'). This is a line of argument that we hope to develop more systematically in the future.

For similar reasons, the same comment may be made about those who major in mathematics at university. There are many reasons for incorporating courses on the history of mathematics for such students, but such courses tend to be inadequate from our point of view, rarely addressing the Eurocentric bias. How many mathematicians know of the cases discussed by Rodrigo Ochigame (2021) such as the development of paraconsistent logic in Brazil in the 1950s or the idea of a non-binary Turing machine based on the Jaina sevenfold system of predication, conceived by scientists in India in the aftermath of its independence?

Final thoughts

What give mathematics and mathematics education the special character that makes it an ideological battleground? It is not controversial to talk about language as a general human faculty *and* particular languages, about *architectures*, in the plural, related to cultural and environmental diversity, about *different* forms of music. Perhaps the answer lies in the importance of mathematics as a powerful means for shaping people as desired by a state? And regardless of the nature of the regime, a common aspect of this is conformity, the following of rules.

George Lakoff (1996) proposed a way of thinking about the differences between conservatives and liberals (his words) in terms of two core metaphors, labelled 'Strict Father' and 'Nurturant Parent', and argues how these help to explain how positions in relation to so many issues that on the surface appear unrelated are, in fact, highly correlated. Foreshadowing the extremes of the contemporary situation in and beyond the United States, he wrote that:

> Conservatives have, at least since the 60s, seen their system of values under attack—from feminism, the gay rights movement, the ecological movement, the sexual revolution, multiculturalism, and many more manifestations of Nurturant Parent morality. (p. 229)

At the most general level, perhaps it is an unavoidable consequence of the complexity of modern societies that there is a conflict between education and governance (Skovsmose, 2022). Munir Fasheh asked:

> Is it possible to teach mathematics effectively – that is, to enhance a critical attitude of one's self, society, and culture; to be an instrument in changing attitudes, convictions, and perspectives; to improve the ability of students to interpret the events of their immediate community, and to serve its needs better –without being attacked by existing authorities whether they are educational, scientific, political, religious, or any other form? (Fasheh, 1982, p. 2)

References

Al-Khalini, J. (2010). *Pathfinders: The golden age of Arabic science.* Penguin.

Asper, M. (2009). The two cultures of mathematics in Ancient Greece. In E. Robson & J. Stedall (Eds.), *The Oxford handbook of the history of mathematics* (pp. 107–132). Oxford University Press.

Boaler, J. (2022, November 3). Educators, you're the real experts. Here's how to defend your profession. *Education Week.* https://www.edweek.org/teaching-learning/opinion-educators-youre-the-real-experts-heres-how-to-defend-your-profession/2022/11

Burkert, W. (1972). *Lore and science in ancient Pythagoreanism.* Harvard University Press. (Original work published 1962)

Civil, M., & Quinteros, B. (2012). Latina mothers' perceptions about the teaching and learning of mathematics: Iplications for parental participation. In B. Greer, S. Mukhopadhyay, A. B. Powell, & S. Nelson-Barber, S. (Eds.), *Culturally responsive mathematics education* (pp. 321–343). Routledge.

D'Ambrosio, U. (2002). *Ethnomathematics.* Sense.

Dayton, A., & Rogoff, B. (2016). Paradigms in arranging for children's learning. In D. S. Guimarães (Ed.), *Amerindian paths: Guiding dialogues with psychology*. Information Age.

Fasheh, M. (1982). Mathematics, culture, and authority. *For the Learning of Mathematics, 3*(2), 2–8.

Ferreira, M. K. L. (2012). Map-making in São Paolo, Southern Brazil: Colonial history, social diversity, and indigenous people's rights. In S. Mukhopadhyay & W-M. Roth (Eds.), *Alternative forms of knowing (in) mathematics* (pp. 115–158). Sense.

Ferreira, M. K. L. (2015). *Mapping time, space and the body: Indigenous knowledge and mathematical thinking in Brazil.* Sense.

Ford, E., Greer, B., Koopman, M., Makadia, B., Mukhopadhyay, S., Wilson, M., & Wolfe, M. (2018). Stories from the trenches: Urban schools in Portland. In S. Crespo, S. Celadó-Pattichis, & M. Civil (Eds.), *Access and equity: Promoting high-quality mathematics, Grades 3–5* (pp. 169–192). National Council of Teachers of Mathematics.

Gay, G. (2009). Preparing culturally responsive mathematics teachers. In B. Greer, S. Mukhopadhyay, A. B. Powell, & S. Nelson-Barber, S. (Eds.), *Culturally responsive mathematics education* (pp. 189–205). Routledge.

Gonzalez, N., Moll, L. C., & Amanti, C. (Eds.). (2005). *Funds of knowledge: Theorizing practices in households, communities, and classrooms.* Erlbaum.

Greer, B. (2021). Learning from history: Jens Høyrup on mathematics, education, and society. In D. Kollosche (Ed.), *Exploring new ways to connect: Proceedings of the Eleventh International Mathematics Education and Society Conference* (Vol. 2, pp. 487–496). Tredition. https://doi.org/10.5281/zenodo.5414119

Greer, B., Mukhopadhyay, S., Powell, A. B., & Nelson-Barber, S. (Eds.). (2009). *Culturally responsive mathematics education.* Routledge.

Gutstein, E. (2006). *Reading and writing the world with mathematics: Toward a pedagogy for social justice.* Routledge.

Gutstein, E. (2012). Mathematics as a weapon in the struggle. In O. Skovsmose & B. Greer (Eds.), *Opening the cage: Critique and politics of mathematics education* (pp. 23–48). Sense.

Hacking, I. (1999). *The social construction of what?* Harvard University Press.

Hersh, R. (Ed.). (2006). *18 unconventional essays on the nature of mathematics.* Springer.

Høyrup, J. (1992). *The formation of a myth: Greek mathematics – our mathematics.* Roskilde Universitetscenter.

Huffman, C. (2018). Pythagoras. In E. N. Zalta (Ed.), *The Stanford Encyclopedia of Philosophy* (Winter 2018 Edition). https://plato.stanford.edu/archives/win2018/entries/pythagoras

Hutchins, E. (2000). *Cognition in the wild.* MIT Press.

Joseph, G. G. (1991). *The crest of the peacock: The non-European roots of mathematics.* Penguin.

Kline, M. (1953). *Mathematics in Western culture.* Oxford University Press.

Koopman, M. (2017). Elementary student t-shirt workers go on strike. *Rethinking Schools, 32*(2).

Lakoff, G. (1996). *Moral politics: What conservatives know that liberals don't.* Chicago University Press.

Lipka, J. (2020). Indigenous knowledge and navigating the rising tides of climate change and other existential threats. *Revista Latinoamericanan de Ethnomathemática, 13*(3), 29–61.

Lipka, J., Yanez, E., Andrew-Ihrke, D., & Adam, S. (2009). A two-way process for developing effective culturally based math: Examples from Math in a Cultural Context. In B. Greer, S. Mukhopadhyay, A. B. Powell, & S. Nelson-Barber (Eds.), *Culturally responsive mathematics education* (pp. 257–280). Routledge.

Lipka, J., Wong, M., Andrew-Ihrke, D., & Yanez, E. (2012). Developing an alternative learning trajectory for rational number reasoning, geometry, and measuring based on Indigenous knowledge. In S. Mukhopadhyay & W-M. Roth (Eds.), *Alternative forms of knowing (in) mathematics* (pp. 159–182). Sense.

McDermott, R. (2012). When is mathematics, and who says so? In B. Bevan, P. Bell, R. Stevens, & A. Razfar (Eds.), *LOST opportunities: Learning in out-of-school time* (pp. 85–91). Springer.

McDermott, R., & Hall, K. D. (2007). Scientifically debased research on learning, 1854-2006. *Anthropology & Educational Quarterly, 38*(1), 9–15.

Memmi, A. (1965). *The colonizer and the colonized.* Orion. (Original work published 1957).

Mukhopadhyay, S. (2009). The decorative impulse: Ethnomathematics and Tlingit basketry. *ZDM Mathematics Education, 41*(1–2), 117–130.

Mukhopadhyay, S. (2009, March). *The appreciation of pattern: Beauty and structure* [Conference presentation]. Wooshteen Kanaxtulaneegí Haa At

Wuskóowu, Sharing our Knowledge: A Conference of Tlingit Tribes and Clans, Juneau, AK.

Mukhopadhyay, S. (2012, November). *Non-literate Indian boat-builders: Vernacular engineering and the rigor of reality* [Conference presentation]. 111th Annual meeting of the American Athropological Association, San Francisco, CA, United States.

Mukhopadhyay, S. (2013, April). *The mathematics of those without power* [Conference presentation]. 7th Mathematics Education and Society conference, Cape Town, South Africa.

Mukhopadhyay, S., & Greer, B. (in press). *The wisdom of others.*

Mukhopadhyay, S., & Roth, W-M. (Eds.). (2012). *Alternative forms of knowing (in) mathematics.* Sense.

Netz, R. (2003). *The shaping of deduction in Greek mathematics: A study in cognitive history.* Cambridge University Press.

Ochigame, R. (2021). *Remodeling rationality: An inquiry into unorthodox modes of logic and computation* [Doctoral dissertation, Massachusetts Institute of Technology].

Paris, S., & Alim, H. S. (Eds.). (2017). *Culturally sustaining pedagogy.* Teachers College Press.

Pinxten, R., van Dooren, I., & Harvey, F. (1983). *The anthropology of space: Explorations into the natural philosophy and semantics of the Navajo.* University of Philadelphia Press.

Powell, A. B., & Frankenstein, M. (1997). *Ethnomathematics: Challenging Eurocentrism in mathematics education.* SUNY Press.

Raju, C. K. (2007). *Cultural foundations of mathematics: The nature of mathematical proof and the transmission of the calculus from India to Europe in the 16th c. CE.* Pearson Longman.

Skovsmose, O. (2022). Concerns of Critical Mathematics Education and of Ethnomathematics. *Revista Colombiana de Educación,* (86), 365–382.

Stevens, R. (2010). Learning as a members' phenomenon: Toward an ethnographically adequate science of learning. *Teachers College Record, 112*(13), 82–97.

Swetz, F. (1994). *From five fingers to infinity: A journey through the history of mathematics.* Open Court.

Tate, W. F. (1995). School mathematics and African American students: Thinking seriously about opportunity-to-learn standards. *Educational Administration Quarterly, 31,* 424–448.

Urton, G. (2012). Mathematics and accounting in the Andes before and after the Spanish Conquest. In S. Mukhopadhyay & W-M. Roth (Eds.), *Alternative forms of knowing (in) mathematics* (pp. 17–32). Sense.

Vithal, R., & Skovsmose, O. (1997). The end of innocence: A critique of 'ethnomathematics'. *Educational Studies in Mathematics, 34,* 131–157.

Washburn, D. K., & Crowe, D. W. (1987). *Symmetries of cultures: Theory and practice of plane pattern analysis.* University of Washington Press.

18. Mathematics education as a racialized field

Christopher C. Jett and Julius Davis

Racism is endemic in society, in education, and in mathematics education. Thus, as a discipline, mathematics education functions as a racialized field. In this chapter, we explore critical race theory (CRT) as a theoretical frame to address this problem. In so doing, we offer offshoots of CRT – BlackCrit, LatCrit, TribalCrit, AsianCrit, and critical Whiteness theory – for mathematics educators to use in their scholarship. We also issue a call to mathematics educators regarding the urgent need to advance race-related work. We conclude the chapter by posing thought-provoking questions for consideration and action.

Introduction

Racism pervades society. It is endemic to the political project called education; it takes specific forms within mathematics education in particular. This phenomenon is readily apparent in the United States of America, referred to as the United States (US), given the racial contention in the broader society and mathematics education.[1] Given this, it must be stated from the outset that this chapter is written primarily in the context of the US, considering our experiences as mathematics education researchers within this country. Of course, there is a great deal of wider significance that can be extracted from US mathematics education

1 Those familiar with the United States context will be aware that mathematics education has been pulled back into the 'culture wars', to a considerable extent because of the efforts of activist researchers and scholars. We do not attempt to address that complex situation within this chapter; the short answer is that it is a manifestation of White supremacy.

©2024 Christopher C. Jett & Julius Davis, CC BY-NC 4.0 https://doi.org/10.11647/OBP.0407.18

regarding racialization. Consequently, examining the racialized nature of mathematics education is important for mitigating racialized barriers that hinder minoritized students' mathematics achievement, persistence, and success.

In Western-dominated countries, and arguably beyond, mathematics education scholarship has propagated and primarily centered Eurocentric perspectives, experiences, knowledge systems, and paradigms (Davis, 2018, 2021; Powell & Frankenstein, 1997; also see Greer, 2021, for a short guide to the immensely important historical/anthropological work of Jens Høyrup). Resultantly, a high proportion of the discipline-specific handbooks, journals, and other mainstream publications have been authored and edited by White men. In the United States, White men have predominantly served as the leaders of mathematics educational organizations. By way of illustration, Julius Davis (2021) argued:

> The Mathematical Association of America (MAA) and National Council of Teachers of Mathematics (NCTM) were founded in 1915 and 1920, respectively, as two predominately White organizations that have and continue to shape the field of mathematics education. White men were the primary founders and leaders of these organizations and were derived historically from White higher education institutions [...] Both organizations have played a significant role in shaping school mathematics, mathematics content, pedagogy, learning, assessment, research, and the future direction of mathematics education. (p. 789)

Mathematics education has largely been constructed through the White male gaze. As a result, racially minoritized groups' perspectives and interests have been largely excluded vis-à-vis disciplinary knowledge systems, policies, and practices. Furthermore, institutional and structural racism persist and, in turn, materialize in the field. It stands to reason, then, that mathematics education functions as a racialized field. Given this reality, mathematics educators must employ race-conscious theoretical frames in their work. That is the goal of the current chapter.

We begin with our positionality to provide some context about who we are as two Black men who conduct mathematics education research through a critical[2] lens (also see Martin & Gholson, 2012, for a compelling discussion on being critical Black scholars in mathematics education).

2 For further related work on critical mathematics education, see, e.g., Frankenstein, 1983; Greer & Skovsmose, 2012; Skovsmose, 2023.

After that, we delve into critical race theory (CRT) and position it as a theoretical frame for consideration in the field. Then, we challenge mathematics education researchers to shift the paradigm, so to speak, and infuse critical theories of race in their work. Afterwards, we share discipline-specific examples and call for mathematics educators to use these theories. We conclude the chapter by briefly summarizing these main ideas and posing some critical questions to advance race-related work in mathematics education.

Our positionality

We are Black men who grew up and completed our mathematics education in the United States, which has a troubled history of racism regarding Black life that persists to this day. As examples, the ongoing Black Lives Matter movement and the murders of George Floyd, Breonna Taylor, and several other Black people at the hands of police have intensified international awareness of racial injustice in an unprecedented way. Because of our racialized identities, we have designed research agendas that interrogate racial issues in mathematics education with respect to Black (male) students, teachers, and communities (Davis, 2014, 2016, 2021, 2022; Jett, 2012, 2019a, 2019b, 2022; Jett et al., 2022; Larnell et al., 2016). As emerging researchers and leaders of CRT scholarship in mathematics education, our professorial experiences have catapulted our desires to contribute race-related work to the field.

The first author has been a race-conscious scholar for as long as he can remember. Stated differently, his critical race journey dates back to his childhood. He is from the predominately Black community of Memphis, Tennessee, which exposed him to racial issues, challenges, and problems. For example, the city of Memphis is where the honorable Dr. Martin Luther King, Jr. – civil rights activist, social justice warrior, and racial equity advocate – was assassinated in 1968. Although this occurred before the first author was born, he learned about the race riots that ensued and has witnessed the racial tensions that still occur within the city. As such, his previous experiences have exposed him to the institutionalized racism endemic in society and propelled him to produce antiracist scholarship, especially work highlighting Black people's strengths.

The second author enters the field as a race-first scholar and uses CRT to interrogate race and racism in mathematics education. He views CRT as the theoretical lens to understand and explain the problems plaguing the global Black community, which include issues within and outside of mathematics education. In prior work, he used CRT to study how race and racism impacted Black people in his community of West Baltimore, a city in Maryland, and better understand how broader social constructions impacted the mathematics education Black people received there. As his knowledge of CRT scholarship has matured, he has begun to use it to critique mathematics education holistically and challenge false notions that White men have been the primary architects of mathematics (Davis, 2018, 2021). Jointly, the authors use CRT to counter White supremacist logics that attempt to devalue the mathematics contributions and accomplishments of Black scholars.

Critical race theory in mathematics education

CRT is the leading theoretical lens employed to examine the racialized experiences of minoritized people in law, education, and mathematics education. As a framework, CRT emanates from critical legal studies (CLS) because of the dissatisfaction with how legal scholars fail to address race and racism in the law (Crenshaw et al., 1995; Tate, 1997). From a CRT perspective, social constructions of race are essential to understanding how racism functions in society, its institutions, and the law (Delgado & Stefancic, 2001). CRT operates from the premise that racism is a deeply rooted, permanent, and ever-changing feature of American society and its institutions to maintain White supremacy and power. CRT recognizes mainstream legal claims of objectivity, neutrality, colorblindness, and meritocracy as disguises for the self-interest of Whites in power. Furthermore, CRT validates the experiential knowledge of minoritized people and recognizes the importance of crossing disciplinary boundaries to understand the racialized experiences of minoritized people. In short, critical race theorists are committed to achieving racial justice.

In the 1990s, CRT began to take shape in education. Even though William Tate (1993) used CRT in mathematics education prior to 1995, the formal introduction of CRT in the broader education research community occurred in 1995 through Gloria Ladson-Billings and Tate's (1995) path-breaking article 'Toward a Critical Race Theory of Education', published in *Teachers College Record.* In it, they ushered this theoretical perspective from CLS into (mathematics) education to center race and challenge mainstream multiculturalism, which largely pointed to equitable access to mathematics education without questioning the nature of what is being accessed. Danny Martin (2019) commented that 'the forms of inclusion offered up in equity-oriented discourses and reforms have typically involved two trajectories: (1) inclusion accompanied by marginalization; and (2) assimilation into the existing cultures of mathematics education' (p. 460). As a result, CRT radically shifted the multiculturalist and colorblind paradigm in education to accentuate race, racism, and other forms of oppression.

Daniel Solórzano and Tara Yosso (2002) assert that there are five defining elements of CRT in education; namely, that it: (1) asserts that race and racism are endemic and permanent fixtures of American society and structures; (2) challenges dominant ideology; (3) ascribes a commitment to social justice; (4) centralizes the experiential knowledge of minoritized people; and (5) uses an interdisciplinary approach to better understand racism, sexism, and classism.

As mentioned earlier, the genealogy of CRT in mathematics education (CRT(ME)) can be traced back to the scholarship of Tate. Tate (1993) published the first CRT(ME) article entitled 'Advocacy Versus Economics: A Critical Race Analysis of the Proposed National Assessment in Mathematics'. That same year, Tate and colleagues published an article merging law, CRT, education, and mathematics education (Tate et al., 1993). Because of this foundational work, scholars have credited Tate as the chief architect of CRT in (mathematics) education (Davis, 2014; Davis & Jett, 2019; Lynn & Adams, 2002). Tate, however, credits Derrick Bell, recognized as the father of CRT, with being the first scholar to use CRT to examine educational issues; see Bell's (1975) article 'Serving Two Masters: Integration Ideals and Client Interests in School Desegregation Litigation' in *The Yale Law Journal* for an education-related critical race analysis. Be that as it may, Tate is a trailblazer regarding the production of scholarship focused on CRT in education that interweaves mathematics education specifically.

Martin (2009), another prominent race scholar in mathematics education, argues that CRT reckons with the historically and socially constructed nature of race and racism. He challenges the racial hierarchies of mathematical ability, participation, and power. Moreover, he asserts that:

> Drawing on analyses of the ways that race and racism are conceptualized and studied outside of mathematics education will help illustrate the need for similar kinds of analyses within mathematics education. I argue that rather than exploiting the usefulness of sociological and critical theory frameworks, the vast majority of mainstream mathematics education research and policy purporting to explain so-called racial achievement gaps between African American, Latino, and Native Americans on one hand, and White and Asian students on the other, continues to rely on inadequate and impoverished approaches to race, racism, and racialized inequality. (p. 297)

In this article, Martin admonishes the mathematics education community for its poor treatment and analysis of race in research and policy. A more comprehensive race-based analysis, he argues, will further illuminate the racial issues present in mathematics education research, elucidate how mathematics education policy operates and is implicated in sustaining racial hierarchies, and lead to more nuanced race-based analyses among mathematics educators.

Building on this firm foundation of race scholarship, many Black mathematics education researchers have used CRT(ME). To illustrate, we co-edited *Critical Race Theory in Mathematics Education* (Davis & Jett, 2019). This edited volume includes chapters written by scholars who extrapolate the tenets of CRT to advance race-conscious analyses in our field. More precisely, scholars merged CRT and mathematics education knowledge to establish stronger connections between the two. We acknowledge the racial atrocities thrust upon Black people and center Black students in mathematics contexts in our work. We also acknowledge that other minoritized groups have experienced mathematics education as a racialized field. Therefore, we call for a paradigm shift – one that explicitly and unapologetically attends to issues of race, racism, and racialization – to the study of these groups to contribute race scholarship to the field.

Shifting the paradigm: Infusing critical theories of race and ethnicity into mathematics education

CRT has a growing influence on education. The CRT literature continues to expand, as CRT-related publications have focused on school-based education, tertiary education, and other (emerging) academic disciplines. Over the last two decades, CRT has incorporated the racialized and ethnicized experiences of minoritized people. Given this emphasis, offshoots of CRT, including BlackCrit, LatCrit, TribalCrit, AsianCrit, now address racial oppression beyond the Black/White binary.[3] Further, White scholars have used critical Whiteness theory to look inward and behind the mirror to expose White privilege and challenge racism (Delgado & Stefancic, 1997).

Here, we expound on the four aforementioned theoretical perspectives that were birthed out of CRT to more systematically examine race, racism, ethnicity, classism, and other forms of oppression applied to diverse racial and ethnic groups in mathematics education as well as the complementary perspective of Whiteness studies. These frameworks can be useful for mathematics educators who explicate race and ethnic-conscious analyses within the field. We acknowledge that, in this overview, we are not attempting to address the complexities of diversity within each of these groupings. Contemporary immigrants from Africa differ in significant ways from the descendants of enslaved Africans; Indigenous people are as diverse as the environments in which they are ecologically embedded, and so on. These issues have added importance because of the tendency of supremacists to essentialize, and, over time, they will evolve as an essential extension of our analysis.

BlackCrit

Since CRT's inception in education, scholars have argued that this theoretical framework has privileged the Black experience, Blackness, and in some regards, African Americans (Dumas & ross, 2016; Phillips, 1998). Michael Dumas and kihana ross argued that CRT in education

3 FaithCrit, QuantCrit, and QueerCrit represent other emerging offshoots of CRT. For related reading, see, e.g., Garcia et al., 2018; Malone & Lachaud, 2022; Valdes, 1998.

moved away from focusing on Blackness because of critiques from non-Black scholars. However, they emphasized that there is a need to incisively analyze the specificity of Blackness and antiBlackness in education and agreed that there is, indeed, an implicit focus on Blackness in CRT in education.

BlackCrit analyses the specificity of Blackness and antiBlackness by explaining how Black people have been marginalized, disdained, disregarded, and excluded in educational spaces (Dumas & ross, 2016). Distinguishing BlackCrit from CRT, Dumas and ross proffer that CRT is not intended to address how antiBlackness informs and influences racist ideology and institutional practice. Rather, BlackCrit is necessary to evaluate 'how blackness matters in our understanding of key tenets related to, for example, the permanence of racism and whiteness as property' (p. 417).

Dumas and ross (2016) offer three foundational ideas of BlackCrit in education: (1) 'AntiBlackness is endemic to, and is central to, how all of us make sense of the social, economic, historical, and cultural dimensions of human life' (p. 429); (2) 'Blackness exists in tension with the neoliberal-multicultural imagination' (p. 430); and (3) 'BlackCrit should create space for Black liberatory fantasy, and resist a revisionist history that supports dangerous majoritarian stories that disappear Whites from a history of racial dominance' (p. 431).

Mathematics educators have begun to use BlackCrit to advance understandings of Black children's and adults' experiences (Martin et al., 2019; Matthews et al., 2021). Martin and colleagues used BlackCrit to examine systemic violence regarding the experiences of Black mathematics learners. They also used BlackCrit to offer a Black Liberatory Mathematics Education, which advances a radical reimagination of mathematics education for Black students. Martin and colleagues recognized that liberatory mathematics education cannot exist within the current educational system rooted in racism (White supremacy) and antiBlackness. In other work, Lou Matthews and colleagues used BlackCrit in a reflective essay to explore possibilities for the Black community to honor families' agency, expand digital equity, and prioritize approaches that support liberatory mathematics education during the COVID-19 global pandemic.

LatCrit

LatCrit examines the positioning of Latinas/os[4] writ large and encompasses unique issues directly related to the Latina/o community, including immigration, language, and phenotype (Bernal, 2002). LatCrit also values the strengths of Latina/os to highlight community wealth, challenges commonly held beliefs about the racial hierarchy of ability, and problematizes the notion of a race-neutral society. Similar to CRT, LatCrit is derived from the following five foundational tenets: (1) the centrality of race and racism intersecting with other forms of oppression; (2) a challenge to dominant discourse and ideology; (3) a profound commitment to social justice; (4) the validation of experiential knowledge; and (5) the incorporation of a transdisciplinary perspective (Bernal, 2002; Fernández, 2002; Solórzano & Yosso, 2001). Drawing from these tenets, scholars have used LatCrit to highlight how their ethnicity, language, immigrant status, and culture have been rejected in classrooms.

While there is a rich literature on Latinos/as and mathematics education (e.g., Tellez, Moschkovitch, & Civil, 2011), there has been limited use of the term 'LatCrit' in mathematics education. Rochelle Gutiérrez (2013) offered it as a theoretical perspective to consider using in the socio-political turn. She recognized that scholarship on LatCrit was scant, but the framework provided evidence about the ability to challenge and dismantle social constructions of race, racism, Whiteness, sexism, classism, and other forms of oppression (e.g., language, immigrant status). Gutiérrez also pointed out that social activism and testimonios were important features for those using LatCrit in mathematics education.

In her study, Maria Zavala (2014) used LatCrit to examine Latina/o students' narratives of learning mathematics in an urban multilingual high school. Her findings revealed that Latina/o students had to grapple with racial stereotypes and linguistic challenges given that American English served as the official language of mathematics

4 Like other critical scholars, we problematize the gender binary present in Latina/o. We use this language here to honor the way it was presented by the cited authors even though we recognize that language continues to shift regarding the Latinx community.

instruction. These complex factors significantly impact Latina/o students' mathematics identities and ultimately influence how they see themselves as mathematics learners. In summary, LatCrit has allowed mathematics education researchers to better understand how Latina/o students' mathematics identities were co-constructed in relation to their ethnic, linguistic, and gendered experiences.

TribalCrit

TribalCrit addresses the racialized challenges thrust upon Indigenous people (Brayboy, 2005). CRT purports that racism is endemic, while TribalCrit purports that colonization is endemic. Regarding theory, many Indigenous scholars view it as a roadmap for their community's continuous survival. Researchers have used TribalCrit to examine the misappropriation of Native mascots, the misrepresentation of cultural symbols, and fraudulent ethnic policies (Castango & Lee, 2007; Marshall, 2018). The interest convergence tenet, in particular, has often been used to demonstrate how these matters have served the interests of White people, institutions, and systems.

Education researchers have used TribalCrit in mathematics education and other science, technology, engineering, and mathematics (STEM) fields (Kokka, 2018; Marshall, 2018; also see Deloria, 1997; Stavrou & Miller, 2017). Samantha Marshall used TribalCrit to produce an Indigenous sovereign Tribal nation in mathematics education. In her qualitative study, she provided insights into the value Indigenous leaders hold for the education of their youths, including cultural congruity, cultural and linguistic sustainment, and sovereignty. Her article illuminated the complexities of Indigenous education and Tribal nation-building in and through the lens of mathematics education. In a different study, mathematics educator, Kari Kokka, used TribalCrit to examine the experiences of four STEM teacher activists who created a social justice STEM organization. She found that firsthand experiences with being oppressed led them to become STEM teacher activists. She also found that being STEM activists became a vehicle of healing for them as they addressed the inequities they experienced or witnessed in their communities.

AsianCrit

As another adaptation of CRT, AsianCrit grounds the experiences, perspectives, and voices of Asian Americans in light of the racism thrust upon them (Iftikar & Museus, 2018; Museus & Iftikar, 2014). Asian American groups often include: 'Bhutanese, Burmese, Cambodian, Chinese, Hmong, Indian, Indonesian, Japanese, Korean, Lao, Pakistani, Taiwanese, Thai, and Vietnamese Americans, in addition to many other ethnic groups' (Iftikar & Museus, 2018, p. 4). White supremacist constructions of these groups have attempted to lump them into a singular category, but they have distinct languages, norms, values, and traditions. Martin (2009) notes that social constructions of 'Asian' in an international context frame them as negative and in direct competition with White students and the global White power structure to invoke strong US nationalism.

AsianCrit proposes seven tenets to better understand and examine the racial and ethnic realities of Asian Americans (Iftikar & Museus, 2018). These tenets elucidate the White supremacist sustenance that attempts to corral Asian Americans into a single group and simultaneously thwart their advancement. They include: (1) '*Asianization* is grounded in the reality that people within the US only become "Asian" because of White supremacy and the racialization processes that it engenders. Specifically, White supremacy and pervasive nativistic racism in the US result in Asian Americans being racialized as perpetual foreigners, threatening yellow perils, model and deviant minorities, and sexually deviant emasculated men and hypersexualized women' (p. 8); (2) *transnational contexts* highlights the global relationships between Asian Americans and White supremacist logics; (3) *(re)constructive history* draws upon the assets, contributions, and voices of Asian Americans to produce an accurate depiction of the group's history; (4) *strategic (anti)essentialism* purports that Asian Americans join forces to gain power and advocate against White supremacist structures and policies; (5) *intersectionality* explores how other constructs such as gender, class, and sexual orientation fuel systematic oppression for this group; (6) the *story, theory, and praxis* tenet espouses that racially minoritized groups' experiential knowledge challenges dominant, deficit narratives about Asian Americans; centers their real experiences; and offers a more holistic orientation to their perspectives; and (7) the commitment to *social justice* tenet advances that AsianCrit seeks 'to eradicate racism, sexism, heterosexism, capitalist exploitation, and other systemic forms of dehumanization and domination' (p. 9).

Scholarship using AsianCrit in mathematics education[5] is in short supply. Kokka and Theodore Chao (2020), two Asian American mathematics educators, used AsianCrit to study Asian American teachers' experiences with racism and conceptualize their conjoining racial, ethnic, and mathematics teacher identities in the wake of the model minority myth that pervasively suggests that Asians are good at math (Shah, 2019). Kokka and Chao's findings indicate that the four Asian American mathematics teachers experienced internalized racism and engaged in stereotype management by distancing themselves from other Asian Americans, avoiding discussions about their own difficulties in mathematics, and intentionally reaching out to develop relationships with Black and Latinx students.

WhiteCrit?

The four cases just described suggest a fifth, namely 'WhiteCrit', yet that word has scarcely appeared in the literature. There are, however, critical studies of Whiteness in mathematics education, influenced by CRT (Battey, 2013; Foste & Irwin, 2020; Nishi, 2021). These studies name and expose the inner workings of White privilege and examine race, racism, and racial identity within mathematics education, revealing how Whiteness functions to maintain racial dominance. Power and oppression are articulated, redefined, and reasserted through individual and institutional practices that privilege Whiteness (Corces-Zimmerman & Guida, 2019). The power of Whiteness intersects with other systems of domination, including but not limited to patriarchy, capitalism, ableism, and genderism.

Jeremy Bohonos (2019) contends that critical Whiteness theory does not have a clear set of tenets that govern it. However, other scholars have identified the following three core elements that guide it: (1) *Thinking Whitely* explains the many ways that Whiteness and White supremacy

---

5 In related work, see Cvencek et al. (2015) for a poignant discussion regarding the development of math–race stereotypes.

influence White people's conscious and unconscious thoughts to maintain racial superiority and dominance (Corces-Zimmerman & Guida, 2019). (2) *Behaving Whitely* includes well-intentioned White people who maintain White domination consciously and unconsciously through actions, White complicity, and White emotionality (Corces-Zimmerman & Guida, 2019; Foste & Irwin, 2020). White complicity includes the unconscious negative beliefs that White people hold about non-White people, which affect their practices and habits centered on Whiteness, and the consequences of those practices and habits. White emotionality underscores the racialized ways that White people experience and act on emotions such as shame, guilt, denial, anger, rage, sadness, discomfort, and defensiveness as a means to protect White fragility (Anderson, 2016; Corces-Zimmerman & Guida, 2019; DiAngelo, 2018). (3) *Speaking Whitely* refers to the discursive and rhetorical strategies, the elusive and deceptive language, White talk, colorblindness, and color-evasiveness that White people use to reinforce the status and privilege of Whiteness (Annamma et al., 2016; Bonilla-Silva, 2006; Corces-Zimmerman & Guida, 2019; McIntyre, 1997).

Martin (2009, 2019) notes how there has not been a systematic study of Whiteness and its relations to mathematics education (i.e., mathematics participation, opportunity to learn, and achievement) even from mathematics education researchers who claim to study race in their analysis of mathematics achievement. Whiteness is a salient aspect of the mathematics education enterprise; most professors, researchers, teachers, and students are White and benefit from the individual, institutional, and structural arrangement of Whiteness (Battey & Leyva, 2016; Davis, 2021; Stinson, 2017). Martin (2009) also notes that Whiteness is a part of the larger system of racism that operates to privilege White mathematical knowledge construction. White male mathematics educators and researchers represent a highly racialized and gendered space that is privileged in the field and has influenced social and policy perspectives about mathematics (Davis, 2021; Martin, 2008, 2009).

BlackCrit, LatCrit, TribalCrit, and AsianCrit are, in general, about how these groups of people are harmed by White domination. What might it mean to postulate WhiteCrit as analysis of the harm done to White people, specifically through mathematics education? At first sight, that might seem a strange suggestion given White privilege; however, James Baldwin did raise the question, pointing out that, for White racists, 'their moral lives have been destroyed by the plague called color' (Buccola, 2019, p. 383).[6] WhiteCrit in this sense could add another dimension.

Discipline-specific examples

Mathematics education researchers have emphasized the racialized nature of mathematics policies, textbooks, and word problems (Martin, 2009; Martin et al., 2019; Tate, 1995). Numerous other examples demonstrate the racialized nature of the field and show how racist practices manifest in systematic and institutional ways. One striking example is tracking, which denies minoritized students access to 'high-quality' mathematics instruction. There are several issues in relation to testing, perhaps starting with the characterization of differences in test scores as 'achievement gaps' (Davis & Martin, 2018; Miller-Jones & Greer, 2009). And the traceable connection between the Eurocentric myth of the development of academic mathematics and intellectual White supremacy is discussed at various points in this volume.

In this section, we provide two discipline-specific examples that brought racial issues to the fore and simultaneously gained national attention in the United States. The first example occurred in the state of Georgia; ironically, the first author currently works at one of the state institutions that prepares mathematics teachers for this particular school district (although it was not revealed where the teacher in this scenario received their teaching credential). Notwithstanding, a third-grade teacher shared the following exercise with students: 'Each tree had 56 oranges. If 8 slaves pick them equally, then how much would each slave pick?' (AFRO Staff, 2012). Another exercise read: 'If Frederick got two beatings per day, how many beatings did he get in 1 week?' Interestingly, another teacher made copies of the assignment. As a result, these word

6 The quotation is from a debate with the avowed White supremacist, William F. Buckley, Jr., at the Cambridge Union in 1965, described in detail, and with interwoven biographies of the protagonists giving historical background, in Buccola's *The Fire Is Upon Us*.

problems, embedded with racist logics, were spread to a racially diverse group of third-grade students in four distinct classrooms.

In a California mathematics classroom, a White woman teacher was placed on leave after a video went viral of her mocking Native Americans. In the video, she was dressed in clothing representative of Native American culture with imaginary tomahawks in her hand. She was repeatedly saying SOH-CAH-TOA, a mnemonic often used to help students remember the trigonometric functions: Sine (opposite over hypotenuse), Cosine (adjacent over hypotenuse), and Tangent (opposite over adjacent). The blatant racism encroached in this mathematics teacher's primitive stereotypes about Native American practices provides another example of how racist ideologies thrive in mathematics classrooms. Taken together, these two discipline-specific examples could use BlackCrit and TribalCrit, respectively, to highlight the dehumanizing aspects of mathematics practices for racialized students and clearly indicate which racial groups win and which ones lose apropos their mathematics education. In so doing, these examples, among others not mentioned here, necessitate a call to the mathematics education community regarding the racialized nature of the field.

A call to mathematics educators

Socially and politically constructed meanings of Black, Latinx, Native, Asian, and White American pervade the racialized and ethnicized experiences of these groups in mathematics education research, policy, and practice. Hence, these constructions perpetuate the racial hierarchy of mathematical ability, participation, and power that remains unchanged and unchallenged in meaningful ways. Martin (2009) argues that, 'rather than questioning and deconstructing the *racialized* nature of this hierarchy, many mainstream math educators accept it as their natural starting point' (p. 316). He also maintains that the racialized nature of students' experiences, research, and policy in mathematics education contributes to the social devaluing of African Americans, Latinx, and Native Americans. Contrastingly, Whiteness or being White occupies a privileged space that does not result in being socially devalued. Earlier, we delved into BlackCrit, LatCrit, TribalCrit, AsianCrit, and WhiteCrit, and our heartfelt charge is for mathematics educators to use these frameworks to critically examine issues of race, racism, ethnicity, and other forms of oppression.

In this vein, we are calling for a racial awakening that expands CRT in mathematics education to address, question, and deconstruct the racialized and ethicized experiences of Black, Latinx, Native, Asian, and White Americans. CRT's offshoots, namely, BlackCrit, LatCrit, TribalCrit, AsianCrit, and WhiteCrit, offer tools for systematic examinations. These manifestations of CRT should be further conceptualized, expanded, and merged with our disciplinary field to properly address, mitigate, and dismantle racist practices. Empowering racially minoritized scholars to explore their own racialized and ethnicized group's mathematics education experiences will unearth additional insights given their requisite knowledge systems, skills, and realities.

Concluding thoughts

In this chapter, we have expounded CRT, which has been central to our work as Black male mathematics education researchers. We have also brought focused attention to BlackCrit, LatCrit, TribalCrit, AsianCrit, and Whiteness studies with the goal of having more mathematics education researchers use them in their work. While these theoretical frames draw from CRT, they can all distinctly address the racialized nature of the field with respect to and across different racial categories. Furthermore, we offer the following questions for consideration:

- In what nuanced ways can mathematics education scholars use these critical theories of race to advance knowledge about the racialized nature of the field and ways to address it?
- In what ways can we conjoin theories of race to build on and extend foundational scholarship in mathematics education (e.g., critical mathematics education) to strengthen race-related work?
- What is the specific role of mathematics education scholars in this racist and xenophobic political climate (i.e., when critical race scholars are being attacked, mathematics curricular materials are being censored, and disciplinary textbooks are being banned)?

- What opportunities exist to facilitate change regarding the racialized nature of mathematics beyond securing grant funding, conducting research studies, presenting at conferences, and writing scholarly publications?
- What can we learn from international scholars, their histories, and their race-oriented epistemologies about mathematics education that could be instructive for the field?

In closing, please know that we have not hit a plateau with race work, as these questions indicate that issues of race, racism, and racialization continue to run amok in mathematics education. It is important to emphasize that Black, Latinx, Native, and Asian American scholars, students, and families should not bear all of the responsibility to address Whiteness and racism in the field. White mathematics educators must take greater responsibility for doing race work with White populations to challenge and dismantle the system of racism and White supremacy. As this chapter demonstrates, there is still much more work that needs to be done to understand the experiences of racially minoritized students in mathematics education, break the image of mathematics as a White male domain, and honor the full humanity of racialized groups in mathematics contexts.

References

AFRO Staff. (2012, January 12). Racist math problems at Ga. school anger parents, NAACP. *AFRO News*. https://afro.com/racist-math-problems-at-ga-school-anger-parents-naacp

Anderson, C. (2016). *White rage: The unspoken truth of our racial divide.* Bloomsbury.

Annamma, S. A., Jackson, D. D., & Morrison, D. (2016). Conceptualizing color-evasiveness: Using dis/ability critical race theory to expand a color-blind ideology in education and society. *Race, Ethnicity and Education*, *20*(2), 147–162. https://doi.org/10.1080/13613324.2016.1248837

Battey, D. (2013). Access to mathematics: 'A possessive investment in whiteness'. *Curriculum Inquiry*, *43*(3), 332–359. https://doi.org/10.1111/curi.12015

Battey, D., & Leyva, L. A. (2016). A framework for understanding whiteness in mathematics education. *Journal of Urban Mathematics Education*, *9*(2), 49–80. https://doi.org/10.21423/jume-v9i2a294

Bell Jr., D. A. (1975). Serving two masters: Integration ideals and client interests in school desegregation litigation. *The Yale Law Journal*, *85*(4), 470–516.

Bernal, D. D. (2002). Critical race theory, Latino critical theory, and critical raced-gendered epistemologies: Recognizing students of color as holders and creators of knowledge. *Qualitative Inquiry*, *8*(1), 105–126. https://doi.org/10.1177/107780040200800107

Bohonos, J. W. (2019). Including critical Whiteness studies in the critical human resource development family: A proposed theoretical framework. *Adult Education Quarterly*, *69*(4), 315–337. https://doi.org/10.1177/0741713619858131

Bonilla-Silva, E. (2006). *Racism without racists: Color-blind racism and the persistence of racial inequality in the United States* (2nd ed.). Rowman & Littlefield.

Brayboy, B. (2005). Towards a Tribal Critical Race Theory in education. *The Urban Review*, *37*(5), 425–446. https://doi.org/10.1007/s11256-005-0018-y

Buccalo, N. (2019). *The fire is upon us: James Baldwin, William F. Buckley Jr., and the debate over race in America*. Princeton University Press.

Castagno, A. E., & Lee. S. J. (2007). Native mascots and ethnic fraud in higher education: Using Tribal Critical Race Theory and the interest convergence principle as an analytical tool. *Equity & Excellence in Education*, *40*(1), 3–13. https://doi.org/10.1080/10665680601057288

Corces-Zimmerman, C., & Guida, T. F. (2019). Toward a critical whiteness methodology: Challenging whiteness through qualitative research. In J. Huisman & M. Tight (Eds.), *Theory and method in higher education research* (Vol. 5, pp. 91–109). Emerald.

Crenshaw, K., Gotanda, N., Peller, G., & Thomas, K. (Eds.). (1995). *Critical race theory: Key writings that formed the movement*. The New Press.

Cvencek, D., Nasir, N. S., O'Connor, K., Wischnia, S., & Meltzoff, A. N. (2015). The development of math–race stereotypes: 'They say Chinese people are the best at math'. *Journal of Research on Adolescence*, *25*(4), 630–637. https://doi.org/10.1111/jora.12151

Davis, J. (2014). The mathematical experiences of Black males in a predominately Black urban middle school and community. *International Journal of Education in Mathematics, Science, and Technology*, *2*(3), 206–222.

Davis, J. (2016). Free to conduct research of race and racism in my West Baltimore community. In R. T. Palmer, L. J. Walker, R. B. Goings, C. Troy, C.

T. Gipson, & F. Commodore (Eds.), *Graduate education at historically Black colleges and universities* (pp. 79–89). Routledge.

Davis, J. (2018). Redefining Black students' success and high achievement in mathematics education: Toward a liberatory paradigm. *Journal of Urban Mathematics Education, 11*(1–2), 69–77. https://doi.org/10.21423/jume-v11i1-2a359

Davis, J. (2021). A liberatory response to antiblackness and racism in the mathematics education enterprise. *Canadian Journal of Science, Mathematics and Technology Education, 21*(4), 783–802. https://doi.org/10.1007/s42330-021-00187-x

Davis, J. (2022). Disrupting research, theory, and pedagogy with critical race theory in mathematics education for Black populations. *Journal of Urban Mathematics Education, 15*(1), 9–30. https://doi.org/10.21423/jume-v15i1

Davis, J., & Jett, C. C. (Eds.). (2019). *Critical race theory in mathematics education.* Routledge.

Davis, J., & Martin, D. B. (2018). Racism, assessment, and instructional practices: Implications for mathematics teachers of African American students. *Journal of Urban Mathematics Education, 11*(1&2), 45–68.

Delgado, R., & Stefancic, J. (Eds.). (1997). *Critical White studies: Looking behind the mirror.* Temple University Press.

Delgado, R., & Stefancic, J. (2001). *Critical race theory: An introduction.* New York University Press.

Deloria, V. (1997). *Red Earth, white lies: Native Americans and the myth of scientific fact*. Fulcrum.

DiAngelo, R. (2018). *White fragility: Why it's so hard for white people to talk about racism.* Beacon.

Dumas, M. J., & ross, k. m. (2016). 'Be real Black for me': Imagining BlackCrit in education. *Urban Education, 51*(4), 415–442. https://doi.org/10.1177/0042085916628611

Fernández, L. (2002). Telling stories about school: Using critical race and Latino critical theories to document Latina/Latino education and resistance. *Qualitative Inquiry, 8*(1), 45–65. https://doi.org/10.1177/107780040200800104

Foste, Z., & Irwin, L. (2020). Applying critical whiteness studies in college student development theory and research. *Journal of College Student Development, 61*(4), 439–455. http://doi.org/10.1353/csd.2020.0050

Frankenstein, M. (1983). Critical mathematics education: An application of Paulo Freire's epistemology. *Journal of Education, 165*(4), 315–339. https://doi.org/10.1177/002205748316500403

Garcia, N. M., Lopéz, N., Vélez, V. N. (2018). QuantCrit: Rectifying quantitative methods through critical race theory. *Race, Ethnicity and Education, 21*(2), 149–157. https://doi.org/10.1080/13613324.2017.1377675

Greer, B. (2021). Learning from history: Jens Høyrup on mathematics, education, and society. In D. Kollosche (Ed.), *Exploring new ways to connect: Proceedings of the Eleventh International Mathematics Education and Society Conference* (Vol. 2, pp. 487–496). Tredition. https://doi.org/10.5281/zenodo.5414119

Greer, B., & Skovsmose, O. (2012). Seeing the cage? The emergence of critical mathematics education. In O. Skovsmose & B. Greer (Eds.), *Opening the cage: Critique and politics of mathematics education* (pp. 1–18). Sense.

Gutiérrez, R. (2013). The sociopolitical turn in mathematics education. *Journal for Research in Mathematics Education, 44*(1), 37–68. https://doi.org/10.5951/jresematheduc.44.1.0037

Iftikar, J. S., & Museus, S. D. (2018). On the utility of Asian critical (AsianCrit) theory in the field of education. *International Journal of Qualitative Studies in Education, 31*(10), 935–949. https://doi.org/10.1080/09518398.2018.1522008

Jett, C. C. (2012). Critical race theory interwoven with mathematics education research. *Journal of Urban Mathematics Education, 5*(1), 21–30. https://doi.org/10.21423/jume-v5i1a163

Jett, C. C. (2019a). Mathematical persistence among four African American male graduate students: A critical race analysis of their experiences. *Journal for Research in Mathematics Education, 50*(3), 311–340. https://doi.org/10.5951/jresematheduc.50.3.0311

Jett, C. C. (2019b). Using personal narratives to elucidate my CRT(ME) journey. In J. Davis & C. C. Jett (Eds.), *Critical race theory in mathematics education* (pp. 164–182). Routledge. https://doi.org/10.4324/9781315121192-10

Jett, C. C. (2022). Racial equity in mathematics: Reflections and recommendations from a Black mathematics educator. *Notices of the AMS, 69*(9), 1566–1569. https://dx.doi.org/10.1090/noti2547

Jett, C. C., Yeh, C., & Zavala, M. (2022). From argumentation to truth-telling: Critical race theory in mathematics teacher education. *Mathematics Teacher Educator, 10*(3), 223–230. https://doi.org/10.5951/MTE.2022.0007

Kokka, K. (2018). Radical STEM teacher activism: Collaborative organizing to sustain social justice pedagogy in STEM fields. *Educational Foundations, 31*, 86–113.

Kokka, K., & Chao, T. (2020). 'How I show up for Brown and Black students': Asian American male mathematics teachers seeking solidarity. *Race Ethnicity and Education, 23*(3), 432–453. https://doi.org/10.1080/13613324.2019.1664002

Ladson-Billings, G., & Tate, W. (1995). Toward a critical race theory in education. *Teachers College Record, 97*(1), 47–68. https://doi.org/10.1177/016146819509700104

Larnell, G. V., Bullock, E. C., & Jett, C. C. (2016). Rethinking teaching and learning mathematics for social justice from a critical race perspective. *Journal of Education, 196*(1), 19–29. https://doi.org/10.1177/002205741619600104

Lynn, M., & Adams, M. (2002). Introductory overview to the special issue Critical Race Theory and Education: Recent Developments in the Field. *Equity & Excellence in Education, 35*(2), 87–92. https://doi.org/10.1080/713845285

Malone, L., & Lachaud, Q. (2022). FaithCrit: Towards a framework of religio-spirituality in critical race theory. *Journal of Critical Race Inquiry, 9*(2), 93–109. https://jcri.ca/index.php/CRI/article/view/15370

Marshall, S. A. (2018). To sustain tribal nations: Striving for Indigenous sovereignty in mathematics education. *The Journal of Educational Foundations, 31*(1&2), 9–37.

Martin, D. B. (2008). E(race)ing race from a national conversation on mathematics teaching and learning: The National Mathematics Advisory Panel as White institutional space. *The Montana Mathematics Enthusiast,* 5(2&3), 387–398. https://doi.org/10.54870/1551-3440.1117

Martin, D. B. (2009). Researching race in mathematics education. *Teachers College Record, 111*(2), 295–338. https://doi.org/10.1177/016146810911100208

Martin, D. B. (2019). Equity, inclusion, and antiblackness in mathematics education. *Race, Ethnicity and Education, 22*(4), 459–478. http://dx.doi.org/10.1080/13613324.2019.1592833

Martin, D. B., & Gholson, M. (2012). On becoming and being a critical Black scholar in mathematics education: The politics of race and identity. In O. Skovsmose & B. Greer (Eds.), *Opening the cage: Critique and politics of mathematics education* (pp. 203–222). Sense. https://doi.org/10.1007/978-94-6091-808-7_10

Martin, D. B., Price, P. G., & Moore, R. (2019). Refusing systemic violence against Black children: Toward a Black liberatory mathematics education. In J. Davis & C. C. Jett (Eds.), *Critical race theory in mathematics education* (pp. 32–55). Routledge. https://doi.org/10.4324/9781315121192-4

Matthews, L. E., Jessup, N. A., & Sears, R. (2021). Looking for 'us': Power reimagined in mathematics learning for Black communities in the pandemic. *Educational Studies in Mathematics, 108*(1), 333–350. https://doi.org/10.1007/s10649-021-10106-4

McIntyre, A. (1997). *Making meaning of whiteness: Exploring racial identity with White teachers.* State University of New York Press.

Miller-Jones, D., & Greer, B. (2009). Conceptions of assessment of mathematical proficiency and their implications for cultural diversity. In B. Greer, S. Mukhopadhyay, S. Nelson-Barber, & A. B. Powell (Eds.), *Culturally responsive mathematics education* (pp. 165–186). Routledge. https://doi.org/10.4324/9780203879948-14

Museus, S. D., & Iftikar, J. (2014). Asian critical theory. In M. Y. Danico (Ed.), *Asian American society: An encyclopedia* (pp. 96–98). Sage. https://dx.doi.org/10.4135/9781452281889.n35

Nishi, N. W. (2021). White hoarders: A portrait of whiteness and resource allocation in college algebra. *The Journal of Higher Education, 92*(7), 1164–1185. https://doi.org/10.1080/00221546.2021.1914495

Phillips, S. L. (1998). Convergence of the critical race theory workshop with LatCrit theory: A history. *University of Miami Law Review, 53*(4), 1247–1256. https://repository.law.miami.edu/umlr/vol53/iss4/37

Powell, A. B., & Frankenstein, M. (1997). *Ethnomathematics: Challenging Eurocentrism in mathematics education.* State University of New York Press.

Shah, N. (2019). 'Asians are good at math is not a compliment': STEM success as a threat to personhood. *Harvard Educational Review, 89*(4), 661–686. https://doi.org/10.17763/1943-5045-89.4.661

Skovsmose, O. (2023). *Critical mathematics education*. Springer. https://doi.org/10.1007/978-3-031-26242-5

Solórzano, D. G., & Yosso, T. J. (2001). Critical race and LatCrit theory and method: Counter-storytelling. *International Journal of Qualitative Studies in Education, 14*(4), 471–495. https://doi.org/10.1080/09518390110063365

Solórzano, D. G., & Yosso, T. J. (2002). Critical race methodology: Counter-storytelling as an analytical framework for educational research. *Qualitative Inquiry, 8*(1), 23–44. https://doi.org/10.1177/107780040200800103

Stavrou, G. S., & Miller, D. (2017). Miscalculations: Decolonizing and anti-oppressive discourses in Indigenous mathematics education. *Canadian Journal of Education, 40*(3), 92–122. https://journals.sfu.ca/cje/index.php/cje-rce/article/view/2382

Stinson, D. W. (2017). Beyond White privilege: Toward White supremacy and settler colonialism in mathematics education. *Journal of Urban Mathematics Education, 10*(2), 1–7. https://doi.org/10.21423/jume-v10i2a348

Tate, W. F. (1993). Advocacy versus economics: A critical race analysis of the proposed national assessment in mathematics. *Thresholds in Education, 19*(1–2), 16–22.

Tate, W. F. (1995). Returning to the root: A culturally relevant approach to mathematics pedagogy. *Theory into Practice, 34*(3), 166–173. https://doi.org/10.1080/00405849509543676

Tate, W. F. (1997). Critical race theory and education: History, theory, and implications. *Review of Research in Education, 22*(1), 195–247. https://doi.org/10.3102/0091732X022001195

Tate, W. F., Ladson-Billings, G. & Grant, C. (1993). The Brown decision revisited: Mathematizing social problems. *Educational Policy, 7*(3), 255–275. https://doi.org/10.1177/0895904893007003002

Tellez, K., Moschkovich, J., & Civil, M. (Eds.). (2011). *Latinos/as and mathematics education: Research on learning and teaching in classrooms and communities*. Information Age.

Valdes, F. (1998). Theorizing outCrit theories: Coalitional method and comparative jurisprudential experience-RaceCrits, QueerCrits and LatCrits. *University of Miami Law Review*, 53(4), 1265–1322.

Zavala, M. D. R. (2014). Latina/o youth's perspectives on race, language, and learning mathematics. *Journal of Urban Mathematics Education, 7*(1), 55–87. https://doi.org/10.21423/jume-v7i1a188

19. Gender, mathematics, and mathematics education

David Kollosche, Daniela Steflitsch, and Kora Maria Deweis-Weidlinger

This chapter approaches the discipline of mathematics from the perspective of gender studies. It provides an overview of the gendering of mathematics and mathematics education based on aspects such as images of mathematics, achievement, representation, biology, cognition, learning preferences, classroom interaction, and belonging. These aspects are then critically addressed from a post-structural perspective on gender and mathematics. Special attention is paid on moral dilemmas in dealing with gender inequality in mathematics and on the question how the perspective of gender studies can enrich our understanding of mathematics.

Introduction

Why bother to look at mathematics and mathematics education from the perspective of gender? At a first glance, the political struggles for the recognition and non-discrimination of different genders seem to be very far away from the presumably objective and logical shores of mathematics. Daring a closer look, we might be surprised that we have entered a rose garden of thorny questions and paradoxes: Why is mathematics commonly regarded as a male domain? Why is there only one female winner of the Fields Medal, the highest decoration for success in mathematics research? Why is there a clear male majority of university professors of mathematics? And why are far fewer women than men pursuing mathematics-intensive careers? These are examples for

©2024 David Kollosche et al., CC BY-NC 4.0 https://doi.org/10.11647/OBP.0407.19

persistent inequalities, while school achievement in mathematics differs only slightly between girls and boys, while some countries even show higher achievements among girls, and while female mathematicians are just as successful as their male counterparts.

These differences between men and women in the field of mathematics demand an explanation for different reasons. Politically, they raise the question whether women in mathematics are being systematically discriminated against. Economically, motivating more women to pursue mathematics-intensive careers is often believed to be beneficial for a country's economic development. Didactically, we might want to ask if girls and boys have different resources and needs in the mathematics classroom and should be taught differently. Even for the sake of mathematics, an investigation into gender aspects of mathematics and mathematics education might be illuminating: Would not an answer to the question why mathematics is commonly regarded a male domain deepen our understanding of mathematics? Might not the hypothetical result that women do mathematics differently but are systematically excluded from higher mathematics call for a different way of doing mathematics? We shall return to these questions.

We know no other field of inquiry into mathematics education which would include as many theoretical perspectives and interest-based positions as the gender-and-mathematics discourse. For an intense experience of that variety, see the forty-two divergent peer commentaries on Camilla Persson Benbow's (1988) contribution to the topic in the journal *Behavioral and Brain Sciences*; and even further perspectives have been developed since. The questions that Betty Johnston (1995) raised give a quick taste of this melange:

> How does the imbalance manifest itself? What 'facts' are we using to help us see it, who collected them, for what purpose, on what evidence? What does 'good at maths' mean and how do we measure it? How do we construct our understanding of the 'facts'? How do we use it? And, finally, why do we care so very much that everyone should do mathematics? (p. 228)

Noteworthy of this research field is the nearly exclusive reliance on a men-versus-women dichotomy. There is hardly any research, especially nothing that we found to be a useful reference, about gender issues in mathematics and mathematics education, that would transcend that dichotomy and focus on non-binary identities. We think that it is an urgent research desideratum to open up the gender concepts used in mathematics education research, but we will not be able to achieve that in this chapter.

The next part of this chapter will deal with the concept of gender, which has proven to be an important concept for explaining and problematising differences between men and women. In the third part of this chapter, we will address differences and explanations regarding mathematics. We dedicate a separate part to post-structural perspectives on gender and mathematics, as we consider them to be extremely powerful but also surprisingly alien to common thinking. In the fifth part, we allow ourselves to place a critique of the common discourses on mathematics and gender by proposing a closer focus on a possible gender bias of mathematics itself. In the last part, we turn to moral questions in asking what could and should be done in mathematics and mathematics education as a consequence of the provided insights.

The concept of gender

The concept of gender has been developed as a counterpart to the concept of sex. The first section of this part of the chapter presents the gender concept and addresses how it can help us understand certain aspects of differences between men and women. The second section problematises the concept of gender from a post-structural perspective.

Sex and gender

The distinction between the concepts *sex* and *gender* has been established to point out that some differences between men and women are biological in nature while many others are social constructions and open to change. The term 'gender' has presumably been introduced to academia in the above sense by John Money (1955). His perspective on the cultural conditions of the differences between men and women became of wider academic and public interest to describe socially created inequalities.

We are usually able to assign the sex of a person with quite a certainty by detecting the external sexual organs, and we start doing so even before a baby is born. However, when it comes to the attribution of gender, it is usually more difficult or less clear. What we can perceive are gendered signs and forms of behaviour, for example a certain stature, shape of a face, haircut, dress code, certain ways of moving and talking, also interests in certain activities, such as in mathematics. In rare cases, we might be puzzled by seemingly conflicting signs and wonder in which pigeonhole to put that person, but in most cases, we easily assign a gender and, by doing that, activate certain expectations or gender-specific role models. These stereotypes can be clearly visible and obvious, as for example in a group photo of the all-male staff of a fire station. They can also be more subliminal, for example in television advertisements for cough syrup, when the image of the caring and nurturing mother and wife is implicitly conveyed in the marketing of medical products that are equally suitable for both sexes.

In the United States, an early influence in academia had been anthropologist Margaret Mead (1949) who showed that different societies ascribed different social roles to the sexes. Michael N. Friedan's (1963) *The Feminine Mystique*, documenting the dissatisfaction of housewives, became a bestseller and made feminism an issue of the general public in the US and beyond. Such contributions opened an intellectual space in which alternative roles for women in society could be envisioned and expressed. The analysis and change of gender roles were at the heart of the subsequent women's right movement in the second half of the twentieth century. This movement included not only political struggle for women's suffrage, equal access to education and professions and sexual autonomy, it also led to a critique of theories that positioned women as inferior to men and motivated research on femininity.

Janet Saltzman Chafetz (2006) provides an impressive overview of the variety of theoretical approaches to gender theory. Among this variety, social learning theories have become widely used to describe gender as a social role one is educated into. As Jennifer Marchbank and Gayle Letherby (2014) put it, social learning theories assume 'that girls and boys learn gender-appropriate behaviour from birth as we are all surrounded by gender socialisation messages from our families, peers and the media', and they have 'been the basis of most sociological work on masculinity and femininity, mainly focused on determining how we learn, internalise, and then recreate gender stereotypical roles' (p.

9). Social learning theories make us aware that, having undergone education in a gender-biased society, we are always-already part of this social structure.

Problematising gender

The gender discourse has been fundamentally criticised from a post-structural perspective. This perspective is closely connected to the French philosopher Michel Foucault, who worked out the discursive constitution of reality in such fields as mental illness (Foucault, 1961), delinquency (Foucault, 1975), sexuality (Foucault, 1976), or patterns of thought more generally (Foucault, 1966). The main line of argumentation is that reality is not just out there in a pre-structured form but constructed by humans in discourse. Reality, therefore, is not objective but ambiguous, constituted differently from different perspectives in different times and places, and a product of interests and power. To take the example of mental illness, Foucault (1961) worked out how the idea of mental illness is the product of a modern discourse which is designed to exclude from society any forms of behaviour deviating from the modern rationalist ideal and tempting us to fall back to a pre-rationalist existence by abandoning our self-discipline. Thereby, the construction of reality does not only include the assertion of a certain discourse, it also includes the organisation of social practices and institutions, the legitimisation of specific arguments, and the validity of a certain body of knowledge. In the case of mental illness, institutions such as asylums, academic discourses such as psychology, the distinction of experts on mental illness, and practices of removing the mentally ill from the public sphere work together in a complex web that constitutes what mental illness means for us today.

In his later work, Foucault (1982) investigated how power is executed on people by discourses. He explained that discourses do not simply direct and forbid, rather they create temptations and design rooms in which to position the self. People then make these discourses their own by filling them out individually within the given boundaries, thus becoming an accomplice of the discourse itself. As an example, consider speaking in your mother tongue: sure enough, you have a distinct style of expressing yourself in it, but there are also certain boundaries you would not cross, certain rules you will stick to, rules you would even demand others to follow. Foucault (2007) stressed the possibility of resistance against discourses that govern us and understood critique as the art 'not the be governed *like that*, by that, in the name of those principles, with such and such an objective in mind and by means of such procedures, not like that, not for that, not by them' (p. 33, original emphasis). Now, from this post-structural perspective, it would seem promising to direct one's attention at the discursive constitution of women, men and gender.

In *Gender Trouble* and her later *Undoing Gender*, Judith Butler (1999, 2004) approached the gender discourse from a post-structural perspective. With traditional feminism she shared the assumption that gender roles are cultural constructs and designed in a way to secure the social privilege of men. However, Butler's view differed from traditional feminism in some crucial points. Foremost, studies asking for the nature of femininity, for example with the goal of a more girls-friendly organisation of education, reproduce the idea that we are *born* into our roles as males and females. In contrast to that assumption, Butler argues that even sex is discursively constructed. As Penelope Eckert and Sally McConnell-Ginet (2003) pointed out, 'there is no obvious point at which sex leaves off and gender begins, partly because there is no single objective criterion for male or female sex' (p. 10). For example, the configuration of genitals is sometimes ambiguous, and controversially discussed medical procedures are systematically being undertaken to adjust the configuration of genitals to social expectations. In the light of the variety of possibly inconsistent biological features used to determine sex, including anatomical, genetical and hormonal features, and in the light of the sometimes ambiguous nature of these features, the decision which sex to assign to a person is ultimately social.

Butler (1999) problematised that traditional feminist studies, by adhering to the dichotomy of the two genders, proved unable to explain cases of third, mixed, and changing gender. Further, neither biology nor social learning theory can explain the different varieties of masculinity and femininity. Neither can they explain instances of individual resistance to gender roles, of testing their boundaries, playing with them, maybe even redefining them.

Post-structural theories redirect our attention from the question what could typically count as masculine or feminine to the question how the necessarily political discourses on gender interact with the constitution of our identities, with the goals we deem approachable, the roles we feel comfortable in, and the expectations directed at us. Our various discourses often include gender roles and direct men and women to specific positions within the discourse, for example to positions of ability or disability. Deviations from such discursively set roles are difficult to accept for others sharing that discourse, but they may also lead to conflict within the respective person. For instance, it may be difficult to position oneself as a loving mother and well-organised housewife and simultaneously as following a professional career. Heather Mendick (2005) calls the inner negotiation of the different discourses 'identity work'.

At the same time, the post-structural perspective points out that we are not only inevitably educated into gender discourses, but, from that position within the discourse, always-already accomplices of gender discourses and 'doing gender' (Butler, 1999, p. 41). This perspective does not present women as passive victims of gender stereotypes but assumes an active role of women in positioning themselves in gender discourses. This shift is not meant to reassign the blame for gender inequalities to women. Instead, it is meant to highlight the ways in which we could do gender differently. Eventually, the benefit of understanding gender as a product of discourse is that gender differences are not inescapable but open to change. From a post-structural perspective, promoting such change begins with a deconstruction of dominant discourses and a search for possibilities to think differently. At least in academia, though not that easily in the social pressures of daily life, we are not reduced to the decision where to position us in given gender discourses but can eventually decide 'not to be governed like that'.

Differences and explanations

How do men and women differ at all concerning mathematics? Before we give a short presentation of gender differences in mathematics, we want to address the question whether investigating such differences bears dangers. In her insightful book *Sex Differences in Cognitive Abilities*, Diane F. Halpern (2012) explained:

> Many psychologists and others are opposed to any comparisons of women and men. Much of the opposition is based on the fear that when differences are found, the data will be interpreted and misused in ways that support a misogynist agenda or unwittingly provide support for the idea that there are 'proper roles' for men and women. (p. 3)

Roberto Ribeiro Baldino (2000) was surprised by the acceptance and tolerance for his conference presentations where he explained that a gene had been identified which allowed for higher mathematical understanding in the first place, and for which 'it has been possible do determine that only 10 to 15 per cent of men and 5 to 8 per cent of women are [...] carriers', explaining among other inequalities 'why the majority of mathematicians are men' (p. 145). The whole theory was bogus, the presentation an experiment conducted with a clueless audience of researchers in mathematics education, but, aside from much else, it showed that seemingly scientific explanations for differences are often too uncritically accepted, maybe even welcomed, and eventually set to use in the legitimisation of inequalities. We assume that any contestation of differences is dialectical in nature, on the one hand assisting in the explanation of inequalities, while on the other hand forming a basis for their legitimisation. We will have to bear in mind this twofold nature of stated differences if we seek to avoid being trapped by explanations that, from a different perspective, turn out to be questionable and problematic.

Paul Ernest (2007) warned against taking up a one-dimensional perspective in which we assume that there is *one* gender problem in mathematics. He found that what is addressed when discussing gender and mathematics is a whole array of different inequalities, and that authors often talk about rather different issues under similar headings. Ernest claimed that what poses a gender-related problem is eventually a question of perspective and interest. We can conclude that talking about *the* gender problem in mathematics is diffusing the discussion and may even be proposing that gender itself is the problem, rather than facing a range of different problems, which shine up from different perspectives and become problematic for people with specific interests. Here, we decided to distinguish the following perspectives on gender differences in mathematics:

1. Gender differences in images of mathematics.
2. Gender differences in achievement.
3. Gender differences in representation.
4. Gender differences in biology.
5. Gender differences in cognition.
6. Gender differences in learning preferences.
7. Gender differences in classroom interaction.
8. Gender differences in belonging.

These areas can serve as an explanatory basis for each other, but they are not easily brought into a linear order of cause and effect. For example, gender differences in learning preferences can explain differences in achievement *and* demand an explanation themselves. Consequently, gender-oriented studies in mathematics education need to navigate through a complex web of interrelating issues, which shine up with different intensity when different interests come to play. Ernest (1995) proposed to understand gender inequality in mathematics as a vicious cycle of mutually reinforcing phenomena. We will address this idea in the section titled 'Where is mathematics?'. On the following pages, we wish to address the various perspectives on gender differences in their own right. Afterwards, we will revisit these discourses more critically from a post-structuralist perspective.

Images of mathematics

For a long time and in many places, mathematics had been considered to be male. Even today, mathematics and mathematics-related domains are still stereotyped as masculine and are therefore difficult to reconcile with female gender roles. Public images are influences by role models, with whom people can identify more or less easily. In mathematics, the names and faces of such role models are predominantly male (and White, one might add). There are several reasons for that. One reason is that the successes of female mathematicians have often been withheld in the history writing of predominantly male historians. The most outstanding ancient example is Hypatia of Alexandria, whose story has only recently become a vivid field of historical study. A more recent example is the controversy around the impact of Albert Einstein's wife Mileva Marić on her spouse's work in theoretical physics.

However, the number of published contributions of women to mathematics have indeed been few compared to those of men. The reason here is not necessarily that women are less interested in mathematics. Teri Perl (2010) examined an annual magazine published in England from 1704 to 1841, which contained enigmas, queries, and mathematical questions and was aimed specifically at women. Its existence over decades suggests that there was a wide market for such publications among women and that women were indeed interested in mathematics. The main explanation rather seems to be that, until fairly recently in human history, patriarchal structures in society have not allowed or at least strongly hindered women to appear as an independent intellectual, to receive the necessary education and formal qualification, to have their voices heard and their work published, and to secure paid positions in mathematics. In this vein, Reuben Hersh and Vera John-Steiner (2011) tell the stories of Sophie Germain, Sofya Kovalevskaya and Emmy Noether. Today, as we shall see in a later section on 'Representation', female mathematicians are no longer the exception but still a minority.

Stereotypical images about mathematics and mathematicians are transported through popular media. Gilah Leder (1995) analysed articles of well-known newspapers in Australia and Canada regarding gender equity. These articles confirmed the prevailing stereotypical views about male-dominated power structures much more often than they questioned them. She concluded that 'it appears that the subtle messages conveyed in the popular press are consistent with small but consistent differences in the ways females and males perceive and value mathematics and related careers as appropriate for themselves' (p. 125).

Mendick (2005) analysed movies presenting mathematics and mathematicians and stated: 'This dominant discourse around mathematicians in popular culture depicts them as boring, obsessed with the irrelevant, socially incompetent, male and unsuccessfully heterosexual' (p. 214). Movies with mathematical contents (e.g., *A Beautiful Mind, Enigma, Good Will Hunting, Pi*) make use of such stereotypical images and at the same time shape our image of a 'typical' mathematician. In all films, the protagonists struggle with mental illness that is directly or indirectly related to mathematics. In these movies, mathematics is presented as something where femininity does not fit in. However, in a later analysis of fictions published after the financial crises of 2008, Mendick (2017) finds that mathematics has been portrayed in more diverse and critical ways. *Hidden Figures*, a movie about a group of female Afro-American mathematicians contributing crucial work to NASA projects, was too new to find its way in Mendick's paper but serves as an outstanding example for that shift.

Stereotypical images about mathematics and mathematicians are also reproduced in and influence education. Natthapoj Vincent Trakulphadetkrai (2017) analysed the representation of girls and women in sixty-four Anglophone mathematical picture books produced for private education and entertainment. He found that girls and women were considerably underrepresented. In a study on images of mathematics in the mathematics classroom, Mary Schatz Koehler (1990) concluded that the image of mathematics as a male domain was reinforced by the portrayal of mostly male mathematicians and the use of mostly masculine context in test questions.

In a mathematics course designed for re-entry into science and technology fields, Zelda Isaacson (1990) explored the reasons why the attending women (all in their twenties and thirties) chose to opt out of mathematics somewhen in their school career. Some explained that as a woman you are considered weird if you like maths. One example of a conversation with colleagues about what course one of these women is in, shows this clearly: 'They look at you in absolute horror, and that's the end of the conversation' (p. 24). Within these conversations, the impact of stereotype-based family pressure and peer influence crystallised. Another woman described mathematics as a subject where competences such as creativity or imagination are not used and thus women who are more likely to show such skills (in her view) cannot connect with mathematics. While we will return later to the question of how women negotiate their identities in the light of such images of mathematics, Isaacson's study showed that images of mathematics do have an impact on women's choices. Other evidence, which we will only mention here, comes from a statistical analysis of gender-science stereotypes and sex differences in science and math achievement in thirty-four countries, which established 'that nation-level implicit stereotypes predicted nation-level sex differences in 8th-grade science and mathematics achievement' (Nosek et al., 2009, p. 10593).

Achievement

Mathematics is often considered a discipline in which men show higher proficiency than women. Statistical data reveals that this is the case in some environments but no general phenomenon. As an illustration, we can look at the results of large-scale assessment regimes such as PISA. In PISA 2012, the latest PISA run with an emphasis on mathematics when we wrote this chapter, boys had significantly higher mean scores than girls in mathematics performance in thirty-seven of the participating sixty-five countries, reaching differences in the mean scores of boys and girls as high as 22 score points in Austria and 25 in Colombia and Chile (OECD, 2014, p. 305). However, there were also six countries in which girls performed significantly better than boys, including Iceland, where girls on average scored 6 points more than boys, and Jordan, where the difference amounted to 21 points. Summing up the data from the thirty-four participating OECD states (the only average values provided by the study), the mean score of boys is 12 points higher than that of girls. What do these numbers tell us? First, the PISA score is normalised with a standard deviation of 100. 12 score points amount to only 12% of that deviation in score points. Second, the mean scores between comparable countries often differ more drastically than between girls and boys in the countries. For example, Belgium scored 20 points higher than neighbouring France, Switzerland scored 25 points higher than neighbouring Austria, and Finland scored 40 points higher than neighbouring Sweden. On average, Finnish girls scored 37 points higher than Swedish boys. Other comparative assessments yield similar patterns (e.g., Hanna 1989, 1994; Ernest 2007). We conclude that differences in mathematics achievement in favour of boys can be detected but appear to be very small and possibly negligible when compared to other variations.

Countries in which girls perform significantly better than boys prove that higher performance by boys is no general phenomenon and indicate that local factors such as culture and school systems must play a crucial role. Obviously, it would be difficult to explain these differences between countries by biological theories, which claim universal validity, alone (Nosek et al., 2009). In a study with 2300 school leavers in England, Geoffrey Driver (1980) compared the mathematical achievement of students with English descent with that of students with West Indian descent. It was not very surprising that boys of English descent slightly outperformed girls of English descent. However, West Indian girls outperformed English boys and West Indian boys performed at the level of English girls. Driver explained these gender differences in performance between students of English and students of West Indian descend by the different social roles of men and women in both societies.

Admittedly, one might contend that the mathematical literacy defined and measured by PISA and school mathematics as assessed in Driver's study are something different than higher secondary or even tertiary mathematics, where gender differences could show differently. This is why we compared the number of enrolments in tertiary education programs (including Bachelor, Master and doctoral studies) in mathematics and statistics with the number of graduations in these programs within the European Union as documented by the European Statistical Office Eurostat. In 2014, the latest year for which we have union-wide data, there were 117064 women and 147901 men enrolled, while 25074 women and 27871 men graduated in these programs (Eurostat, 2019a, 2019b). That yields 0.214 female graduates per female enrolments, whereas the ratio is only 0.188 for men. Although the statistics include no information about the obtained grades, women appear to be more successful students of mathematics than men. We conclude that there is no evidence that men would generally achieve better in tertiary mathematics either.

Representation

As the Eurostat data presented in the last paragraph documented, considerably less women than men enrol in tertiary education in mathematics, and slightly less women than men graduate from such programmes. Such an underrepresentation of women can be found in many forms. We already addressed the fact that women are underrepresented in depictions of mathematics in popular media, but underrepresentation can also be found in schools and in research.

When we look at academic positions, the proportion of women, depending on the ascending level of qualification, ranges from 35% (lower level) to 15% (highest level) in STEM fields compared to 46% and 24% across all scientific fields in the European Union (European Commission, 2019). In the US, an average of 30% of PhDs in mathematics were awarded to women between 2005 and 2008 and in 2005 only 9% of all full-time tenured professors in doctoral-level mathematics departments were women (Popejoy & Leboy, 2012). Even more severe underrepresentation of women is reported from African countries (Gerdes, 2006; Masanja, 2010). Apart from some exceptions, this list could easily be extended further. In summary, both horizontal and vertical segregation can be observed. This means that not only the proportion of women is significantly lower in mathematics-related areas than in other areas, but also that women are less frequently represented in higher status areas than in lower status areas.

The underrepresentation of women in mathematics can partly be explained historically. Isolde Kinski (1994) studied the history of the underrepresentation of women in mathematics from a German perspective. Until the last century, education was reserved for only a few people and a privilege of men. The distribution of tasks between the sexes and the societal roles associated to the sexes restricted women to the domestic sphere and thus excluded them from higher education. It was argued that dealing with science or mathematics was against women's nature. Even when, from around 1820 onwards, bourgeois daughter schools became more common in the German-speaking world and elsewhere, these were limited to teach girls only elementary arithmetic necessary for keeping the household.

At the beginning of the twentieth century, there were initial discussions about girls' access to school-leaving examinations and thus access to universities. Even though women were officially admitted to study in German countries from 1900 onwards, they had to fight different obstacles for decades. In those days it was unthinkable for many universities to award doctorates to women. For example, Christine Ladd, who was the first American woman to fulfil all formal requirements for a PhD in 1883, was not officially recognised until forty-three years later (Green, 2001). Even in 1981, only about 6% of the thousand speakers at the meetings of the American Mathematical Society were women

(Kenschaft, 1982). Nevertheless, with the admission of women to the universities, the contents of the higher girls' schools changed. Curricula for girls' and boys' schools were unified and coeducation was introduced in the course of the twentieth century. Since then, representational inequalities have decreased but they still exist to a considerable extent.

Sapna Cheryan (2012) suggested that 'seeking out math-related careers is still a gender role violation for women' (p. 184, without original emphasis). She explained that the public image of mathematics as a male domain and activity remained an obstacle for the perusal of mathematical careers by women. However, there are other possible explanations for the underrepresentation of women in mathematics which we will address in the following sections.

Biology

Biological explanations for gender differences in mathematics often focus on abilities in spatial visualisation. Three different approaches are repeatedly found in the literature: a cerebral explanation, a hormonal explanation, and a genetic explanation. The cerebral explanation refers to the neuropsychological effect of lateralisation. This describes the unequal distribution of individual functions between the two halves of the brain. Studies show that an asymmetrical organisation of the brain is more pronounced among men (Halpern, 2012). Therefore, women tend to use both brain hemispheres when solving exercises requiring spatial abilities while men particularly use the right brain half, to which skills such as spatial thinking and numerical reasoning are attributed. However, there is no evidence that one way of using the brain for spatial ability tasks is more successful than the other. Other studies show that the level of certain hormones might have an impact on the performance in spatial ability tests (Hampson, 1990; Hampson & Rovet, 2015). Approaches which linked spatial ability to specific genes have been found unconvincing (Boles, 1980), but the assumption that men and women might have different genetical dispositions for spatial ability as a result of their archaic roles as hunters and fighters persists among some scholars (Geary, 1998; Halpern et al., 2007).

Despite these attempts to explain sex-related differences biologically, different meta-studies on biological differences between men and women concluded that, in most areas, differences are minimal, if not negligible. That includes fine motor skills, mental rotations, spatial perception and visualisation, mathematical ability, computational skills and understanding of mathematical concepts (Hines, 2010) but also general intelligence (Colom et al., 2000) as well as language skills, communication behaviour, computer use, self-esteem, aggression, helpfulness, leadership skills and sexual behaviour (Hyde, 2005). Consequently, Janet Shibley Hyde (2005) proposed 'that males and females are similar on most, but not all, psychological variables' (p. 581). Obviously, biology appears not to explain much.

Cognition

Some scholars argue that gender differences in academia derive from the fact that men and women think differently. One of the most provocative studies in this direction is Mary Field Belenky, Blythe McVicker Clinchy, Nancy Rule Goldberger, and Mattuck Jull Tarule's *Women's Ways of Knowing* (for a similar study resulting in a somewhat different categorisation see Magolda, 1992). Belenky and colleagues (1997) argued that women think differently than men and that academia is usually organised in a way that values only masculine ways of thinking with the effect of silencing women. The psychologists conducted interviews with a wide variety of US-American women to learn about 'the way they perceived themselves and the world around them' (p. 4):

> What is truth? What is authority? To whom do I listen? What counts for me as evidence? How do I know what I know? Yet to ask ourselves these questions and to reflect on our answers is more than an intellectual exercise, for our basic assumptions about the nature of truth and reality and the origins of knowledge shape the way we see the world and ourselves as participants in it. (p. 3)

Belenky and colleagues (1997) described *received knowledge*, which relied solely on authorities and was the main form of knowledge organisation for many women before they developed *subjective knowledge*. The latter form of knowledge is the first step towards building confidence in oneself and includes a drastic refusal of authoritarian truth claims:

> Subjectivist women distrust logic, analysis, abstraction, and even language itself. [...] The fervor with which subjectivist women draw

> sharp lines between intuitive knowledge and what they assume to be the impersonality of abstract thought harks back to the dogmatism and either/or thinking characteristic of the women we described in earlier chapters. It is not that these women have become familiar with logic and theory as tools for knowing and have chosen to reject them; they have only vague and untested prejudices against a mode of thought that they sense is unfeminine and inhuman and may be detrimental to their capacity for feeling. (p. 71)

Some women in the study felt a need to abandon purely subjectivist positions for the sake of universal perspectives which would, for example, fulfil the requirement of academic or other interpersonal debate. However, the way of knowing adopted by women differed from that of men in what Belenky and colleagues (1997) termed *procedural knowledge*. Procedural knowledge focused not on how things are but on how something can be done, including an appreciation of different perspectives on situations. Here, objectivity was gained not by assuming that there is one true way but by leaving subjective positions and opening up for a variety of perspectives. However, the authors stressed that the adopted procedures are most often man-made and might bear in them a gender bias, which allows for answering questions relevant for men rather than questions relevant for women. Such a bias might be found in a very specific example, but it was also identified on a very general level: Belenky and colleagues reported men to strive for *separate knowing*, which positions the learner in a distance to the object of learning and looks for procedures for its manipulation, whereas women are found to strive for *connected knowing*, which is a very personal endeavour to find ever new procedures to understand the object in question. Women, especially those engaged in academia, were found to also perform separate knowing, but they often reported to find it meaningless or to have turned to connected knowing later. A last way of knowing, termed *constructed knowledge*, assumes the post-structuralist position that knowledge is but a construction. Constructers of their own knowledge embark on the mission to unify valuable input and subjective positions into a narrative that personally makes sense and still meets the requirements of successful communication with others.

Discussing the relevance of women's ways of knowing to mathematics education, Olive Chapman (1993) argued that a positivist 'view of mathematics tends to dehumanize or objectify it, thus limiting it to a framework with characteristics that are more compatible with separate knowing than connected knowing' and that 'all of the circumstances that could facilitate connected knowing are stripped away' in traditional teaching settings with the result of silencing connected and mostly female knowers (pp. 208–209). She demanded 'that mathematics be reconceptualized to reflect its humane features and mathematics classroom processes revised to facilitate the characteristic ways of knowing of both males and females' (p. 209), a demand that we will return to. We recommend Joanne Rossi Becker's (1995) more detailed account of what it might mean to allow for connected knowing in the classroom.

The work of German mathematics educator Inge Schwank departs from different theoretical frameworks but could be interpreted as an application of the insights presented above. In her analysis of problem-solving strategies in Information Technology studies and mathematics, Schwank (2002) identified two typical cognitive approaches, which she terms *predicative* and *functional* thinking:

> The label *predicative* was used to characterize a problem solving behaviour highly orientated at and sensible for features, relations and judgements, whereas the label *functional* was used to characterize a problem solving behaviour highly orientated at and sensible for courses, modes of actions and effects. (p. 489, original emphasis)

We find it striking that predicative thinking as described by Schwank aims at understanding very much in the sense of what Belenky and colleagues (1997) called procedural knowledge, while functional thinking as described by Schwank aims at manipulation very much in the sense of what Belenky et al. called separate knowing. Given that analogy, it does not come as a surprise that Schwank (1994) reported that women tend to think mostly in a predicative way, while the majority of men seems to apply a functional way of thinking. A small proportion of men was found to prefer predicative thinking but functionally thinking girls seem to be a real exception. However, there are people who cannot be assigned to either one or the other way of thinking.

Through the analysis of problem-solving behaviour, electroencephalogram (EEG) patterns, and eye movement in experiments, Schwank (1999, 2002) showed that the way of solving a task within logical reasoning differs greatly with the method of thinking. There are tasks where one method of thinking is more successful than the other, depending on the way the task is presented. Broken down to mathematics classes, it also seems to be decisive how the teacher articulates help when students have problems in solving tasks. In contrast to the functional thinker, who likes to elaborate solutions step by step with active trying, it will not be quite helpful for a predicative thinker to tell him or her to 'just give it a try' (Bischof-Köhler, 2002/2011). Students who are using a predicative strategy tend to think about the whole problem with all its aspects before they start working on it. Teachers largely attributed this approach of solving problems to the female students in their class and therefore often characterised the girls as being insecure. Hence, we can imagine that the way of thinking might also explain why girls and boys differently participate in class. While boys as a part of their way of thinking just try to find solutions to the problem and accept wrong answers along the way without being discouraged, girls might need some time to grasp a problem in its entirety and to build up the relations between single elements. The reason why boys and girls act differently in mathematics lessons could, therefore, be partly due to different ways of thinking.

Halpern and colleagues (2007) suggested that there might be sex differences within the underlying cognitive processes, as there seem to be specific tasks where females perform better than males and *vice versa*. Examples where women tend to excel men are language production, reading, writing, and algebra. These are all tasks requiring fast retrieval of information stored in long-term memory and the use of language to create connections. Males, on the other hand, seem to use strategies focusing on the maintenance and manipulation of mental representation in working memory, letting them outperform females in mathematical problem solving, mental rotation or spatial perception tasks. This approach seems to largely coincide with that of Schwank, suggesting that girls and boys might just use different strategies for solving problems. However, it is not possible to say with certainty to what extent these findings really influence performance in mathematics class. The step from cognitive psychology to mathematics teaching and learning does not seem to have been sufficiently explored and one can only speculate to what extent the different cognitive processes lead to gender differences in mathematics achievement or cause different degrees of interest in mathematics.

Learning preferences

The idea that people use different styles of thinking and that these styles are used with different frequency among men and women, quickly leads to the question of how far classrooms are organised in a way that welcomes both styles. Jo Boaler (1997) took up that question. She followed the cohorts moving from school years 9 to 11 in two schools over the course of three years and used ethnographic methods, including classroom observations as well as questionnaires and interviews with students and teachers, to learn how students experience mathematics education.

One issue Boaler pointed out is that addressing a 'lack of understanding of the mathematics they encountered in class [...] was particularly acute for the girls, not because they understood less than the boys, but because they appeared to be less willing to relinquish their desire for understanding' (p. 112). For example, in an interview, a student called Marsha explained that 'you have to work it out and you get the right answers but you don't know what you did, you don't know how you got them' (p. 114), whereas a student called Gary explained that 'once you know how to do it, you're away' (p. 115). Apparently, students such as Marsha strive for what Belenky and colleagues (1997) called connected knowing, while students such as Gary strive for separate knowing.

Boaler (1997) pointed out that the use of different ways of knowing interacts with teaching styles. Textbook work, which offers only one perspective on the mathematical content and usually one approach to performing procedures, is usually preferred by boys. In contrast to that, girls preferred working with individualised booklets and in groups:[1]

> The reasons that the girls liked these approaches were related to the freedom they experienced to use their own ideas, work as a group or

1 In a similar study in Germany, female students were reported to prefer to work with the textbook as it provided orientation, whereas male students preferred to work without the textbook (Jahnke-Klein, 2001, p. 119), which proposes that reoccurring and singular gender preferences have yet to be discriminated.

> work at their own pace. All these practices, the girls claimed, gave them access to a depth of understanding that textbook work denied them. (p. 114)

Boaler also discussed the desire expressed by many boys to solve problems quickly. This competitive desire to excel in speed and number, or at least not to fall behind, reduced the need for knowledge to knowing *how* to solve the involved problems and stands in opposition to a wish for understanding. Teachers quickly pacing through the mathematical contents are then clearly meeting the desires of separate knowers and disregarding the needs of connected knowers.

Some of Boaler's findings had already been reported by Jacquelynne Eccles (1989), who stated that competitive activities, drills, and practices are attributes of classrooms that produce high sex differences, while classrooms with low sex differences in mathematics tend to be more co-operative and supportive. In her course for re-entrants, which was attended only by women, Isaacson (1990) worked mainly in the form of group work, which most women reported to be the decisive difference to mathematics lessons at school. The participants acknowledged the importance of this form for building deeper understanding through explaining contents to other group members and discussing their answers until all agree. Boaler's study coincided with a German study by Sylvia Jahnke-Klein (2001). She obtained similar results and posed the question whether boys, who seek technical understanding and want to move on faster, appear more gifted or higher achieving than girls, who want to dwell on contents to understand them thoroughly.

It becomes apparent that different styles of teaching are to a different extent able to meet the desire for learning of connected knowers. Boaler (1997) reported a number of cases where girls with a high potential in mathematics hardly participate in mathematics lessons for reasons closely associated with unfitting ways of knowing. As this 'disparity [...] was most acute for the highest ability girls' (p. 123), Boaler raised the question if unfitting teaching styles could be responsible for unequal achievement and representation of girls in high-ability environments.

On the other hand, Boaler reported of boys in reform-oriented classrooms with open teaching approaches, who complained that 'they wanted more structure in their work, they wanted someone to tell them what to do' (p. 120). Obviously, there are also teaching styles that do not meet the wishes of separated knowers. However, Boaler added that these boys were able to adjust to the different teaching styles within one- or two-years' time.

Classroom interaction

There are numerous studies on the behaviour of both teachers and pupils in mathematics lessons. Many of these studies show that teachers treat boys more favourably than girls and that boys act differently from girls. This may not only lead to the conclusion that girls are disadvantaged by the interactions, structure, and climate in the classroom, but might also have an impact on the performance, the choice of course, and also on the motivation and self-image of students.

While different treatment can already be found in interactions with children who do not yet attend school (Olson et al., 2010), most studies have been conducted with students in secondary school education. Becker (1981) reported that classroom environment seems to be more supportive of males than of females both academically and emotionally. Interactions on a non-academic level such as joking are much more common between teachers and male students. Females do not seem to have an active role during class and seem to react to teachers' bonding with boys by becoming even more passive. Becker observed that teachers seemed to reinforce the traditional view of mathematics as a male domain, for example through language or examples used for explanations. Therefore, the identified class environment was not considered as a positive way to stimulate girls to continue their study in mathematics. In contrast, it seemed to have led girls to behave in ways that reinforce teachers' expectations of male superiority.

Koehler (1990) confirmed that boys receive more help from teachers than girls, they are more often involved in conversation with teachers, they receive more feedback on their behaviour from teachers, and they are provided more critical feedback on their work outcomes by teachers. Joachim Tiedemann (2002) documented that teachers attributed higher mathematical abilities and higher effort-resources to boys than to girls. Elizabeth Fennema, Penelope L. Peterson, Thomas P. Carpenter, and Cheryl A. Lubinski (1990) argued that the potential in mathematics of boys is usually overestimated, while that of girls is rather underestimated.

By examining teacher perceptions about their students' mathematical ability by letting them classify if they consider maths classes as too easy or too hard for them, Catherine Riegle-Crumb and Melissa Humphries (2012) found that especially on mid-level classes, White females are more likely to be judged as being in a course too difficult for them.

Helga Jungwirth (1991) investigated how boys and girls interact in teacher-centred classroom conversations, which are still widely found in everyday school life in the German-speaking part of the world. Jungwirth does not refer to the work of Belenky and colleagues (1997), but analogies in the results are obvious. She found that boys are more likely to respond to open and ambiguous questions or to give fragmentary answers, on which this teaching method basically thrives, while girls hardly respond at all to ambiguous questions or try to answer questions completely. This tends to disrupt the structure of this method and might, therefore, let them appear less competent. Furthermore, boys tend to hide their lack of knowing by dismissing it as a little mistake, taking up the teacher's advice and corrections, while girls tend to emphasise their lack of knowing by sticking to their solution in a desire for thorough understanding. However, this mismatch of conversational expectations might give the impression that girls are less competent than boys. Jungwirth stressed that girls adjust to such classroom cultures, so that changing classroom culture without addressing the role expectations of girls might be futile.

It is conceivable that certain aspects mentioned above show other patterns in single-sex schools, whose effects have been investigated in several studies (Becker, 2001; Delon, 1995; Hiddleston, 1995; Lee & Anderson, 2015; Morrow & Morrow, 1995; Prendergast & O'Donoghue, 2014; Thompson, 1995). Although the research interests and approaches in the various studies are quite different, most studies show that girls benefit from single-sex schooling in mathematics education. The question is if this benefit stands even when girls engage with mathematics together with boys in future situations.

Jessica Brooke Ernest, Daniel L. Reinholz, and Niral Shah (2019) put on record that men appear more competent in mathematics than women also in student-centred post-secondary education in mathematics. They were able to show that women prove their mathematical competence in small-group discussions and in side talk and that they participate in these discussions almost equally as men. However, in many cases their ideas do not find their way into public plenary discussions, which means that these discussions are not only dominated by men, but also that women's mathematical competence remains invisible.

Regarding the gender-specific treatment of women and men in higher education in mathematics, studies by Irene Pieper-Seier (2009) not only showed that women report more experiences of discrimination as their level of qualification increases. Moreover, personal support within the academic community is found to be particularly decisive for success but more often offered to men than to women, possibly due to a culture of self-sex support.

Belonging

Experiences based on the dimensions of gender and mathematics discussed above add up to very different feelings of belonging to mathematics, often to the disadvantage of girls and women. Catherine Good, Aneeta Rattan, and Carol S. Dweck (2012) pointed out that the sense of belonging acts as a contributory factor when it comes to choosing maths courses or a career in this field. They found that the sense of belonging to maths is a strong predictor of the intent to pursue maths and to remain in the discipline. Feelings of belonging are more than changing the perspective from the structural effects of the dimensions of gender and mathematics discussed above to individual manifestations of gender inequalities. They are very personal answers to situations in which mathematics, gender and the self interact in complex ways.

Women's sense of belonging seems to be quite vulnerable to the perceptions of their academic environment. Stereotype threat – the effect of explicitly or implicitly believing a negative stereotype about a social group – seems to trigger psychological processes leading to a weaker performance. Steven J. Spencer, Claude M. Steele, and Diane M. Quinn (1999) found that when female students believe the stereotype that they are not as able as males to do good at maths, their test performances decline. Sian L. Beilock, Elizabeth A. Gunderson, Gerardo Ramirez, and Susan C. Levine (2010) showed that female teachers' maths anxiety correlated with negative self-concepts and low achievement among female primary students. These finding might be just the tip of the iceberg, and stereotype threat might influence women's feelings of belonging to mathematics more widely. Consequently, this could well lead to less interest to become a member of this community, even among high-achieving women. This is especially likely to be the case if the surrounding environment of women underlays the fixed-ability concept. Good, Rattan, and Dweck (2012) showed that, in environments with malleable ability attitudes, even high gender stereotypes did not harm women's sense of belonging to maths. Therefore, it appears to be crucial if the surrounding community of women holds mathematical ability to be fixed or developable, and how it sees women's maths ability relative to the ones of men.

A special research focus has been laid on the interplay of feelings of belonging and the personal attribution of success and failure. Bettina Hannover (1991) documented that female students of German grammar schools are likely to assume less competence and expect less success than male students with comparable mathematical achievement – an effect that did not show for the subject of German language. Their self-assessment of their mathematical abilities is therefore much less favourable than their actual performance (Ludwig, 2010). Alternatively, one could speak of overestimation among boys, as they tend to judge their mathematical achievements higher than their grades are.

The issue of confidence in the own mathematical ability also shows up on university level. In a project on students' experiences of undergraduate mathematics, Melissa Rodd and Hannah Bartholomew (2006) conducted interviews to ask female students for their experiences studying mathematics. As at earlier educational levels, the women showed patterns of doubting their own mathematical abilities. Two keywords came up again and again during these interviews: specialness and invisibility. The stories of the women drew a picture of still being special when studying mathematics as a woman. In their observations of a lecture, Rodd and Bartholomew found that women showed a higher attendance rate than men, were the higher attaining group but mostly men were participating in it. Even high attaining women felt uncomfortable when they were asked to contribute. Consistent with the studies on the different roles of boys and girls in mathematics lessons, it seems that women choose different ways to acquire knowledge and that traditional forms of university teaching do not meet women's learning preferences. Jillian M. Knowles (2010) took up these insights when she noticed that her female mathematics students made considerably less use of tutoring than male students, identified underlying organisational structures as obstacles for the female students, and introduced a gender-friendly support system for students of mathematics. In a study at a US university, Abbe H. Herzig (2004) found that even female mathematics staff were troubled by feelings of not belonging and chose to abandon mathematics on these grounds.

Dweck and N. Dickon Reppucci (1973) introduced the concept of learned helplessness to explain the effect of the way in which success and failure are dealt with on a learner's motivation and self-image. Men often attribute failure externally to certain circumstances or other person's fault or simply talk about having had bad luck. On the other hand, they usually attribute success to personal skill. This means that even in the case of failure there is no reduction in self-confidence, while self-confidence is boosted in the light of success. Women often show exactly the opposite pattern. They blame themselves for failure and attribute success to external factors. This tends to lead to a lower level of self-confidence (Dweck et al., 1978). Leder (1980) suggested that learned helplessness at least partly explains gender differences in mathematics education, as they stand in the way of feelings of belonging.

Another explanation is that girls deny any belonging to mathematics by underestimating their abilities for tactical reasons (Bischof-Köhler, 2002/2011). Matina Horner (1968) caused some controversy with her fear-of-success theory, according to which women are reluctant to prove their abilities because they fear that performing well within male domains will cause them to be rejected by society. Therefore, lower self-confidence within this area might act as a kind of self-protection to meet the stereotypical expectations of society. Although not all subsequent studies came to the same results, some did reach the same conclusion: women in non-traditional careers were the least popular within both sexes (Pfost & Fiore, 1990). Men still seem to tend to react critically when women enter traditionally male domains, and women reflect this attitude. So, the fear of success is not without reason, even though it may not be as severe a problem today as it was in the days of Horner's study.

A different line of explanation locates the problem rather in environmental conditions than in the individual. Dweck (2007) argued that the combination of the mindsets that success in mathematics required talent rather than work and that this talent was more common among men than among women led to a loss of confidence when challenges occur. Especially in mathematics, it adds to the problem that effort is not valued as high as giftedness. She stated:

> So if you believe in a math gift and your environment tells you that your group does not have it, then that can be disheartening. However, if, instead, you believe that math ability can be cultivated through your efforts, then the stereotype is less credible. (pp. 49–50)

A talent-focused mindset is therefore especially harming girls, as they might easily believe the negative stereotype that boys show more mathematical ability and hence are more likely to not deepen their mathematical knowledge. A change to mindsets in which mathematical ability is not a fixed thing but a consequence of work might protect women from such stereotypical views and might increase their confidence within mathematical tasks. Studies showed that the gender gap in mathematical achievement nearly disappeared when only looking at the students with such more beneficial mindsets (Dweck, 2007). It seems like girls who believe that mathematical abilities can be developed and are not something unchangeable are doing just as good as boys do. This would indicate the need for some change within cultural values in the educational environment. Instead of believing that one is smart only if things come easily, a greater belief in the power of effort and the appreciation of it will lead to more confidence among all students.

Problematising the discourse on mathematics and gender

Earlier, we introduced a post-structural perspective, which assumes that our reality is constituted through discourse, and that such discourse is necessarily ambiguous and interest-driven. We stressed that we are necessarily part of such discourses and reproduce them, positioning us as accomplices of such discourses. But we also stressed that we can step out of such discourses and deconstruct them by analysing the way the discourses present reality, how they allow to interpret what is happening, which social roles they arrange, which forms of conduct they demand. Here, we will first attempt to deconstruct the discourses on mathematics and gender presented above. Then we will outline how a post-structural approach to mathematics and gender can help us build new ways of understanding.

Deconstructing the gendering of mathematics

Deconstructing these discourses does not mean to discard their validity but to contextualise and relativise the claims made. It aims at revealing the underlying assumptions and interests manifested in certain discourses and at opening spaces to think differently. In this spirit, we revisit the perspectives addressed above:

- The idea that different *images of mathematics* compete, and that they represent the interests of different groups, is no post-structural insight. From a post-structural perspective, we have however become aware that images of mathematics are closely intertwined with more general ideas of rationality, objectivity, and government, and that they usually direct at deprivileged positions for women. Valerie Walkerdine and the Girls and Mathematics Unit (1989) draw on Foucault (1975) to argue that rationalism has constructed modern academia as a truth-seeking and masculine enterprise, relegating women to household work. This idea is deeply rooted in contemporary discourses on how and by whom academic work should be approached and creates struggles for women who want to unite mathematical success and femininity in their identity work.
- Much effort is being laid in the measurement and comparison of *achievement* in mathematics. From a post-structural perspective, objective achievement does not exist and the very idea of achievement and differences in achievement are inseparably linked to the practices through which discourses on achievement are constituted. Most assessment programs on the basis of which performance is compared by gender

use batteries of short and one-ended tasks. As shown in our discussion on cognition, learners differ in how quickly and technically instead of carefully and holistically they want to understand issues, in how much they enjoy racing through competitive situations, and in whether they share the style of thinking in which a problem is presented. Pamela L. Paek (2010) tested a group of 122 Californian high school students once with a timed pen-and-paper test and then with an untimed online computer system. She found that boys showed better achievements in the timed pen-and-paper test while girls performed better in the untimed online test. How far can the small differences in national and international assessments between boys and girls then be said to indicate differences in mathematical ability, and how far do they only indicate that the assessment instruments serve rather male than female expectations and strengths?

- As the often-documented *underrepresentation* of girls and women in mathematics addresses limited educational options and limited access to socio-economically prestigious careers, such underrepresentation is usually considered a problem that has to be challenged. However, we propose to also critically address the social environment in which such underrepresentation can be constituted as a problem. For example, from a perspective in which uncritical use of mathematics and science is held responsible for the technological devastation of our planet, the overrepresentation of boys and men in mathematics might be seen as a social problem. Such a change of perspective does not change the socio-economic consequences of underrepresentation, but it might help to understand and question the system in which women deciding against mathematics come to be seen as a problem.
- *Biological* attempts to explain inequalities between men and women in mathematics depart from an uncritical understanding of biology. From a post-structural perspective, biology itself is not objective, but necessarily a political discourse. This becomes obvious when revisiting historical

biological theories that were intended to prove the superiority of a specific gender, ethnicity or race, and it is still effective today as becomes obvious in the often-practiced dichotomic definition of gender or the assumption that body and mind are widely separated entities. Alternative positions, which are just as scientifically sound, propose that body and mind are closely interwoven, that our experiences and discourses can influence the composition of the body, and that bodily differences between the sexes, at least those deemed to affect our relationship to mathematics, may well be produced discursively. For example, visuospatial skills are not biological invariants but can be improved through training (Marulis et al., 2007; Sorby & Baartmans, 2000).

- *Cognitive* explanations suggest that there are innate differences in the ways men and women think. Studies such as *Women's Ways of Knowing* (Belenky et al. 1986/1997) had a historical function in claiming a distinct female identity instead of regarding femininity merely as an inferior version of masculinity, but they also carry the message that there are ways of knowing closely connected with being a woman in a biological sense. From a post-structural perspective, we contend that there are different meta-discourses that regulate how knowledge is valued, connected, structured and communicated. Again, such meta-discourses are ambiguous and interest-driven. Cognitive studies show that different styles of thinking are not used exclusively by one gender but cross gender boundaries. Gender differences appear because a majority of men prefers a certain style amongst the alternatives they have, while most women prefer a different style. Such preferences need not be natural; instead, they can be assumed to result from other gender-biased cultural influences. Therefore, it would be short-sighted to uncritically take different ways of knowing as a departure of a differentiated organisation of mathematics education. Instead, the different cognitive approaches deserve closer analysis as to how they organise our relationship to the world. Such an analysis might come to the conclusion that education in a specific way of knowing is crucial for our society and that it is legitimate to form students' thinking accordingly. Such an analysis might also come to the conclusion that it is paramount to introduce all students to a plurality of ways of knowing, and to critically address the position of mathematics within this cognitive field.

- Just like ways of knowing, *learning preferences* clearly differ between men and women, but do not allow to draw clear lines between the genders. Boaler (1997) documented that a few girls have the same preferences as the majority of boys, that some boys have the same preferences as the majority of girls, and that some preferences are shared by both genders. Anna Llewellyn (2012) raises the question if such preferences are formed by the discourses of what it means for girls and boys to be a good student. Again, it would be short-sighted to take the discourses which say that women need to be taught in one way and men in another way for granted. These attempts to change *classroom interaction*, which has already been identified as a source of gender inequality in mathematics education, can easily produce new inequalities and reinforce the assumption that boys and girls are different species in the mathematics classroom.
- While it has been established that feelings of *belonging* and *not belonging* influence students' educational and vocational choices, it would be dangerous to search for the reasons for different feelings in the individuals or in a specific gender alone. Indeed, psychological work on stereotype threat, learned helplessness, and fear of success depart from an analysis of the individual. Interventions based on such theories usually attempt to change the mindset of the individual. However, all these theories already acknowledge that feelings of not belonging result from an interplay of individuals and their social environment. Consequently, the social environment and the interplay mechanisms, which allow for structural exclusion by feelings of not belonging in the first place, should receive just as much attention as the psyche of the individual.

Understanding the gendering of mathematics

Post-structural analysis investigates how gender is incorporated when learners construct discourses that explain their relationship to mathematics. Thereby, gender is seen as a problematic discourse itself, rather than as the safe grounds from which to engage in further analysis. Walkerdine and the Girls and Mathematics Unit (1989) stated that dominant views tend to present possible causes of such differences as something real and true. Contrary to that, their way of dealing with the gender problem is 'one which treats truth not as something easily empirically verifiable but as slippery stuff created out of fantasies and fictions which have been made to operate as fact' (p. 19).

They pointed out that there is the trap of thinking mathematics and gender in patriarchal patterns formed by society, trying to prove the assumption that there is a gap between girls and boys and then finding ways to put right what was found wrong. They argued that research about gender tends to focus on searching for differences, whereas similarities are often neglected and seen as a failure to show significant differences. The interpretation of results, they wrote, often indicated that the approach of the study was to show that girls lack something that boys have (e.g., competence, confidence, spatial abilities) or to show that girls are different. The problem with this approach, however, is that with a search for deficits, one group is always portrayed as a problem and bears the blame. Walkerdine and the Girls and Mathematics Unit questioned if such a perspective can help women in any way.

According to them, rationality and mathematical thinking are still viewed as closely linked to the cultural definition of masculinity within society. Such patriarchally shaped societies are affected by the myth that women and mathematics are not inherently compatible and therefore differences between males and females are seen even when there are none. Since we are all part of society, we are quite likely to adopt this view, even if we are not aware of it. Thus, whenever participating in society we tend to confirm and reproduce this view.

Just as much as gender discourses influence the discourses we relate to when making sense of us in relation to mathematics, our relation to mathematics influences how we constitute ourselves as gendered people. Thereby, identity work on gender cannot be reduced to deciding whether we are male or female. Rather, identity work on gender requires a positioning in discourses that describe gender roles differently, in combining different discourses, maybe in rejecting and reauthoring them. In this sense, although we have evidence of women who do not combine discourses of mathematical success and femininity and consequently abandon mathematics when they have the chance to (Herzig, 2004), mathematical success can also be the source of a new kind of femininity (Foyn et al., 2018, addressed below).

In her analysis of case studies, Mendick (2006) documented how identity work negotiates among divergent discourses on gender and mathematics. She found that the students positioned themselves and others within binary oppositions such as talent vs. hard work or real understanding versus rote learning. Thereby, both sides of the oppositions are unequally valued with higher-valued sides associated with masculinity and lower-valued sides associated with femininity. This is the discursive minefield in which girls and women have to build an identity as a learner of mathematics.

After Mendick (2006) had documented the gendered identity work of school students in mathematics, further case studies resulted in similar findings and deeper insights, also in different environments. Trine Foyn, Yvette Solomon, and Hans Jørgen Braathe (2018) presented case studies of high-achieving girls in mathematics and presented how they identified as a 'nerd' and how they had to renegotiate their social roles. Elizabeth de Freitas (2008) authored a fictional biography of a girl obsessed with mathematics and illuminated possible connections between the epistemology of mathematics and gender stereotypes. Jennifer Hall (2010) reported how female high school and university students of mathematics dealt with the feeling of not belonging to mathematics. A longitudinal study by Fiona Walls (2010) traced the identity work of both male and female students of mathematics from primary to secondary school and illustrated how identity work in mathematics relates to the gendered discourses that adolescents are subjected to. Yvette Solomon, Darinka Radovic, and Laura Black (2016) presented a retrospective on the identity work of a female mathematician within a field of experienced contradictions.

Where is mathematics?

It is astonishing how little reference is made to mathematical contents and methods when gender differences in mathematics and mathematics education are discussed. Our argument here is that an important perspective on gender and mathematics is missing in the current literature. We will show this shortcoming in the discussion of a text by Ernest, only to later express some initial thoughts on how mathematics might come into play.

The strange absence of mathematics

One would assume that the discourse called mathematics would have an impact on gender differences in mathematics and mathematics education. How then is mathematics addressed in gender-oriented research in mathematics education? We already saw that the relationship between gender and mathematics can be located in different dimensions, but none of these address mathematics as a discourse in itself.

Public images of mathematics might be the closest to an analysis of the discourse of mathematics, as the suitability of such images is usually evaluated by their suitability to certain philosophies of mathematics. Ernest (1995) presented one of the most profound discussions of images of mathematics in connection to the philosophy of mathematics. He indicated analogies between absolutist philosophies of mathematics, traditional teaching styles, and a masculine style of thought on the one hand and fallibilist philosophies of mathematics, reform teaching styles, and a feminine style of thought. On this basis, which would itself be worthy of discussion, he postulated that 'such values, stereotypes and beliefs end up as a vicious cycle denying women equal opportunities' (p. 456).

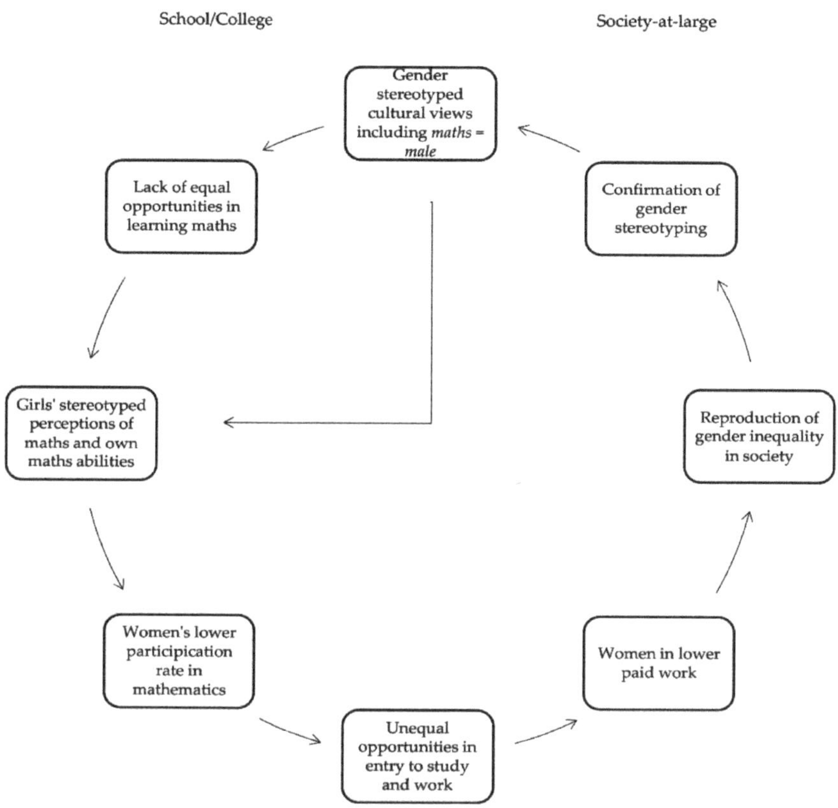

Fig. 19.1 The reproductive cycle of gender inequality in mathematics education (after P. Ernest, 1995, p. 457).

Ernest's vicious cycle, presented in Figure 19.1, is powerful indeed. However, what role does mathematics play in this explanation? Where in the cycle does the specificity of mathematics come into play? What is special about mathematics so that this vicious cycle works for mathematics as it would for chemistry but not for psychology? If we do not want to make ourselves believe that gender inequalities in mathematics are a mere coincidence or historical artefact, we will have to look for the reasons *in* mathematics.

Ernest's argument that a fallibilistic philosophy of mathematics is associated with reformist teaching styles and that it approaches mathematics in line with feminine ways of thinking assumes that mathematics itself has no gender-bias so that the image of mathematics could be changed towards a more gender-inclusive form. This assumption is shared by many, both in studies on gender and mathematics (e.g., Chapman, 1993) and beyond (e.g., Skovsmose, 2011). In his *Invitation to Critical Mathematics Education*, Ole Skovsmose (2011) sees 'mathematics education as being undetermined', 'without "essence"', able to 'be acted out in many different ways and come to serve a grand variety of social, political, and economic functions and interests' (p. 2). In this spirit, Leone Burton (1995) proposed 'that the perceived male-ness of mathematics is equally an artefact of its production and its producers' (p. 215). Critical and feminist research in mathematics education is then supposed to find a way of teaching mathematics that allows for equality. Here also, critique is reduced to the teaching methods and does not cover the discourse of mathematics.

We deem it possible that the social turn in the philosophy of mathematics, which allowed to understand mathematics as a social construction, left the impression that this construction was fully open to change and could be adjusted to social interests. Pat Rogers and Gabriele Kaiser (1995) assumed that a gender-inclusive form of mathematics 'will involve a fundamental shift in what we value in mathematics, in how we teach it, in how mathematics is used, and in the relationship of mathematics to the world around us' (p. 9). Obviously, such a change of mathematics might mean that mathematics would no longer be able to play the role it plays in society today, that we create something completely different. Here, we do not talk about what we approach as mathematics today, but about a fiction called mathematics, without any idea of how much both ideas would overlap.

From a post-structural perspective, contemporary mathematics is a collection of practices, knowledge, beliefs, and applications whose meaning is constituted discursively. As a discourse, mathematics is necessarily ambiguous, interest-driven, and inseparably connected to the other discourses it is defined by. Mathematics might not have any metaphysical 'essence', but it has meaning which is well-demarcated through its use in discourses. We propose to investigate that meaning in order to analyse in what sense mathematics can be understood as a gendered activity.

Mathematics as separated knowing

With reference to the distinction between separate and connected knowing presented by Belenky and colleagues (1997), Chapman (1993) warned us that 'the traditional view of mathematics as a value-free, purely cognitive endeavour [...] tends to dehumanize or objectify it, thus limiting it to a framework with characteristics that are more compatible with separate knowing than connected knowing' (p. 208). She demanded that 'mathematics be reconceptualised to reflect its humane features and mathematics classroom processes revised to facilitate the characteristic ways of knowing of both males and females' (p. 209). In 'Moving Towards a Feminist Epistemology of Mathematics', Burton (1995) drew on conceptualisations of mathematics as a social practice, refused the myth of objectivity, and stressed that mathematics is open to error, never fixed, and requires different perspectives and exchange – all in line with connected knowing.

We argue that this discourse still does not advance to the issue of mathematics. Admittedly, these philosophical attitudes try to make sense of mathematics and can align more to separate or connected ways of knowing. However, this tells us more about these attitudes than about the social practice of mathematics. Ian Hacking (2014) argued that usually mathematicians neither participate in these philosophical discussions nor do they find them enlightening of their practice. The work of Davis and Hersh (1980) was composed in this spirit. They dedicated only one chapter of their *The Mathematical Experience* to these philosophical debates and far more to reflections of doing mathematics. They further noted 'that the typical working mathematician is a Platonist [assuming mathematics to be objective and truth to be recognisable] on weekdays and a formalist [assuming mathematics to be ambiguous constructions and knowledge related to such constructions] on Sundays' (p. 321). It is telling that such a view constitutes a paradox for the philosopher but does not hinder the work of the mathematician. We propose that, while it is clear that mathematical work is partly subjective, open to debate, and in need of different perspectives, objectivity is an *ideal* that mathematicians strive for. On weekdays they have a Platonist attitude when working towards this ideal, while on Sundays they can lean back and admit that mathematics stays messy and the ideal has not been reached. Platonist attitudes can then be considered one form of describing that ideal, while alternative attitudes, which argue that such an ideal cannot be fully realised, do not hinder anyone on holding these ideals.

We also propose that this ideal permeates the whole mathematical discourse. David Kollosche, in Chapter 5 of this volume, argues that calculation and logic, two practices present in any area of mathematics, are explicitly directed at such objectification, at separating mathematics from us rather than connecting it to a plurality of meanings. This disconnectedness of mathematics is what constitutes it as a discipline in its own right and establishes it in opposition to applied disciplines such as physics. From this perspective, mathematics is not waiting to be humanised or to open up for connective knowing. Rather, separated knowing is the very ideal of mathematics. Changing that ideal would require mathematics to become something very different, something that might not be assigned the same cultural value. We think that before asking mathematics to change, we should direct our attention more closely to the social situatedness of mathematics and its connection to separated knowing. That does not mean that the teaching and learning of mathematics cannot be changed to more gender-inclusive forms, nor do we think that the separated nature of mathematical knowledge necessarily excludes women. Instead, we should endeavour towards a transparent, explicit, and unbiased discussion of the importance, potential, limits, and dangers of mathematical forms of knowledge. Making the gendering of mathematics and connected styles of thinking explicit should constitute a vital step towards rejecting presupposed gender-roles and finding new ways of identifying with mathematics.

Mathematics as a patriarchal project

Another feature of the discourse of mathematics which is closely related to gender issues is the analogy between logical hierarchies and patriarchal social systems as discussed in detail by Kollosche in Chapter 5 of this volume. While communities before did not know fatherhood and private property, patriarchal societies connected the ideas of property and father-son relationship through the idea of inheritance. In this logic of inheritance, one's position in the net of relationships is static as you always stay your father's son. Further, you either are somebody's son or you are not – this dichotomy does not allow for mixed or third identities. Eventually, you are defined by whose household you will inherit. This logic of inheritance can be found again in the organisation of the military and in the organisation of the world of polytheistic gods. Eventually, a logic of thought has been developed in patriarchal Ancient Greece, which shows the same organisation: the meaning of its objects is static through definition, its subsumption under other objects is either true or false, and its features can be deduced from objects under which it is subsumed.

We find it hard to disregard the structural analogy between the patriarchal organisation of society and the logic that some Ancient Greek philosophers and rationalists celebrated as its purest expression. It seems unlikely that the development of logic as a very distinct organisation of thought in Ancient Greece was completely independent from the development of patriarchy in the same culture only a few centuries earlier. We propose reconsidering to what extent the ideas that meaning is static, can be expressed through dichotomies, and is inherited through directed bipolar relationships represent gender-biased assumptions. These assumptions, we argue, lie at the heart of mathematics and present it as a patriarchal project for making sense of our world. Thereby, understanding mathematics as a patriarchal project does not mean that women are excluded from mathematics or less able to do mathematics. Instead, it means that mathematics mirrors a social organisation which reproduces male hegemony, thereby possibly reinforcing the latter and provoking sentiments differing by gender.

Moral dilemmas instead of a conclusion

The documentation of inequalities in the education system and beyond is often accompanied by demands to take action. But, as might be guessed from the deconstruction of gender, what might seem to be a good intervention from one perspective may raise serious concerns from a different perspective on mathematics and gender. Sue Willis (1996) differentiated between four perspectives on mathematics and gender, identifying either women, classroom practices, the mathematics curriculum, or social inequality itself as the main problem. We will address some of the moral paradoxes invoked by these perspectives in this last part of our chapter with the hope of further elaborating the grounds on what can be done.

Changing women

Since the 1960s, programs have been installed around the globe to positively influence the achievements, career choices and self-images of girls and women in mathematics. Burton (1990), Rogers and Kaiser (1995), as well as Lynda R. Wiest (2010) included reports about various such programs. However, many of these programs have been criticised, as they tend to ignore feminist approaches to mathematics and seek to adjust feminine intellect, choices, and attitudes to a masculine norm, thus installing masculine approaches to mathematics as normality and portraying feminine approaches as inferior. For example, Olive Fullerton (1995), who taught mathematics methodology for prospective primary school teachers at university, stated:

> One of the barriers to understanding mentioned frequently by the teacher candidates working with me, was that, for them, mathematics was neither relevant nor meaningful. They did not appreciate that mathematics permeated their lives, that their every action was in some way connected to mathematics, that the beauty and harmony of their world was due in large measure to mathematics. (pp. 44–45)

In her statement, Fullerton assumed that there is a truth about the role mathematics plays and should play in our world, and that her female students require assistance to recognise this truth. As Willis (1996) framed it from a deficit perspective, 'the problem lies with the children, who, because of their gender, race, ethnicity, social class or disability, lack the knowledge, skills or motivations necessary for access to, and success in, school mathematics' (p. 44). Boaler (1997) criticised how such a perspective is not interested in the girls' experiences and in the reasons for their action; it is only interested in changing the girls so that they can reach the achievements and representation of boys.

There are, however, many programs that opened up for feminine approaches to mathematics. For instance, Charlene Morrow and James Morrow (1995) repeatedly organised summer schools for female high-school students in the United States. These included classes in problem solving, in fundamental mathematics concepts, and in programming. The organisers wrote that they laid special foci on allowing for women's ways of knowing, on increasing self-confidence in mathematics, and on developing a voice in mathematics. From the student feedback presented by the organisers, it becomes clear that such opportunities can have an empowering effect for the participants. Such feedback is reassuring and documents that this approach is worth considering.

While programs that take place independently from school can be assumed to usually meet the interests of the girls and women who choose to participate, programs interacting with schools, for example by working in schools, inviting classes, or providing teaching materials, approach girls who might have no desire to renegotiate their relationship to mathematics. No matter how well-intentioned, sensitive, and open such obligatory programs might be, they are still forms of manipulation. Is it really emancipation to encourage women to pursue mathematics, even when they have valid reasons not to? If the environment in which women learnt not to identify with mathematics does not change, would not fostering enthusiasm for mathematics, especially in secure environments, only create a deeper conflict of identity for these women when they return to their original environments? May there even be good reasons not to be too enthusiastic about mathematics as a way to approach our world? It seems fair to conclude that programs, which offer women a safe space to investigate their relationship to mathematics, are important. However, they should be open to the possibility that women might have good reasons not to pursue mathematics, and they should be accompanied by gender-inclusive changes of the mathematics classroom and the mathematics curriculum.

Besides, it is an interesting thought experiment to consider whether boys, rather than girls, should change their relationship to mathematics. This is not to suggest that boys should aim for lower achievements in mathematics to achieve equal outcomes. However, given the critique that assessment instruments usually benefit a masculine approach to mathematics, instruments favouring women might put men in a deficit position and demand them to change in order to fill that gap. Concerning the didactical shift from mastering mathematical procedures, which now can be outsourced to computers, to a critical understanding of mathematics, connected knowers might indeed be advantaged over separated knowers. Eventually, it might turn out that a separated and therewith uncritical approach to mathematics laid the basis for devastating applications in modernity. Such a perspective might explicitly position mathematics as a problem of masculinity.

Changing classrooms

Changing the teaching and learning of mathematics in order to challenge gender inequality in mathematics leads to moral dilemmas as well. The work of Boaler (1997) shows clearly that the preferred ways of teaching and learning differ between boys and girls. Does that mean that the organisation of teaching and learning should be adjusted to meet the wishes of all learners? Is that even possible? Or is the implication that students should be allowed to choose from mathematics courses that differ in their styles of teaching and learning, even if that led to something close to single-sex education in mathematics? If we followed this latter plan to approach mathematics in such safe havens, what would that mean for our students' abilities to approach mathematics once they leave our safe havens and engage with mathematics in their future lives? Do we risk that our students fall back into old patterns because they had no chance to learn how to navigate in an environment with divergent approaches to mathematics? In the end, does meeting students' gender-sensitive learning preferences actually reproduce gender inequalities?

A more general question is whether teaching styles should be chosen on the basis of students' learning preferences at all. Educational theory might present well-founded arguments on why to teach mathematics in specific ways. For example, if mathematics education was meant to produce technocrats who are able to technically master mathematics without asking many why questions, would it not be only natural to rely on exposition and individual exercise, irrespective of how boys and girls reacted to this orientation? If, on the other hand, mathematics education was meant to allow for a critical and multifaceted perspective on mathematics, would it not be only natural to allow for diverse approaches to mathematics and to facilitate discussions on these approaches, again irrespective of how boys and girls reacted to that orientation? What causes gender inequalities here is that specific ways of teaching and learning are gendered and not equally accessible for or valued by all genders.

We conclude that there are no clear answers on how to change classroom practice to combat gender inequalities in mathematics education. The overall problem is that the teaching of mathematics is embedded in a complex social system. Even if we know that specific measures have the potential to fight gender inequalities, they might produce gender inequalities of a different kind and further problems or radical changes regarding the teaching and learning of mathematics. A pragmatic approach here, in order to avoid getting stuck in the midst of these paradoxes, is to promote changes to the best of our knowledge and to deal with upcoming problems as they arise. The awareness that there are no easy answers for classroom interventions may help us to see such problems coming and to meet them with open eyes.

Changing contents

Willis (1996) reminded us that 'the choices made in developing school mathematics curricula will reflect the values, priorities and lifestyles of the dominant culture' (p. 45). She continues that, with respect to gender, school mathematics could privilege characteristics of mathematics which are identified more closely with the masculine over characteristics more closely identified with the feminine, such as the logical over the intuitive, the context-free over the context-bounded, the rational and abstract over the personal and social, the unambiguous over the ambiguous, or the absolute over the relative (ibid.).

Indeed, many of the curriculum reforms undertaken in many countries in the last decades can be interpreted as pointing in this direction. The mastery of standard techniques for calculation and proof, together with a desire to accumulate truths, which were dominant in traditional mathematics curricula, are well aligned to what Belenky and colleagues (1997) termed the more masculine separate knowing. The corresponding activities made room for more connected knowing in the sense of a multi-faceted understanding: individual approaches in problem solving and modelling, appreciation of different voices in group work and classroom discussions, productive ways of dealing with seemingly wrong ideas, even critical reflections on the use of mathematics. Aside from questioning how far this shift from viewing mathematics as a product to seeing it as a social activity (see Chapter 1 in this volume) has affected the mathematics classrooms, could it be that the prescribed mathematics curriculum has already opened up for more diverse and especially connected approaches to the subject? We see two problems here.

First, it is unclear how far such reforms indeed affect schools. Kollosche (2018) proposes that many reform initiatives in mathematics education fail because they disregard the social expectations directed at schools. Contributions in the so-called math-war discussions in the US, where some mathematicians expressed their interest in preserving a conservative conception of mathematics (summarised in J. R. Brown, 2008, pp. 207–217), lay bare such expectations, but we can expect that this is just the tip of the iceberg and that most expectations are communicated through more subtle channels. Tony Brown and Olwen McNamara (2005) captured how the growing industry of standard testing with its focus on test items, which can be answered unambiguously and quickly, has caused teachers to direct their teaching away from the reform curricula and back to teaching mathematical techniques and truths. Consequently, we wonder how far curriculum reforms that would open up mathematics for various ways of knowing is still nothing but an aspiration.

Second, it is unclear how far the mathematics curriculum can open up at all. Curriculum change has addressed the methods through which the mathematical contents are approached. We agree that it is possible to allow for other ways of knowing here. However, the mathematical contents remain widely untouched. The triangle stays a triangle; the Pythagoras theorem stays the Pythagoras theorem. As we concluded earlier in this chapter, mathematics as a body of knowledge is closely related to a distinct way of knowing, and it is questionable how far it can be altered at all.

Changing us

Socio-critical scholars proposed that gender inequalities in mathematics are an active part of a larger problem, and that they should be addressed as such. Paul Dowling (1991) stressed that the inequalities are actually functional in reproducing social advantages of men. The mathematics classroom then appears as a mere part in a larger system which aims at securing and reproducing discourses that position men in superior social positions. From such a perspective, it seems questionable how far gender equity in mathematics education can be achieved while the discourses reproducing inequality survive throughout society. In short, does it make sense to fight gender inequalities in the mathematics classroom, or would it be necessary to become a gender activist with a general agenda?

We would argue that change has to start somewhere and in many places before social systems can be said to have changed as a whole. It would be unethical to tolerate structural disadvantages for girls and women while waiting for global answers to gender inequality. Pending further research and without a global solution, change has to be promoted by us, mathematics educators or not, concerning the ways mathematics is publicly perceived, the ways we understand mathematical achievement, the ways of knowing that we accept for approaching mathematics, the styles in which we teach and learn, how we treat each other as teachers and learners of mathematics, and the stories we tell about our and others' relations to mathematics. These goals are unlikely to be achieved through a few interventions alone. We assume that they will provide challenges for generations of teachers and students of mathematics to come. Eventually, these changes may alter our understanding of mathematics and ourselves as mathematical beings in ways that cannot yet be imagined.

References

Baldino, R. R. (2000). Neurone-Z, philosophy of the mind and symptom. In J. F. Matos & M. Santos (Eds.), *Proceedings of the 2nd international mathematics education and society conference* (pp. 143–157). Universidade de Lisboa.

Becker, J. R. (1981). Differential treatment of females and males in mathematics classes. *Journal for Research in Mathematics Education, 12*(1), 40–53. https://doi.org/10.2307/748657

Becker, J. R. (1995). Women's ways of knowing to mathematics. In P. Rogers & G. Kaiser (Eds.), *Equity in mathematics education: influences of feminism and culture* (pp. 164–175). Falmer. https://doi.org/10.4324/9780203990087-29

Becker, J. R. (2001). Single-gender schooling in the public sector in California: Promise and practice. In B. Atweh, H. Forgasz, & B. Nebres (Eds.), *Sociocultural research on mathematics education: an international perspective* (pp. 367–378). Lawrence Erlbaum. https://doi.org/10.4324/9781410600042-31

Beilock, S. L., Gunderson, E. A., Ramirez, G., & Levine, S. C. (2010). Female teachers' math anxiety affects girls' math achievement. *Proceedings of the*

National Academy of Sciences of the United States of America, 107(5), 1860–1863. https://doi.org/10.1073/pnas.0910967107

Belenky, M. F., Clinchy, B. M., Goldberger, N. R., & Tarule, M. J. (1997). *Women's ways of knowing: The development of self, voice, and mind*. Basic Books. (Original work published 1986)

Benbow, C. P [Camilla Persson] (1988). Sex differences in mathematical reasoning ability in intellectually talented preadolescents: Their nature, effects, and possible causes. *Behavioral and Brain Sciences, 11*(2), 169–183. https://doi.org/10.1017/S0140525X00049244

Bischof-Köhler, D. (2011). *Von Natur aus anders: Die Psychologie der Geschlechtsunterschiede* [Different by nature: The psychology of gender differences]. Kohlhammer. (Original work published 2002)

Boaler, J. (1997). *Experiencing school mathematics: Teaching styles, sex, and setting*. Open University.

Boles, D. B. (1980). X-linkage of spatial ability: A critical review. *Child development, 51*(3), 625–635. https://doi.org/10.2307/1129448

Brown, J. R. (2008). *Philosophy of mathematics: A contemporary introduction to the world of proofs and pictures*. Routledge.

Brown, T., & McNamara, O. (2005). *New teacher identity and regulative government: The discursive formation of primary mathematics teacher education*. Springer. https://doi.org/10.1007/b104107

Burton, L. (Ed.). (1990). *Gender and mathematics: An international perspective*. Cassell.

Burton, L. (1995). Moving towards a feminist epistemology of mathematics. In P. Rogers & G. Kaiser (Eds.), *Equity in mathematics education: Influences of feminism and culture* (pp. 211–227). Falmer.

Butler, J. (1999). *Gender trouble: Feminism and the subversion of identity*. Routledge.

Butler, J. (2004). *Undoing gender*. Routledge.

Chafetz, J. S. (2006). The varieties of gender theory in sociology. In J. S. Chafetz (Ed.), *Handbook of the sociology of gender* (pp. 3–23). Springer. https://doi.org/10.1007/0-387-36218-5_1

Chapman, O. (1993). Women's voice and the learning of mathematics. *Journal of Gender Studies*, 2(2), 206–222. https://doi.org/10.1080/09589236.1993.9960539

Cheryan, S. (2012). Understanding the paradox in math-related fields: Why do some gender gaps remain while others do not? *Sex Roles, 66*(3-4), 184–190. https://doi.org/10.1007/s11199-011-0060-z

Colom, R., Juan-Espinosa, M., Abad, F., & García, L. F. (2000). Negligible sex differences in general intelligence. *Intelligence, 28*(1), 57–68. https://doi.org/10.1016/S0160-2896(99)00035-5

Davis, P. J., & Hersh, R. (1980). *The mathematical experience.* Birkhäuser.

Delon, F. (1995). The French experience: The effects of de-segration. In P. Rogers & G. Kaiser (Eds.), *Equity in mathematics education: influences of feminism and culture* (pp. 142–153). Falmer.

Dowling, P. (1991). Gender, class, and subjecivity in mathematics: A critique of Humtpy Dumpty. *For the Learning of Mathematics, 11*(1), 2–8.

Driver, G. (1980). *Beyond underachievement: Case studies of English, West Indian, and Asian school-leavers at sixteen plus.* Commission for Racial Equality.

Dweck, C. S. (2007). Is math a gift? Beliefs that put females at risk. In S. J. Ceci & W. M. Williams (Eds.), *Why aren't more women in science? Top researchers debate the evidence* (pp. 47–55). American Psychological Association.

Dweck, C. S., Davidson, W., Nelson, S., & Enna, B. (1978). Sex differences in learned helplessness: II. The contingencies of evaluative feedback in the classroom and III. An experimental analysis. *Developmental Psychology, 14*(3), 268–276. https://doi.org/10.1037/0012-1649.14.3.268

Dweck, C. S., & Reppucci, N. D. (1973). Learned helplessness and reinforcement responsibility in children. *Journal of Personality and Social Psychology, 25*(1), 109–116. https://doi.org/10.1037/h0034248

Eccles, J. S. (1989). Bringing young women to math and science. In M. Crawford & M. Gentry (Eds.), *Gender and thought: Psychological perspectives* (pp. 36–58). Springer. https://doi.org/10.1007/978-1-4612-3588-0_3

Eckert, P., & McConnell-Ginet, S. (2003). *Language and gender.* Cambridge University Press.

Ernest, J. B., Reinholz, D. L., & Shah, N. (2019). Hidden competence: Women's mathematical participation in public and private classroom spaces. *Educational Studies in Mathematics, 102*(2), 153–172. https://doi.org/10.1007/s10649-019-09910-w

Ernest, P. (1995). Values, gender and images of mathematics: A philosophical perspective. *International Journal of Mathematical Education in Science and Technology, 26*(3), 449–462. https://doi.org/10.1080/0020739950260313

Ernest, P. (2007). Questioning the gender problem in mathematics. *Philosophy of Mathematics Education, 20.* https://www.exeter.ac.uk/research/groups/education/pmej/pome20/index.htm

European Commission. (2019). *She figures 2018.* Publications Office of the European Union.

Eurostat. (2019a). Graduates by education level, programme orientation, sex and field of education. *Eurostat*. https://ec.europa.eu/eurostat/databrowser/view/educ_uoe_grad02/default/table?lang=en

Eurostat. (2019b). Students enrolled in tertiary education by education level, programme orientation, sex and field of education. *Eurostat*. https://data.europa.eu/data/datasets/tvoyfp236pvctmgfqyudca?locale=en

Fennema, E., Peterson, P. L., Carpenter, T. P., & Lubinski, C. A. (1990). Teachers' attributions and beliefs about girls, boys, and mathematics. *Educational Studies in Mathematics*, *21*(1), 55–69. https://doi.org/10.1007/BF00311015

Foucault, M. (1961). *Histoire de la folie à l'âge classique: Folie et déraison* [Madness and civilization: A history of insanity in the Age of Reason]. Plon.

Foucault, M. (1966). *Les mots et les choses: Une archéologie des sciences humaines* [The order of things: An archaeology of the human sciences]. Gallimard.

Foucault, M. (1975). *Surveiller et punir: Naissance de la prison* [Discipline and punish: The birth of the prison]. Gallimard.

Foucault, M. (1976). *Histoire de la sexualité: La volonté de savoir* [The history of sexuality: The will to knowledge]. Gallimard.

Foucault, M. (1982). How is power exercised? In H. L. Dreyfus & P. Rabinow (Eds.), *Michel foucault: beyond structuralism and hermeneutics* (pp. 216–226). Harvester.

Foucault, M. (2007). What is Critique? In S. Lotringer (Ed.), *The politics of truth* (pp. 41–81). Semiotext(e). (Original work published 1990)

Foyn, T., Solomon, Y., & Braathe, H. J. (2018). Clever girls' stories: The girl they call a nerd. *Educational Studies in Mathematics*, *5*(1), 77–93. https://doi.org/10.1007/s10649-017-9801-4

Freitas, E. de (2008). Mathematics and its other: (Dis)locating the feminine. *Gender and Education*, *20*(3), 281–290. https://doi.org/10.1080/09540250801964189

Friedan, B. (1963). *The feminine mystique*. Norton.

Fullerton, O. (1995). Who wants to feel stupid all of the time? In P. Rogers & G. Kaiser (Eds.), *Equity in mathematics education: Influences of feminism and culture* (pp. 38–49). Falmer.

Geary, D. C. (1998). *Male, female: The evolution of human sex differences*. American Psychological Association.

Gerdes, P. (2006). African women with a doctorate in mathematics. *Newsletter of the African Mathematical Union Commission on the History of Mathematics in Africa*, 33. http://africanwomeninmath.org/women-in-math/report/african-women-doctorate-in-mathematics

Good, C., Rattan, A., & Dweck, C. S. (2012). Why do women opt out? Sense of belonging and women's representation in mathematics. *Journal of Personality and Social Psychology, 102*(4), 700–717. https://doi.org/10.1037/a0026659

Green, J. (2001). How many women mathematicians can you name? *Math Horizons, 9*(2), 9–14. https://doi.org/10.1080/10724117.2001.12021856

Hacking, I. (2014). *Why is there philosophy of mathematics at all?* Cambridge University Press. https://doi.org/10.1017/CBO9781107279346

Hall, J. (2010). The influence of high school and university experiences on women's pursuit of undergraduate mathematics degrees in Canada. In H. J. Forgasz, J. R. Becker, K.-H. Lee, & O. B. Steinthorsdottir (Eds.), *International perspectives on gender and mathematics education* (pp. 365–390). Information Age.

Halpern, D. F. (2012). *Sex differences in cognitive abilities*. Psychology Press.

Halpern, D. F., Benbow, C. P., Geary, D. C., Gur, R. C., Hyde, J. S., & Gernsbacher, M. A. (2007). The science of sex differences in science and mathematics. *Psychological Science in the Public Interest, 8*(1), 1–51. https://doi.org/10.1111/j.1529-1006.2007.00032.x

Hampson, E. (1990). Variations in sex-related cognitive abilities across the menstrual cycle. *Brain and Cognition, 14*(1), 26–43. https://doi.org/10.1016/0278-2626(90)90058-v

Hampson, E., & Rovet, J. F. (2015). Spatial function in adolescents and young adults with congenital adrenal hyperplasia: clinical phenotype and implications for the androgen hypothesis. *Psychoneuroendocrinology, 54*, 60–70. https://doi.org/10.1016/j.psyneuen.2015.01.022

Hannover, B. (1991). Zur Unterrepräsentanz von Mädchen in Naturwissenschaften und Technik: Psychologische Prädiktoren der Fach- und Berufswahl [On the underrepresentation of girls in science and technology: Psychological predictors of subject and career choice]. *Zeitschrift* für *Pädagogische Psychologie, 5*, 169–186.

Hersh, R., & John-Steiner, V. (2011). *Loving and hating mathematics: Challenging the myths of mathematical life*. Princeton University Press.

Herzig, A. H. (2004). 'Slaughtering this beautiful math': Graduate women choosing and leaving mathematics. *Gender and Education, 16*(3), 379–395. https://doi.org/10.1080/09540250042000251506

Hiddleston, P. (1995). The contribution of girls-only schools to mathematics and science education in Malawi. In P. Rogers & G. Kaiser (Eds.), *Equity in mathematics education: influences of feminism and culture* (pp. 148–153). Falmer.

Hines, M. (2010). Sex-related variation in human behavior and the brain. *Trends in Cognitive Science, 14*(10), 448–456. https://doi.org/10.1016/j.tics.2010.07.005

Horner, M. (1968). *Sex differences in achievement motivation and performance in competitive and non-competitive situations* [Doctoral dissertation, University of Michigan].

Hyde, J. S. (2005). The gender similarities hypothesis. *The American psychologist, 60*(6), 581–592. https://doi.org/10.1037/0003-066X.60.6.581

Isaacson, Z. (1990). 'They look at you in absolute horror': Women writing and talking about mathematics. In L. Burton (Ed.), *Gender and mathematics: an international perspective* (pp. 20–28). Cassell.

Jahnke-Klein, S. (2001). *Sinnstiftender Mathematikunterricht* für *Mädchen und Jungen* [Meaningful mathematics education for girls and boys]. Schneider.

Johnston, B. (1995). Mathematics: An abstracted discourse. In P. Rogers & G. Kaiser (Eds.), *Equity in mathematics education: Influences of feminism and culture* (pp. 228–236). Falmer.

Jungwirth, H. (1991). Interaction and gender: Findings of a microethnographical approach to classroom discourse. *Educational Studies in Mathematics, 22*(3), 263–284. https://doi.org/10.1007/BF00368341

Kenschaft, P. C. (1982). Women in mathematics around 1900. *Journal of Women in Culture and Society, 7*(4), 906–909.

Kinski, I. (1994). Mädchenmathematik kontra Knabenmathematik [Girls' mathematics versus boys' mathematics]. *Mitteilungen der Deutschen Mathematiker-Vereinigung, 2*(3), 13–24. https://doi.org/10.1515/dmvm-1994-0308

Knowles, J. M. (2010). Recognizing gender in mathematics relationships: A relational counseling approach helps teachers and students overcome damaging perceptions. In H. J. Forgasz, J. R. Becker, K.-H. Lee, & O. B. Steinthorsdottir (Eds.), *International perspectives on gender and mathematics education* (pp. 421–450). Information Age.

Koehler, M. S. (1990). Classrooms, teacher and gender differences in mathematics. In E. Fennema & G. C. Leder (Eds.), *Mathematics and gender* (pp. 128–148). Teachers College, Columbia University.

Kollosche, D. (2018). Social functions of mathematics education: A framework for socio-political studies. *Educational Studies in Mathematics, 98*(3), 287–303. https://doi.org/10.1007/s10649-018-9818-3

Leder, G. (1980). Bright girls, mathematics and fear of success. *Educational Studies in Mathematics, 11*(4), 411–422. https://doi.org/10.1007/BF00231214

Leder, G. (1995). Equal or different? Cultural influences on learning mathematics. In P. Rogers & G. Kaiser (Eds.), *Equity in mathematics education: Influences of feminism and culture* (pp. 118–126). Falmer.

Lee, K., & Anderson, J. (2015). Gender differences in mathematics attitudes in coeducational and single sex secondary education. In M. Marshman, V. Geiger, & A. Bennison (Eds.), *Mathematics education in the margins: proceedings of the 38th annual conference of the mathematics education research group of australasia* (pp. 357–364). MERGA.

Llewellyn, A. (2012). Unpacking understanding: The (re)search for the Holy Grail of mathematics education. *Educational Studies in Mathematics, 81*(3), 385–399. https://doi.org/10.1007/s10649-012-9409-7

Ludwig, P. H. (2010). Schulische Erfolgserwartungen und Begabungsselbstbilder bei Mädchen: Strategien ihrer Veränderung [Expectations of success at school and self-image of talent in girls: Strategies for change]. In M. Matzner & I. Wyrobnik (Eds.), *Handbuch Mädchen-Pädagogik* (pp. 145–158). Beltz.

Magolda, M. B. B. (1992). *Knowing and reasoning in college: Gender-related patterns in students' intellectual development*. Jossey-Bass.

Marchbank, J., & Letherby, G. (2014). *Introduction to gender: Social science perspectives*. Pearson.

Marulis, L. M., Lui, L. L., Warren, C. M., Uttal, D. H., & Newcombe, N. S. (2007). *Effects of training or experience on spatial cognition in children and adults: A meta-analysis*. Annual Meeting of the American Educational Research Association, Chicago, IL.

Masanja, V. G. (2010). *Increasing women's participation in science, mathematics and technology education and employment in Africa*. https://www.un.org/womenwatch/daw/egm/gst_2010/Masanja-EP.8-EGM-ST.pdf

Mead, M. (1949). *Male and female: A study of the sexes in a changing world*. Morrow.

Mendick, H. (2005). A beautiful myth? The gendering of being/doing 'good at maths'. *Gender and Education, 17*(2), 203–219. https://doi.org/10.1080/0954025042000301465

Mendick, H. (2006). *Masculinities in mathematics*. Open University.

Mendick, H. (2017). Mathematical futures: Discourses of mathematics in fictions of the post-2008 financial crisis. In A. Chronaki (Ed.), *Mathematics education and life at times of crisis* (Vol. 1, pp. 74–89). University of Thessaly Press.

Money, J. (1955). Hermaphroditism, gender and precocity in hyperadrenocorticism: Psychologic findings. *Bulletin of the Johns Hopkins Hospital, 96*(6), 253–264.

Morrow, C., & Morrow, J. (1995). Connecting women with mathematics. In P. Rogers & G. Kaiser (Eds.), *Equity in mathematics education: Influences of feminism and culture* (pp. 13–27). Falmer.

Nosek, B. A., Smyth, F. L., Sriram, N., Lindner, N. M., Devos, T., Ayala, A., Bar-Anan, Y., Bergh, R., Cai, H., Gonsalkorale, K., Kesebir, S., Maliszewski, N., Neto, F., Olli, E., Park, J., Schnabel, K., Shiomura, K., Tulbure, B. T., Wiers, R. W., … Greenwald, A. G. (2009). National differences in gender-science stereotypes predict national sex differences in science and math achievement. *Proceedings of the National Academy of Sciences of the United States of America, 106*(26), 10593–10597. https://doi.org/10.1073/pnas.0809921106

OECD. (2014). *PISA 2012 results: What students know and can do: Student performance in mathematics, reading and science*. OECD Publishing.

Olson, M., Olson, J., Okazaki, C., & La, T. (2010). Conversations of parents and children working on mathematics. In H. J. Forgasz, J. R. Becker, K.-H. Lee, & O. B. Steinthorsdottir (Eds.), *International perspectives on gender and mathematics education* (pp. 33–54). Information Age.

Peak, P. L. (2010). Factors contributing to gender differences in mathematics performance of United States high school students. In H. J. Forgasz, J. R. Becker, K.-H. Lee, & O. B. Steinthorsdottir (Eds.), *International perspectives on gender and mathematics education* (pp. 203–224). Information Age.

Perl, T. (2010). The *Ladies' Diary or Woman's Almanack*, 1704–1841. In H. J. Forgasz, J. R. Becker, K.-H. Lee, & O. B. Steinthorsdottir (Eds.), *International perspectives on gender and mathematics education* (pp. 15–32). Information Age.

Pfost, K. S., & Fiore, M. (1990). Pursuit of nontraditional occupations: Fear of success or fear of not being chosen? *Sex Roles, 23*(1–2), 15–24. https://doi.org/10.1007/BF00289875

Pieper-Seier, I. (2009). Studentinnen und Professorinnen in der Mathematik [Female students and female professors of mathematics]. *Gender, 1*(1), 59–72. http://dx.doi.org/10.25595/76

Popejoy, A. B., & Leboy, P. S. (2012). Is math still just a man's world? *Journal of Mathematics and System Science, 2*, 292–298.

Prendergast, M., & O'Donoghue, J. (2014). Influence of gender, single-sex and co-educational schooling on students' enjoyment and achievement in mathematics. *International Journal of Mathematical Education in Science and Technology, 45*(8), 1115–1130. https://doi.org/10.1080/0020739X.2014.904530

Riegle-Crumb, C., & Humphries, M. (2012). Exploring bias in math teachers' perceptions of students' ability by gender and race/ethnicity. *Gender & Society, 26*(2). https://doi.org/10.1177/0891243211434614

Rodd, M., & Bartholomew, H. (2006). Invisible and special: Young women's experiences as undergraduate mathematics students. *Gender and Education, 18*(1), 35–50. https://doi.org/10.1080/09540250500195093

Rogers, P., & Kaiser, G. (Eds.). (1995). *Equity in mathematics education: Influences of feminism and culture*. Falmer.

Schwank, I. (1994). Zur Analyse kognitiver Mechanismen mathematischer Begriffsbildung unter geschlechtsspezifischem Aspekt [On the analysis of cognitive mechanisms of mathematical concept formation from a gender-specific perspective]. *Zentralblatt* für *Didaktik der Mathematik*, *26*(2), 31–40.

Schwank, I. (1999). On predicative versus functional cognitive structures. In I. Schwank (Ed.), *European research in mathematics education: proceedings of the first conference of the European society in mathematics education* (Vol. 2, pp. 84–96). Forschungsinstitut für Mathematikdidaktik.

Schwank, I. (2002). Analysis of eye-movements during functional versus predicative problem solving. In J. Novotná (Ed.), *European Research in Mathematics Education II: Proceedings* (pp. 489–498). Charles University.

Skovsmose, O. (2011). *An invitation to critical mathematics education*. Sense. https://doi.org/10.1007/978-94-6091-442-3

Solomon, Y., Radovic, D., & Black, L. (2016). 'I can actually be very feminine here': Contradiction and hybridity in becoming a female mathematician. *Educational Studies in Mathematics*, *91*(1), 55–71. https://doi.org/10.1007/s10649-015-9649-4

Sorby, S. A., & Baartmans, B. J. (2000). The development and assessment of a course for enhancing the 3-D spatial visualization skills of first year engineering students. *Journal of Engineering Education*, *89*(3), 301–307. https://doi.org/10.1002/j.2168-9830.2000.tb00529.x

Spencer, S. J., Steele, C. M., & Quinn, D. M. (1999). Stereotype threat and women's math performance. *Journal of Experimental Social Psychology*, *35*(1), 4–28. https://doi.org/10.1006/jesp.1998.1373

Thompson, D. R. (1995). The METRO achievement program: Helping inner-city girls excel. In P. Rogers & G. Kaiser (Eds.), *Equity in mathematics education: influences of feminism and culture* (pp. 28–37). Falmer.

Tiedemann, J. (2002). Teachers' gender stereotypes as determinants of teacher perceptions in elementary school mathematics. *Educational Studies in Mathematics*, *50*(1), 49–62. https://doi.org/10.1023/A:1020518104346

Trakulphadetkrai, N. V. (2017). Where are the girls and women in mathematical picture books? *Mathematics Teaching*, 258, 23–25.

Walkerdine, V., & The Girls and Mathematics Unit. (1989). *Counting girls out*. Virago.

Walls, F. (2010). Freedom to choose? Girls, mathemaitcs and the gendered construction of mathematical identity. In H. J. Forgasz, J. R. Becker, K.-H. Lee, & O. B. Steinthorsdottir (Eds.), *International perspectives on gender and mathematics education* (pp. 87–110). Information Age.

Wiest, L. R. (2010). Out-of-school-time (OST) programs as mathematics support for females. In H. J. Forgasz, J. R. Becker, K.-H. Lee, & O. B. Steinthorsdottir (Eds.), *International perspectives on gender and mathematics education* (pp. 55–86). Information Age.

Willis, S. (1996). Gender justice and the mathematics curriculum: Four perspectives. In L. H. Parker, L. J. Rennie, & B. J. Fraser (Eds.), *Gender, science and mathematics: shortening the shadow* (pp. 41–51). Springer. https://doi.org/10.1007/978-94-011-0143-1_4

20. Societal perceptions of mathematics and mathematics education

Melissa Andrade-Molina and Alex Montecino

That 'people are naturally bad at mathematics' or that 'mathematics is reserved only for people with higher intellect' are naturalised discourses rooted in the image and belief about mathematics and mathematics education. This chapter focuses on mapping societal perceptions of mathematics and mathematics education. These perceptions are tracked within naturalised discourses circulating on social networks, such as YouTube and Twitter, and in the media, such as newspapers and TV shows. We unpack the ways of thinking and understanding mathematics and mathematics education in peoples' comments based on their daily experiences as humans navigating modern society and news websites that have published articles related to mathematics and mathematics education in order to map and take a critical position on societal perceptions circulating about mathematics and mathematics education among the public.

Introduction

> Now, look, we're gonna be dealing with some real serious stuff today. You might have heard of it. It's called math! And without it, none of us would even exist, so let's jump right in. (Mr Goldenfold in *Rick and Morty*, 'Pilot', season 1, episode 1)

Nowadays, it is possible to find many references to mathematics and mathematics education on social networks – YouTube, Facebook, Twitter (now X) – and in the media – newspapers, TV shows, and so on. For

©2024 Melissa Andrade-Molina et al., CC BY-NC 4.0 https://doi.org/10.11647/OBP.0407.20

example, in the sitcom *The Big Bang Theory*, mathematics appears in the form of complex formulas located on whiteboards, whether in the offices of Caltech University or the apartments of the characters Leonard and Sheldon. The formulas are drawn from applied and theoretical physics, which puts mathematics at an unreachable level in which even the physicists themselves are not sure about what they are doing – for example when Leonard needs Sheldon's help and draws a symbol that ends up being Charlie Brown's hair instead of a mathematical notation. Ordinary people, such as the character Penny, an 'average' American, are typically portrayed as people that find mathematics hard and rely on the rest of the characters to help them. This issue is not restricted to gender bias. The character Stuart, an art major from the Rhode Island School of Design and owner of the comic bookstore, The Comic Center, also struggles with complex calculations. Being a scientist, or having a PhD in a STEM (Science, Technology, Engineering, Mathematics) field, seems to guarantee a character's high intellect and interests bordering on geekiness, exemplified by many characters from *The Big Bang Theory* or Ross Geller from *Friends*. Well, except if the character is a woman, apparently. *Futurama* is an animated science fiction sitcom in which the character Amy Wong, a Martian, has a PhD in engineering from Mars University. Her interests are more related to social and fashion topics. She more rarely contributes to her field or engages in scientific discussions than her male peers. However, there are a few exceptions where this pattern does not happen (see, for example, the animated sitcom *The Simpsons*).

That mathematics is challenging and that only smart people can understand its working seem to be common assumptions. 'I'm not a mathematician' is a recurring response when someone is asked to perform any type of calculation that requires even the basic operations of addition, subtraction, multiplication, and division. An example of this phenomenon occurs in *Friends* when the character Joey faces a situation that unfolds as follows:

Joey: Full name.

Cliff: Clifford Burnett.

Joey: Date of birth?

Cliff: November 16, 1968.

Joey: Age?

Cliff: Can't you figure that out based on my date of birth?

Joey: I'm a doctor, Cliff, not a mathematician. (*Friends*, season 8, episode 23)

In movies and TV series, mathematics is portrayed as a highly complex subject. Mathematics is beyond reach, as, for example, in the film *A Beautiful Mind* about the mathematician John Nash. And this is not only for mathematics but extends to mathematics education as well. Even superheroes are nonplussed by mathematics. In the movie *Incredibles 2*, Dash must do his mathematics homework and his father Bob (Mr. Incredible) starts helping him. Dash corrects Bob by saying 'That's not the way you're supposed to do it, dad. They want us to do it this way'. Here Bob replies with a reaction most parents could relate to nowadays: 'I don't know that way. Why would they change math? Math is math. Math is math!' From examples like this, styles of teaching school mathematics have become a common subject of discussion. After watching several movies and TV shows, it becomes natural to begin wondering if mathematics is restricted only to brilliant minds.

When it comes to understanding the struggle with mathematics, it is possible to posit several hypotheses. For example, Neil deGrasse Tyson, a contemporary physicist, elaborates on why most people are bad at mathematics (Cosmology Today, 2017). He asserts that 'I've looked at how much trouble people have with mathematics typically because any one subject that the most people say, "I was never good at..." concerning a topic, it's gonna be math'. His hypothesis deals with how our brains work: 'I say to myself: If our brain were wired for logical thinking, then math would be the most, easiest subject, everything else would be harder. So, I am kinda forced to conclude that our brain is not wired for logic'. Here the apparent fictional reality of movies and TV series becomes an actual hypothetical phenomenon of generalised low academic performance in school mathematics. Being not good at mathematics is, apparently, a shared experience for most people. A Google search on 'why math is hard' gives 667000000 results. The majority are articles addressing why people struggle with mathematics, with titles like 'Why so many students struggle with math', 'Top 6 reasons math is hard to learn', 'Why is math so hard for some children', 'Why so many students hate math (and how to fix it)', and so on. A Google search on 'Why is math easy' gives 593000000 results, within whose titles there is still the stigma of mathematics being hard, such as 'How to learn mathematics easily', '3 ways to make math easy', 'How I rewired my brain to become fluent in math', 'If you can't learn math, maybe it's not your fault', 'How to succeed in math', and so on.

The reaction of people to mathematics is not that distant from what deGrasse Tyson stated. If 'mathematics' is searched on YouTube, the most-watched video with, almost fifty million views, is '15-year-old Yaashwin Sarawanan is a human calculator!' (Asia's Got Talent, 2019). On this video, one of the most 'liked' comments is: 'Q. What is 4+5? Me: I think that should be 9. Pulls out a calculator just to ensure' (22000 likes and 133 replies). At least twenty-two thousand people can relate to this situation, not trusting themselves in performing simple mathematical tasks. On the video 'The surprising beauty of mathematics' (TEDx Talks, 2014), with over six million views, one of the comments with more replies is: 'I used to be great at math until I was taught the "proper" way to do it. Now I pretty much hate math' (375 likes and 61 replies). One reaction for the previous comment refers to the style of teaching his teacher had when he was a student and how it influenced his thoughts about mathematics:

> I had the same problem all my life. I used to be really bad at math, at least I thought so, because of the reasons you gave, teachers wouldn't accept my work either, and worse, sometimes they would say that I just played with numbers without making sense, and copy the result from somebody else, especially, when I didn't know how to explain, how I got my results. And I used to feel so lousy, and a total idiot.

These discourses seem to be part of the common sense of people. For instance, that mathematical problems have only one right answer or that mathematics is so abstract that it is not useful in day-to-day life and impractical due to rote memorisation of formulas (Colagrossi,

2018). Such discourses elevate mathematics to a supra-level of Platonic ideas far from the reach of mundane, average people, not being part of everyday life, and belonging only to mathematical geniuses. A parody of such discourses can be seen in *The Simpsons* (season 14, episode 7) when a math teacher teaches teenage gang members that 'differential equations are more powerful than bullets'.

Mathematics is often portrayed as a highly relevant subject: Without mathematics, humankind, and life as we know it would not exist, but how many people do see mathematics as important? Apparently not so many. For a long time, research on the teaching and learning of mathematics and educational policies have tried to include 'all' people within the practices of school mathematics by arguing that people fully use it in their daily lives and that it should be enjoyable to work with (Pais, 2018). Some dominant narratives around mathematics have become highly recurrent within society, dealing with the value of mathematics (Pais, 2013). For Stephen Lerman (2014), such perceived value performs a role that has resulted in putting mathematics in a privileged position.

In this chapter, we seek to map and take a critical position on societal perceptions circulating about mathematics and mathematics education among the public. The study is relevant because it may critically rethink the role the media have in constructing meaning and the circulation of naturalised discourses about mathematics. The aim is to trace what has been said outside academia, often tropes that propagandise mathematics, offer salvation narratives, and (re)produce myths about, and dehumanise, mathematics. Therefore, the attention is not going to be on the places where mathematics education is taken as a scientific discipline, namely academia, but where the teaching and learning of mathematics stop being virtual.[1] By placing the attention outside the field, we are aiming to break the predominant tendency within mathematics education research, which is to focus on internal issues of the teaching and learning of mathematics (Lerman, 2014), to (re)produce successful experiences (Gutiérrez, 2013), and to promote solutions (Pais, 2012).

1 'Virtual' in the sense of Deleuze (2007).

Mapping social media's discourses

That 'people are naturally bad at mathematics', or that 'mathematics is reserved only for people with higher intellect' are common assumptions found in social discourse. School mathematics has been one of the leading causes of anxiety (math anxiety) and fear (math phobia). How people approach mathematics depends on several factors coming from their previous or present experiences and preconceptions.

As humans, we encounter a plethora of social stimuli in our environments every day and utilise a series of highly adaptive systems (e.g., attention, perception, memory) to make sense of incoming information. Together, these systems systematically alter the information in order to make it interpretable. Ultimately, the aim of using these systems and, more generally, of social information processing is to allow us to make attributions about others. Such attribution-making processes vary interpersonally in that two people may perceive the same event yet conjure up two different attributions. The bases of individual differences in perception are preconceptions, or 'schemas', that are acquired with experience as people encounter people, objects, or events (Garrido, 2020, p. 5071).

Perception becomes an adaptive process through which people sometimes make inaccurate interpretations of the social world (Garrido, 2020). According to David Dunning (2001), research in the field of social perception has been focused on three key aspects: (i) debates regarding the types of information people pay attention to, (ii) tracing the cognitive processes people follow when receiving this type of information, and (iii) the types of judgments people can reach. Our interest in this chapter is how ordinary life's language reveals the circulating and dominant narratives around mathematics and mathematics education. As Philippe Chassy (2014) contends, language provides a systematically biased perception of what is taken to be 'reality'. In this regard, 'the perceptual filters superimposed by language on social realities bias how individuals build a representation of the situation' (p. 36). Along these lines, we approach mathematics through language, which is submerged into linguistic relativity (Sapir, 1929). He states that humans are 'very much at the mercy of the particular language which has become the medium of expression for their society' (p. 210). We cannot free ourselves from bias about mathematics and mathematics education within social activity.

This chapter focuses on mapping societal perceptions of mathematics and mathematics education. Within mathematic education, perception has been addressed mainly in relation to students or teachers and their relationship with some mathematical topics or roles of mathematics in society (see Chan & Wong, 2014; Ikeda, 2018; Leung & Lee, 2013). Toshikazu Ikeda (2018) asserts that how students perceive the role of mathematics in society 'should be distinguished from student appreciation of mathematics' utility in society (p. 261). This is because 'some students might not appreciate the utility of mathematics in society, even though they can recognize its roles'. Aiming to unpack the ways of thinking and understanding mathematics and mathematics education, we search for circulating and naturalised discourses on online news and within expressions of people's opinions. The search involves places where people comment about mathematics and mathematics education in their daily experiences as humans living in modern society and news websites that have published articles related to mathematics and mathematics education.

We set some parameters to delimit the data. We started with sites where people verbalise their opinions without being asked to do so, such as a YouTube video comment box. There are plenty of YouTube channels related to the teaching and learning of mathematics and scientific topics engaged with mathematics. Therefore, social media sites, YouTube, and Twitter in particular, are the first places to look for societal perceptions. Facebook is not part of the sample since many pages dedicated to mathematics or mathematics education are not public. The second source of data is online news in English through the search engine provided by Google at news.google.com. After data were collected, they were further analysed using the big data software Nvivo, and then the analysed results were compiled and graphed using Gephi. The gathered data consist of 1500 YouTube comments, 1250 tweets, and 846 pieces of online news.

On the one hand, social media study is based on: (i) YouTube video comments, and (ii) Twitter posts. Using the YouTube search tool, we identified the five most popular channels regarding mathematics – including mathematics teaching or dissemination of specific mathematical topics. These five channels are the result of triangulating a YouTube search with the keywords: 'math' and 'mathematics' and the filters 'Channel' and 'View count'. Additionally, we considered only the channels that enable posted comments (some channels have disabled the comment section for the most-viewed videos; these were therefore not part of the sample). The channels included: *Numberphile*, *Khan Academy*, *Professor Leonard*, *PatrickJMT*, and *3Blue1Brown*. We selected only the comments from the five most-watched videos of each channel. Using the Chrome extension Twitter Archiver, we collected Twitter posts referring to math, mathematics, and mathematics education. This extension – the free version – enables saving up to one hundred tweets per hour by keywords. The tweets collected are a sample of a significant number of tweets published around the world. We selected only tweets posted in English.

On the other hand, the study from online news is based on a search at Google news. We gathered news regarding mathematics and mathematics education published between 2015 and 2019. We know this methodology can be affected by the search algorithm from Google, that prioritises some news over others. Multiple factors – a lot of these are beyond our control – are key for showing what is more 'appropriate' for each user. However, it is plausible to see a sort of saturation regarding the published online news despite this bias.

The strategy for studying YouTube video comments and Twitter posts is based on saturation of discourses around mathematics and mathematics education. We decided to approach the sample with this strategy given the nature of the data and the number of possible comments and posts we could gather. Saturation becomes helpful since it 'is a term used to describe the point where you have heard the range of ideas and aren't getting new information' (Krueger & Casey, 2014, p, 64). In other words, at some point, the selected samples should become repetitive in their content and purpose. And although some readers may consider this review superficial or general due to the nature of the sample and its limited volume, we may cite social network analytics' recurrences.

The media, societal perception, and mathematics education

It is not rare to encounter news headlines such as 'The myth of being "bad" at maths' (Wen, 2012), published by *BBC News Online*. This article reveals that, according to a study by Silke Luttenberger, Sigrid Wimmer, and Manuela Paechter (2018), around 93% of US adults experience some level of math anxiety. Thus, people unable to understand mortgage interest payments are advised not to feel alone anymore. The article also elaborates on how mathematics has become the school subject of the one-right-answer or the right-or-wrong conundrum, so teachers should emphasise to students that mistakes are part of mathematics learning. And so, math anxiety is naturalised as a cause of bad performance (or, at least, an essential part). This means that only 7% of US adults do not experience math anxiety; such people are not part of the myth of being 'bad' at mathematics. As for young children, they are not exempt from math anxiety. The news article '"Maths anxiety" causing fear and despair in children as young as six' (Weale, 2019) released by *The Guardian* exposes math anxiety as a possible cause of physical symptoms and behavioural problems in class, leading students into a cycle of despair and suffering, harming their mathematics performance. The article asserts that this phenomenon may be contributing to a growing mathematics crisis in the United Kingdom, given that there is a general sense of mathematics being hard compared to other school subjects, which is implicated in students losing their confidence. Here, the way the media decide to present some studies by researchers and international organisations (such as the Organisation for Economic Co-operation and Development, OECD, or the United Nations Educational, Scientific and Cultural Organization, UNESCO) in laypersons' terms plays an important role in circulating certain dominant narratives. The media's power and impact have been an interesting topic for research in modern societies, specifically on consumer goods, services, and election campaigns (see Udanor, Aneke, & Ogbuokiri, 2016). The media – such as online news, social networks, non-Internet-based written news such as newspapers – have enabled a larger and quicker distribution of information, providing the possibility of reacting in real time to events happening anywhere in the world.

According to the data gathered for this analysis, one of the main kinds of circulating and naturalised discourses around mathematics and mathematics education on social media consists of biased opinions based on people's particular experiences, mostly with school mathematics or encounters, whether good or bad, with their school mathematics teachers. Statements such as: 'After watching this, I am convinced that school is not necessary' (YouTube comment), 'My current math teacher makes me worried. You restore my confidence. You have no idea what that means to me' (YouTube comment), or 'Why is math so hard ... maybe its just me being dumb' (Twitter post) reveal the prejudices people have about mathematics and its teaching. For example, someone decided to post on a YouTube video's comment section that the number of dislikes was due to mathematics instructors/teachers feeling deficient while watching the video: 'The dislikes must have been from calculus instructors, who feel inadequate after watching this' (YouTube comment).

Another kind of circulating discourse is about math phobia, the fear of mathematics: 'As a kid I really struggled with math. I just had a flashback of my mom trying to help me with math homework [...] I can't imagine being a kid right now having to do school from home. Me: about to say the wrong answer again Mom' (Twitter post). Another type is built around failure narratives. For instance, 'This test is so hard that I didn't even understand the question' (YouTube comment). Also, about the lack of utility of mathematics in day-to-day life. For example, on a YouTube video that explains how to cut a cake scientifically, a person posts: 'Science needs to know its place and stay away from cake baking and eating [...] I've baked plenty of cakes in the 90s with my Kenwood mixer and without needless advice'. In the same vein, on a video that explains probabilities, a person commented: 'The probability i am ever using probability is 0%'.

It is possible to find opinions stating that people have learned much more from a YouTube video than from their formal classes in school: 'You're a god. I've learned more from you in the last few hours than I have in my math class during the last few months' (YouTube comment). Other comments imply that mathematics is generally beyond human understanding. When this happens – to have learned something considered impossible to learn or to have understood a particular topic or problem – it deserves a celebration: 'AT 35:28: Me screaming...;

roommate: did you win the lottery?; Me: no, I just did my first calculus problem' (YouTube comment).

Figure 20.1 shows frequencies of the categories found within all the YouTube comments collected from the saturation strategy used to approach the gathered data. The statements found on YouTube are based on, and related to, a particular watched video. In this light, videos become a source that evokes past or present experiences with mathematics or mathematics teachers and personal feelings and opinions that help reveal people's perception of math and math education. Likewise, Figure 20.2 shows the frequencies of the categories found within all the Twitter posts collected. Here, opinions are not based on, or related to, a specific video; instead, they are based on people's willingness to post a comment. Although the source that provokes posting a Tweet is not revealed in most of the cases, the comments can also reveal societal perceptions of math and math education.

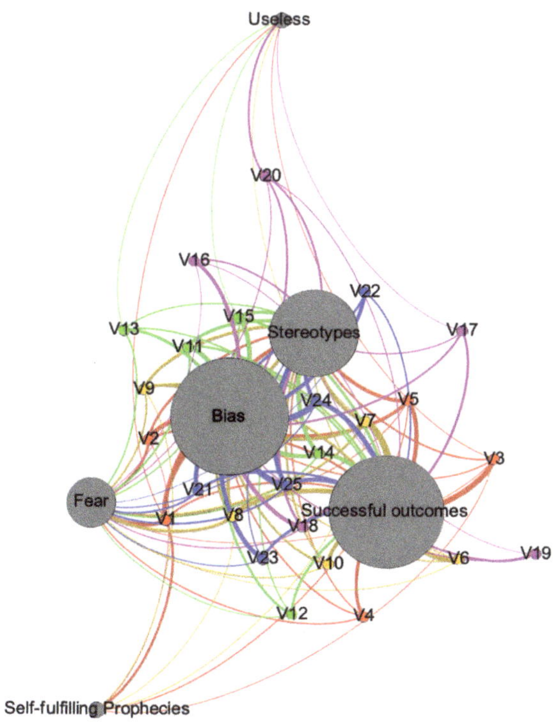

Fig. 20.1 Frequencies of YouTube comment categories. Figure created by authors, using Gephi software.

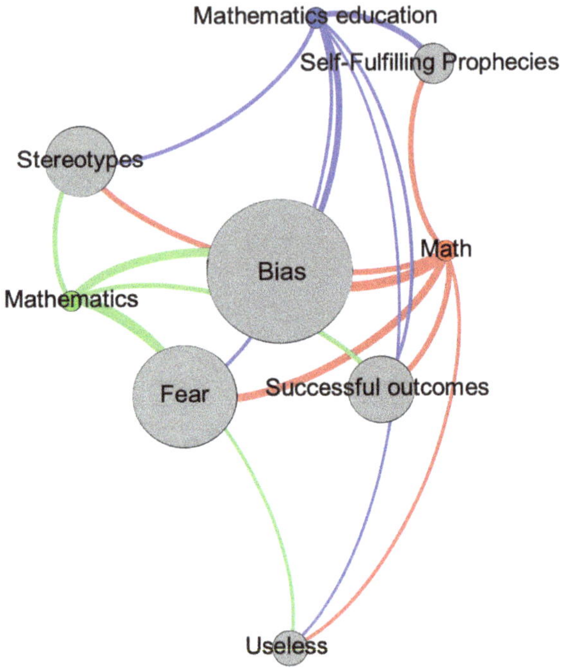

Fig. 20.2 Frequencies of comment categories in Twitter posts. Figure created by authors, using Gephi software.

As previously mentioned, the first category of comments gathered from YouTube and Twitter comprises *biased opinions* regarding school mathematics practices. These comments heighten the narrative that mathematical concepts taught in school are pointless, that teaching styles need attention, and that mathematics education, in general, should be revised.

> This kind of math is the one needed to be taught in classroom. so fun, entertain. (YouTube comment)

> I could not imagine a better way for learning and loving Math <3 Thank u So much for that! (YouTube comment)

> Talking about math is essential. It's what mathematicians do. While students may think that mathematicians simply sit around working out computation problems, that's not at all an accurate picture. Mathematicians ask big questions, come up with ideas, ... (Twitter post)

> my math is so bad 😭😭 10 weeks to get both of these to 160 at the very lowest though we got this (Twitter post)

> I have a question Is it just the topic or math in general that was the problem? (Twitter posts)

A second category comprises comments built around what we take to be *successful outcomes* from mathematics teaching and learning. These comments reveal the difficulties people have while learning mathematics. Certain events – such as watching a video – have a positive impact on their understanding of mathematics or their mathematics tests. Also, overcoming these difficulties is something worth sharing.

> I will now put 'knows how to solve the hardest problem on the hardest test' on my resume (YouTube comment)

> This is legit the only time I've ever enjoyed math in my entire life, ur works beautiful bro (YouTube comment)

> Thank you Bro! I got B+ on exam! You helped me!; I wish you all the best! Sory for my bad english=) thanks from Kyrgyzstan! (YouTube comment)

> i overthink everything but at least i got the answers to da math test now🐢🐢 (Twitter post)

> i have a b+ in math call me the math queen i love algebra (Twitter post)

A third category is composed of comments dealing with *stereotypical* visions of mathematics and its teaching and learning. This understanding of mathematics is based on people's unfortunate experiences of school mathematics that have developed into particular stereotypes of believing that mathematics occurs only in one form. It also includes the beliefs that mathematics is not for all, and that only a certain type of person is mathematically competent – drawing on harmful distinctions regarding race, gender, nationality, etc.

> me: I like math, not my favourite but it's cool; these kind of problems: *appear *; me: wow I'm going to study philosophy and never look at a number ever again (YouTube comment)

> Teacher: there will only be 3 questions; Me: thank god; Test: 1,1A,1B,1C, 2, 2A, 2B, 2C, 3, 3A, 3B, 3C (YouTube comment)

> Asian parents when their kids finish this test: 'That was an improvement, but it's not hard to improve on garbage. *DO IT AGAIN*' (YouTube comment)

> Normal people: im going to major in neuroscience/biology/engineering/business/etc everyone: oh!! thats so cool!! me: im gonna double major in physics and math. everyone: oh . .. . thats .. . . cool. (Twitter post)

> Rewatched Jurassic Park and i've decided the only acceptable way someone can flirt with me is by using mathematic theories (Twitter post)

A fourth category is composed of narratives about *fear and hate.* These narratives are built on people disliking specific experiences they had or still have with mathematics and notions of math phobia. It also includes jokes people make as a way of mocking particular aspects of mathematics. Most comments entail narratives of failure in the practices of learning mathematics and of encountering not-so-good experiences with different teachers and styles of teaching.

> This test is so hard that I didn't even understand the question (YouTube comment)

> Me: not understanding anything; Also me: yes, big brain indeed (YouTube comment)

> I try to solve the tasks; My Brain: don't even think about it (YouTube comment)

> Math keeps making my brain hurt in ways I never thought it was possible. thanks for the video! I freaking love this channel. (YouTube comment)

> oh god, Im still just as lost, I'm screwed! thanks though (Youtube comment)

> apparently i lost my mathematic skills someone please teach me (Twitter post)

> i suck at math woop😔😔 (Twitter post)

> im doing math work but i dont know how to do any of it im going to fail school cause of this virus (Twitter post)

A fifth category is composed of comments about the *useless* nature of mathematics. Most comments within this category entail narratives of perceiving mathematics as lacking meaning and coherence. For many people, mathematics is pointless and without any practical use outside the boundaries of school. It is possible to evidence discourses around the lack of sense school mathematics has in daily-life practices. These discourses come from particular experiences people have when engaging with day-to-day tasks without using the formal mathematics they learnt at school.

> notice how everyone is learning this because they use it at school but no adults that are using this at work (YouTube comment)

> useless as always; Thx you for wasting my time (YouTube comment)

> Does Mathematic still needed? (Twitter post)

> We especially need imagination in science. It is not all mathematics, nor all logic, but it is somewhat beauty and poetry. (Twitter post)

> The best math you can learn is how to calculate the future cost of current decisions (Twitter posts)

A sixth category is built around notions of *self-fulfilling prophecies*. These commentaries reveal how certain beliefs people have regarding mathematics and its teaching become materialised, such as a mathematics test grade. Most comments deal with a low expectation that people will perform well at mathematics tests, or a belief that people will accumulate bad experiences when encountering something related to mathematics. These narratives intertwine with auto-exclusion from mathematics practices.

> Me: thinking I studied well for a test; Putnam: so you've chosen death (YouTube comment)

> I failed Calculus, I not surprising... from now on, I'm watching these videos (YouTube comment)

> Sometimes I wonder why I felt so bold and decided to choose a science major,,,, I ain't good at science nor math and I have to take so many courses for the two uGH. (Twitter post)

> LMAOOOO i definitely failed my math exam: D.D. (Twitter post)

From the frequency graphs for YouTube comments and Twitter posts,[1] it is possible to notice that: (i) Twitter Posts' Frequency Network, TPFN (see Figure 20.2), is more compact in comparison with YouTube Comments' Frequency Network, YCFN (see Figure 20.1). This compactness shows that the peripheral vertices (nodes) – namely the categories of *self-fulfilling*

1 The algorithm Force Atlas 2 from Gephi was applied to visualise both graphs.

prophecies and the *useless* nature of mathematics – in TPFN are most distant from the network. The compactness of a network indicates that discourses within its categories are more balanced. Thus, the dispersion of ideas and discourses about mathematics and mathematics education is less in TPFN than in YCFN. We also notice that (ii) YCFN shows a concentration of discourses built around two categories: *bias* and *stereotypes*. This exhibits the dominant narratives circulating on YouTube comments. *Successful outcomes* and *fear* form the second largest group of categories within the spectrum of dominant narratives. And (iii) the dominant discourses in TPFW occur in a different distribution than those in YCFN. Most Twitter posts are built around narratives of *bias*, this category being the most prominent. Narratives of *fear* and *successful outcomes* are the second-largest circulating discourses. The differences between YCFN and TPFN could be explained by each platform's nature and use. YouTube has become a platform from which people seek virtual help (see, e.g., Aguilar & Puga, 2020). This is not the case for Twitter, which people might use for informative purposes and to post their opinions on diverse subjects.

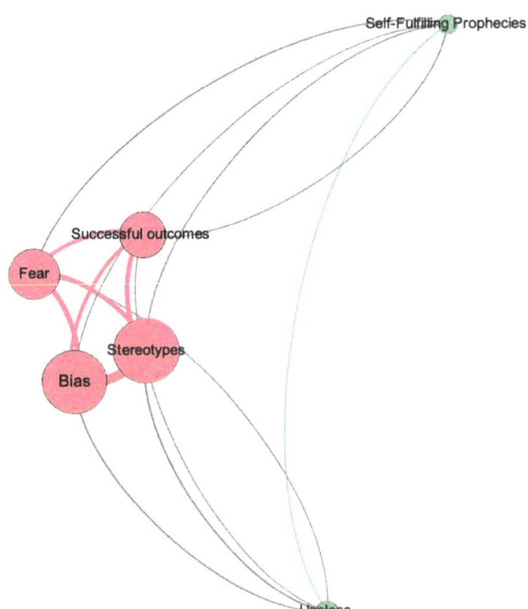

Fig. 20.3 YouTube comments' co-occurrences. Figure created by authors, using Gephi software.

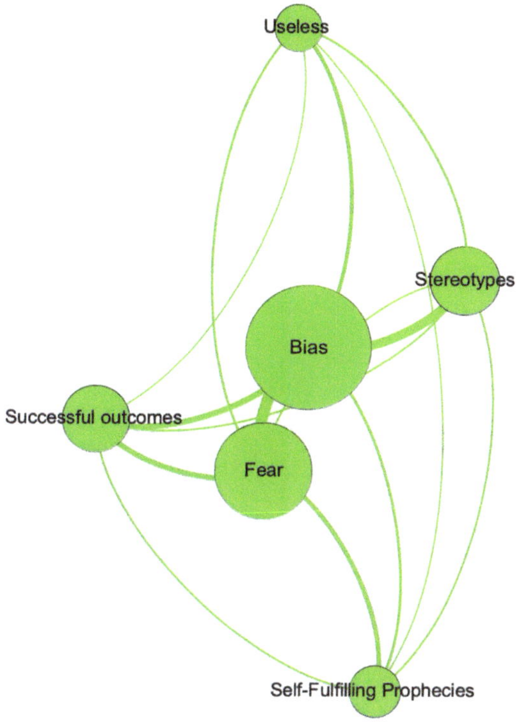

Fig. 20.4 Twitter posts' co-occurrences. Figure created by authors, using Gephi software.

When observing the co-occurrences of such categories, it is possible to note how they relate. From YouTube comments (Figure 20.3), narratives about *bias* intertwine with *stereotypes, fear and hate*, and *successful outcomes* – building a community[2] – more than with the *useless(ness) of mathematics* and *self-fulfilling prophecies.* Communities are visible according to the colour of the nodes. *Useless(ness)* and *self-fulfilling prophecies* are in a different colour than the rest of the categories. This exhibits how comments intertwine amongst these four categories to (re)produce dominant narratives about mathematics and mathematics education. For example,

2 The tool Modularity helps in detecting communities. These are displayed by the use of different colours for each community. Nodes that are connected with higher density belong to the same community.

> Calc II is the last thing standing between me and my degree (It's already been 4 loooong years). You are the only reason I'm graduating. I can't thank you enough! (YouTube comment)

> professor, you are my hero. I started learning with you since my calculus 1. Now I am in calculus 3!!! (YouTube comment)

> The class: A minute has 60 seconds; The homework: calculate how many minutes has the day.; The exam: Calculate the exact amount of seconds there are till the Sun dies, taking into account the seconds you lost reading this question, and then calculate the equivalent hours. (YouTube comment)

Within the dominant narratives circulating in YouTube comments, it is possible to evidence a closer connection – proximity – between *bias* and *stereotypes* categories. This is composed of comments such as:

> It would take me the rest of my life to understand this level of maths (YouTube comment)

> When you understand the main idea of Fourier Transformation finally in a youtube video instead of in a whole semester in university. (YouTube comment)

> Oh.. My.. God; This thing just taught me more math that 2 yrs of Engineering lectures couldn't teach me (YouTube comment)

When it comes to Twitter posts (Figure 20.4), all categories built one large community – this is visible given that all the nodes are displayed in the same colour. However, some connections between categories are denser. The density is visible through the weights of the edges – some lines connecting two categories are visibly thicker than others. Here, it is possible to evidence the dominant narratives that arise from what people decide to share and showcase when posting their *bias* and *fear* related opinions regarding their past or present experiences with mathematics and mathematics education. Within such posts, plenty of statements circulate around people's low expectations for their mathematical skills and performance. Often, people consider themselves as 'bad at math' or 'not good enough or are not meant to do and engage with mathematics'. Opinions such as:

> Why is math so hard... maybe it's just me being dumb. (Twitter post)

> When I read this for the first time in the textbook my mind exploded (YouTube comment)

> Now I have a grudge against that math teacher for life, no offense but I hope he goes to hell or something. He singlehandedly made the bullying worse and gave me anxiety n an irrational fear of people (Twitter post)

> I couldn't even do math in class, how the fuck they think ima be able to do it online? (Twitter post)

> I've been going through finals, failed my math final, and am having an existential crisis about whether or not I wanna be an engineer anymore. but I'm trying. After tomorrow, I should be on more for plotting and writing. (Twitter post)

These narratives shed light on societal perceptions of mathematics and mathematics education as a disciplinary field inserted in school practices. Most people's views on mathematics are entangled with the useless nature of this knowledge in practical daily-life activities. It seems that the apparent impracticality of mathematics instantiates feelings of anger, anxiety, frustration, and discomfort.

Dominant narratives from international organisations, educational policies, and stakeholders acknowledge mathematics as key knowledge for modern society. The salvation themes arising from such discourses involve notions of 'mathematics for all' (see Pais, 2018), in which everyone needs mathematics proficiency to secure their future. Then, people's perceptions of mathematics align with elaborations produced by the media, of mathematics being difficult and inaccessible for people with average IQ. For example, *The Conversation* published an article entitled 'Maths: six ways to help your child love it' (Johnston-Wilder & Penazzi, 2018), which deals with two types of engagement with mathematics: one by mathematicians and the other by everybody else.

> There is a widespread perception that mathematics is inaccessible, and ultimately boring. Just mentioning it can cause a negative reaction in people ... For many people, school maths lessons are the time when any interest in the subject turns into disaffection. And eventually maths becomes a topic many people don't want to engage with for the rest of their lives... At the opposite end of the spectrum, professional mathematicians see mathematics as fun, engaging, challenging, and creative.

Although there have been multiple efforts to make mathematics more accessible to students (including the variety of theories, methods, and materials produced by mathematics educators worldwide), school mathematics seems to be bounded by the alchemic process that produces the mathematics curriculum (see Popkewitz, 2004). This alchemy of the school mathematics curriculum might explain the different nature of the mathematics of professional mathematicians and the mathematics of school practices. Within the media, the apparent spotlight is on the decreasing performance of students. On the one hand, research engages with making mathematics accessible to all to achieve certain standards. In contrast, the media highlight that mathematics-learning standards are not close to being achieved. For example, the French news site *The Connexion* uses the phrase 'catastrophic maths' to describe students' performance (Connexion journalist, 2018): 'A new report that found the level of mathematics among French students today to be "catastrophic" has recommended 21 possible solutions, including the so-called "Singapore method"'.

Another example comes from *The Conversation*, an article titled 'Challenging the status quo in mathematics: Teaching for understanding' (Rakes, 2017). This news story outlines the minimal changes achieved by the national reforms in the United States intended to improve students' performance in mathematics: 'Despite decades of reform efforts, mathematics teaching in the U.S. has changed little in the last century. As a result, it seems, American students have been left behind, now ranking 40th in the world in math literacy'.

The media have also covered stories highlighting the emotions students experience as a result of the impact that teaching and learning school mathematics has on them. These deal with various negative side-effects that arise from engaging with mathematical tasks, such as anxiety, fear, and even physical symptoms (see Weale, 2019; Henry-Nickie, 2018). Within this type of news, mathematics is presented as damaging and scary, with statements even asserting that not all students should be obliged to learn this school subject.

> Children as young as six feel fear, rage, and despair as a result of 'mathematics anxiety', a condition which can cause physical symptoms and behaviour problems in class, according to a study [...] Pupils in both primary and secondary school can find themselves locked in a cycle of despair, suffering from anxiety which harms their maths performance, which in turn leads to increased anxiety. (Weale, 2019)

> The mathematics discipline usually strikes fear into the hearts of most students and working-age adults in the U.S. (Henry-Nickie, 2018)

Along similar lines, there is also a cluster of news stories that deals with disseminating mathematics updates. These news types are focused on the accomplishments in the field, for example, awards or improvements on students' performance worldwide. *ScienceAlert* emphasises the recognition of the first woman awarded the Abel Prize, Karen Uhlenbeck (Starr, 2019):

> For the first time, a woman has been awarded the prestigious Abel Prize. Karen Keskulla Uhlenbeck of the University of Texas at Austin will receive the annual prize for her tremendous contributions to the field... The news is notable because historically, most of the mathematics and science prizes have been awarded primarily to male recipients. Of the 904 individual Nobel laureates, only 51 have been women. (Starr, 2019)

The *Baltic News Network* (Baltic News Network, 2019) elaborates on the positive results of Latvian students in PISA (Program for International Student Assessment), within which low performance has decreased in mathematics by 2.6 points since 2012. *The Telegraph* (Gurney-Read, 2016) highlights the countries with high achievement in mathematics international tests, such as TIMSS (Trends in International Mathematics and Science Study):

> The average results of Latvian pupils were scored 496, which is statistically higher than the average OECD level – 489. Authors of the study say this is the highest result in mathematics Latvian pupils have achieved in seven cycles of the Programme for International Student Assessment (PISA). (Baltic News Network, 2019)

> Singapore, Hong Kong, Korea, Chinese Taipei, and Japan continue to dominate international rankings for maths and science, the latest league tables have revealed. (Gurney-Read, 2016)

Mathematics is perceived as a school subject in which students struggle the most. Sometimes, it is presented as scientific approximations to situations for ordinary life that need more assistance, for example, in finding love. The *Daily Mail* (Gray, 2015) discusses how a mathematician calculated a formula for finding true love. There is also a mathematics formula, according to *Greek Reporter* (Kokkinidis, 2019), to select the most beautiful woman on the planet.

> Mathematics is probably not a subject that many people find sexy, but it could hold the key to finding true love [...] Mathematicians have developed a series of theories that can help people find the perfect partner. (Gray, 2015)

> British scientists have recently determined exactly who is the most beautiful woman on the planet – according to a mathematical formula used by ancient Greeks. (Kokkinidis, 2019)

The media also gather news that focuses on debunking myths around mathematics and mathematics education, such as gender disparities in students' performance. The use of research to build such arguments is a common factor of this type of news. *Genetic Engineering & Biotechnology News* (Genetic Engineering & Biotechnology News, 2019) highlights one piece of neuroscientific research that found no differences between boys' and girls' neural development.

> A comprehensive examination of neural development in boys and girls has now effectively refuted this myth, and demonstrated that neural functioning is similar in both sexes. Findings from the study indicate that, at a neurological level, there should be no reason why girls would have less aptitude for maths than boys. (Genetic Engineering & Biotechnology News, 2019)

And even though gender bias and stereotypes did not appear in YouTube comments or Twitter posts, there are various places where the gender gap is addressed, mostly related to prejudices of women performing poorly within STEM fields. It is often visible in sitcoms and animated series, such as *The Simpsons*, in the episode 'Girls just want to have sums':

> Principal Skinner: You know, Juliana, it's no surprise you became such a success. You always got straight As in school.
>
> Juliana: Well, I remember getting a B or two in math.
>
> Principal Skinner: Well, of course you did. You are a girl.
>
> [Audience gasps.]
>
> Principal Skinner: All I meant was, from what I've seen, boys are better at math, science, the real subjects.
>
> Juliana: [To audience] Calm down, calm down. I'm sure Principal Skinner didn't mean girls are inherently inferior.
>
> Principal Skinner: No, of course not. I don't know why girls are worse. (*The Simpsons*, season 17, episode 19)

Discussion

Can't believe they gave me a math assignment in dental school (Twitter post)

Recalling Chassy (2014), perception of what is taken as real is strongly influenced by languages – that are relativistic, according to Edward Sapir (1929). From social science and social psychological research, it is clear that beliefs and expectations of people can distort perception in various forms, which influence the way of viewing objective 'social reality' (Jussim, 2012). Wendy Wood (2000) explores how social identity theorists have studied group influence as well as other aspects of group behaviour:

> When people categorize themselves as an in-group members, the in-group serves as a reference for social comparison, and people adopt the prototypical in-group attitudes and beliefs as their own [...] agreement from others categorized as similar to self enhances one's subjective certainty and suggests that the shared attitudes reflect external reality and the objective truth of the issue. (p. 557)

Within this taken-as-real context, the media and social networks contribute to fabricating an image for mathematics and shape how people decide to adopt specific beliefs about mathematics and mathematics education. Although there is little research on how the media influence the societal perception of mathematics and mathematics education, there exist some efforts, for example, regarding the learning of mathematics with the use of digital media (see Kynigos, 2008). The social media matrix has become a powerful 'magic wand' able to determine the structure of society as it 'forms a basis for polarization and dissolutions and also ensures mergers and agreements' (Gündüz, 2017, p. 91). And so, social media can reveal, reflect, and even shape, how society views mathematics, whereby stereotypical representations of mathematicians can actually be discouraging to a group of subjects.

Trying to read societal perceptions from a linguistic perspective implies understanding discourse analysis, building on Norman Fairclough's work (1995). However, the focus here is not on using signifiers to critically analyse YouTube comments and Twitter posts. Instead, it is on mapping the dominant narratives surrounding mathematics and mathematics education in order to unfold the particular spatiotemporal conditions that enable a certain type of rationality – a system of reason – that shapes certain kinds of people. From this, the 'taken-as-truth' statements about mathematics and mathematics education circulate to evolve into societal perceptions about what a mathematician is believed to be. These 'taken-as-truth' statements become discursive formations (Jørgensen & Phillips, 2002) that have been produced and reproduced by the interaction of different spheres of modern life (Foucault, 1972) and are entangled in the practices of everyday life. Within this realm of societal perceptions, discourses regarding the relevance of mathematics for society and day-to-day life should be denaturalised. For instance, discourses that link, mathematical proficiency with intellect, enabling the belief that people not performing well in mathematics are cognitively deficient; and discourses that link mathematics proficiency with being male, socially inadequate, not good at sports, poor sight, and so on. Or discourses that link mathematics proficiency with success. Here, discursive frameworks shape the boundaries within which people can negotiate what it means to be good or bad at mathematics, to have or not have mathematical abilities, to enjoy or hate mathematics etc. When people take a position regarding mathematics or school mathematics, their perceptions based on dominant narratives about success, usefulness, and value of mathematics play a determinant role. In this light, circulating discourses about mathematics and mathematics education shape people's perceptions and predispositions towards mathematics, given that:

> Our perception of objects is formed within the limits of discursive constraints [Discourse] causes a narrowing of one's field of vision, to exclude a wide range of phenomena from being considered as real or as worthy of attention, or as even existing; thus, delimiting a field is the first stage in establishing a set of discursive practices. (Mills, 2001, p. 51)

People respond to mathematics and school mathematics in diverse ways. How people perceive mathematics and mathematics education will depend on their frame of reference, values, beliefs, experiences, interests, etc. These alter not only our perceptions of social reality but also of reality itself. Societal perception could be understood as a portrait of reality shaped by our beliefs, experiences, or interests; but also, as a way of constructing our reality and social order. Johann

Engelbrecht, Salvador Llinares, and Marcelo Borba (2020) assert that at the same time that humans develop and build new media, the media themselves transform and 'construct' a new kind of human. The media complex does more than represent or describe the world and its relations; it produces and reproduces narratives about mathematics and mathematics education. It seems natural to think that how people conceive of mathematics is determined and situated. However, subjects are not passive entities within a system. In this context, the media might be seen as having agency (see Butler, 1997).

Mathematics undergoes an alchemic process (Popkewitz, 2004) deeply rooted in spatiotemporal conditions that transform the mathematics curriculum. As Yip-Cheung Chan and Ngai-Ying Wong (2014) contend, 'social, political, and even economic and cultural/religious backgrounds of a period generate a social mood that affects the curriculum worldview' (p. 274). Under this light of 'social mood', mathematics becomes the epitome of modernism, whether by the intrinsic characteristics of mathematics itself (logical and rational), or the capacity of promoting the development of valuable competencies for humankind (such as problem solving and the capability of changing reality). School mathematics is structured to reach as much of the population as possible, materialised in political agendas such as 'no child left behind' or 'mathematics for all'. Educational policies are entangled within a system of reason in which mathematics becomes a powerful knowledge for securing people's future and, thus, it is framed as something that should be enjoyable to learn (see Pais, 2018). However, societal perceptions about mathematics and mathematics education, rooted in the same system of reason from which political agendas portray mathematics as useful and vital, differ entirely from educational aspirations and desires. This finding problematises the teaching and learning of mathematics as a means of self-segregation, by constructing categories of 'normal' and 'smart' citizens in contrast to 'abnormal' and 'undesirable' citizens.

References

Aguilar, M. S., & Puga, D. S. E. (2020). Mathematical help-seeking: Observing how undergraduate students use the Internet to cope with a mathematical task. *ZDM Mathematics Education*, *52*(5), 1003–1016. http://doi.org/10.1007/s11858-019-01120-1

Asia's Got Talent. (2019, March 7). *15 year old Yaashwin Sarawanan is a human calculator!* YouTube. https://www.youtube.com/watch?v=kvymoFdjuHw

Baltic News Network. (2019, December 4). Latvian pupils' mathematics score above average among OECD member states. *Baltic News Network*. https://bnn-news.com/latvian-pupils-mathematics-score-above-average-among-oecd-member-states-208092

Butler, J. (1997). *The psychic life of power.* Stanford University Press.

Chassy, P. (2014). How language shapes social perception. In D. Evans (Ed.), *Language identity: Discourse in the world* (pp. 36–54). Bloomsbury.

Chan, Y., & Wong, N. (2014). Worldviews, religions, and beliefs about teaching and learning: Perception of mathematics teachers with different religious backgrounds. *Educational Studies in Mathematics*, *87*(3), 251–277. https://doi.org/10.1007/s10649-014-9555-1

Colagrossi, M. (2018). Why America is bad at math. *Big Think*. https://bigthink.com/technology-innovation/why-america-is-bad-math

Connexion journalist. (2018, February 13). Catastrophic maths report advises Singapore method. *The Connexion*. https://web.archive.org/web/20210507181454/https://www.connexionfrance.com/French-news/Catastrophic-maths-report-advises-Singapore-method

Cosmology Today. (2017, December 24). *Why most people are bad at mathematics - Neil deGrasse Tyson asks Richard Dawkins*. YouTube. https://www.youtube.com/watch?v=fbrQ8F-LQNs

Deleuze, G. (2007). The actual and the virtual. In G. Deleuze & C. Parnet (Eds.), *Dialogues II* (pp. 148–152). Columbia University Press.

Dunning, D. (2001). What is the word on self-motives and social perception: Introduction to the special issue. *Motivation and Emotion*, *25*(1), 1–6. https://doi.org/10.1023/A:1010627820999

Engelbrecht, J., Llinares, S., & Borba, M. C. (2020). Transformation of the mathematics classroom with the internet. *ZDM Mathematics Education*, *52*(5), 825–841. https://doi.org/10.1007/s11858-020-01176-4

Fairclough, N. (1995). *Critical discourse analysis.* Longman.

Foucault, M. (1972). *'The archaeology of knowledge' and 'The discourse on language'*. Pantheon.

Garrido, C. (2020). Social information processing. In V. Zeigler-Hill & T. K. Shackelford (Eds.), *Encyclopedia of personality and individual differences* (pp. 5070–5073). Springer.

Genetic Engineering & Biotechnology News. (2019, November 11). Brain function study suggests boys and girls have equal aptitude for mathematics. *GEN Genetic Engineering & Biotechnology News*. https://www.genengnews.com/news/brain-function-study-suggests-boys-and-girls-have-equal-aptitude-for-mathematics/

Gray, R. (2015, Mar 27). Choose ugly friends, highlight your flaws and don't settle down before the age of 22: Mathematician reveals the formula for finding true love. *Daily Mail*. https://www.dailymail.co.uk/sciencetech/article-3014452/Choose-ugly-friends-highlight-flaws-don-t-settle-age-22-Mathematician-reveals-formulas-finding-true-love.html

Gurney-Read, J. (2016, November 29). Revealed: World pupil rankings in science and maths – TIMSS results in full. *The Telegraph*. https://www.telegraph.co.uk/education/2016/11/29/revealed-world-pupil-rankings-science-maths-timss-results/

Gutiérrez, R. (2013). The sociopolitical turn in mathematics education. *Journal for Research in Mathematics Education*, *44*(1), 37–68. https://doi.org/10.5951/jresematheduc.44.1.0037

Gündüz, U. (2017). The effect of social media on identity construction. *Mediterranean Journal of Social Sciences*, *8*(5), 85–92. https://doi.org/10.1515/mjss-2017-0026

Henry-Nickie, M. (2018, September 11). The 21st century digital workplace makes mathematics inescapable. *Brookings*. https://www.brookings.edu/blog/techtank/2018/09/11/the-21st-century-digital-workplace-makes-mathematics-inescapable/

Ikeda, T. (2018). Evaluating student perceptions of the roles of mathematics in society following an experimental teaching program. *ZDM Mathematics Education*, *50*(1–2), 259–271. https://doi.org/10.1007/s11858-018-0927-3

Johnston-Wilder, S., & Penazzi, D. (2018, August 17). Maths: Six ways to help your child love it. *The Conversation*. https://theconversation.com/maths-six-ways-to-help-your-child-love-it-96441

Jørgensen, M. W., & Phillips, L. J. (2002). *Discourse analysis as theory and method*. Sage.

Jussim, L. (2012). *Social perception and social reality: Why accuracy dominates bias and self-fulfilling prophecy*. Oxford University Press.

Kokkinidis, T. (2019, October 16). Greek mathematics reveal most beautiful woman on the planet. *Greek Reporter*. https://greece.greekreporter.com/2019/10/16/greek-mathematics-reveal-most-beautiful-woman-on-the-planet/

Krueger, R. A., & Casey, M. A. (2014). *Focus groups: A practical guide for applied research*. Sage.

Kynigos, C. (2008). Theories, context and values to understand learning with digital media: Book review of 'Humans-with-media and the reorganization of mathematical thinking', by M. Borba and M. Villareal. *ZDM Mathematics Education, 40*(5), 909–911. https://doi.org/10.1007/s11858-008-0145-5

Lerman, S. (2014). Mapping the effects of policy on mathematics teacher education. *Educational Studies in Mathematics, 87*(2), 187–201. https://doi.org/10.1007/s10649-012-9423-9

Leung, A., & Lee, A. M. S. (2013). Students' geometrical perception on a task-based dynamic geometry platform. *Educational Studies in Mathematics, 82*(3), 361–377. https://doi.org/10.1007/s10649-012-9433-7

Luttenberger, S., Wimmer, S., & Paechter, M. (2018). Spotlight on math anxiety. *Psychology Research and Behaviour Management, 11*, 311–322. https://doi.org/10.2147/PRBM.S141421

Mills, S. (2001). *Discourse*. Routledge.

Pais, A. (2012). A critical approach to equity. In O. Skovsmose & B. Greer (Eds.), *Opening the cage: Critique and politics of mathematics education* (pp. 49–91). Sense.

Pais, A. (2013). An ideology critique of the use-value of mathematics. *Educational Studies in Mathematics, 84*(1), 15–34. https://doi.org/10.1007/s10649-013-9484-4

Pais, A. (2018). Truth, power, and capitalist accumulation in mathematics education. In M. Jurdak & R. Vithal (Eds.), *Sociopolitical dimensions of mathematics education* (pp. 95–109). Springer.

Popkewitz, T. (2004). The alchemy of the mathematics curriculum: Inscriptions and the fabrication of the child. *American Educational Research Journal, 41*(1), 3–34. https://doi.org/10.3102/00028312041001003

Rakes, C. (2017, Jun 21). Challenging the status quo in mathematics: teaching for understanding. *The Conversation*. https://theconversation.com/challenging-the-status-quo-in-mathematics-teaching-for-understanding-78660

Sapir, E. (1929). The status of linguistics as a science. *Language, 5*(4), 207–214. https://doi.org/10.2307/409588

Starr, M. (2019, March 19). For the first time ever, a woman has won the prestigious Abel Prize for mathematics. *ScienceAlert*. https://www.sciencealert.com/for-the-first-time-a-woman-has-won-the-abel-prize-for-mathematics

TEDx Talks. (2014, August 9). *The surprising beauty of mathematics | Jonathan Matte | TEDxGreensFarmsAcademy*. YouTube. https://www.youtube.com/watch?v=SEiSloE1r-A

Udanor, C., Aneke, S., & Ogbuokiri, B. O. (2016). Determining social media impact on the politics of developing countries using social network analytics. *Program: Electronic Library and Information Systems*, *50*(4), 481–507. https://doi.org/10.1108/PROG-02-2016-0011

Weale, S. (2019, March 14). 'Maths anxiety' causing fear and despair in children as young as six. *The Guardian*. https://www.theguardian.com/education/2019/mar/14/maths-anxiety-causing-fear-and-despair-in-children-as-young-as-six

Wen, T. (2012, May 12). The myth of being 'bad' at maths. *BBC*. https://www.bbc.com/worklife/article/20200506-how-to-tackle-your-anxiety-about-maths

Wood, W. (2000). Attitude change: Persuasion and social influence. *Annual Review of Psychology*, *51*, 539–570. https://doi.org/10.1146/annurev.psych.51.1.539

21. Beginning again

Brian Greer, David Kollosche, and Ole Skovsmose

In this book, we started from the position that the doing of mathematics and mathematics education are human activities, with all that that implies. As the book was developed, the notion of 'images' of mathematics and mathematics education, both influencing, and being propagated by, human actors became salient. We suggest that an analysis of such images may shed important light on the many acknowledged discontents of mathematics education. For too many people, their actual and remembered engagement with mathematics in schools is unnecessarily alienating rather than the enlightening and empowering experience that it could be.

The title of this volume refers to breaking images of mathematics and mathematics education. We began by advocating, in alignment with new developments in what qualifies consensually among those in the field as 'philosophy of mathematics', a radical shift from the chimerical quest to define mathematics as some kind of entity. In particular, we reject what is called the Platonic view that the entity of mathematics exists, in some way, independently of human beings. Instead it is argued throughout the book that the appropriate framing is to describe what people do when they 'do mathematics'. Further, what counts as mathematics is socially negotiated and these negotiations are historically contingent and subject to cultural diversity. Accordingly, such a shift in perspective necessitates historical and cultural lenses. As stated at the outset, mathematics and mathematics education are situated within historical, cultural, social, ethical, and political – in short, human – contexts.

We argue that mathematics education and its discontents cannot be adequately addressed internally, by yet another empirical study of

©2024 Brian Greer et al., CC BY-NC 4.0 https://doi.org/10.11647/OBP.0407.21

children struggling with fractions and the Holy Grails of the teacher-proof textbook, the perfect curriculum, the all-revealing test. Nor is there some complete architecture of cognitive development which, when fully developed, will render mathematics education straightforward. The discontents highlighted throughout this volume demand consideration of external contexts; the problems are human problems, and cannot be solved by technical means alone.

The aspiration to disassociate mathematics from the perceived contamination of human limitations has a long heritage, within which Platonism has played a dominant role. There is the image from early in the modern scientific era of the universe operating like a clockwork mechanism, suggesting that if one possessed total knowledge of every aspect of the universe at any given time, the future could be predicted. The desire to eliminate human imperfection can be seen in the progressive waves of formalism in mathematics, logical positivism in science, behaviourism in psychology. Unwarranted power of numbers is pervasive; psychology is heavily implicated through the conflation between 'X' and 'an inescapably imperfect measure of X'; the ignoring of that conflation underlies failure to acknowledge the limitations of psychometrics. An obvious example, with massive implications, is the idea that intelligence can be measured as a single number.

Similar issues of (de)humanisation arise in relation to mathematical modelling, whereby the relative precision of models of physical phenomena may be projected onto models of phenomena involving human complexities. Hence the formatting of our lives, accelerated by information technology, by models to which few have access, and over which even fewer have any control. The above considerations have been complexified by developments in Artificial Intelligence and the phenomena of the post-truth era, in which constructed images of alternative realities can dominate.

Over the five-year gestation of this book, in the creating of its diverse chapters, the phrase 'image of mathematics' has become increasingly salient. This phrase is necessarily nebulous but speaks to very real phenomena. The images of mathematics that people acquire through schooling and social life, and the images that people project in furtherance of ideological aims, have extraordinary power in both school mathematics and in the control of societies by state apparatuses. While explicitly or implicitly touched upon by many of the contributors in this volume, the intimate relationship between mathematics education and capitalism in its many forms remains to be thoroughly explicated. The reader of this book will have picked up multiple resonances of how these factors play out in practice. One theme that has been stressed is that the writing of the history of academic mathematics by the winners has contributed, to a significant and consequential extent, in the creation of an image underpinning intellectual White supremacy.

Our collective objective of questioning accepted wisdom about mathematics and mathematics education may be served by describing and interrogating such images. The word 'image' has numerous connotations, including: pictorial representation; idol, object of veneration; the conscious attempt to create a positive impression of a person or object, idea or picture in the mind. Obviously, there are many pictorial representations that reflect images of mathematics. That topic merits a series of books in itself. There is also the fascinating field of the fictional and non-fictional portrayal of mathematics and mathematicians in books, plays, and films.

Then there is the use of 'image' to refer to an object of veneration. In relation to mathematics, we have highlighted two aspects at various points in the book. Common among mathematicians, and uncritically accepted by many non-mathematicians, are what we consider inflated notions of the intellectual superiority of mathematics compared to other intellectual achievements. Among all groups in society, there is an associated public image of mathematical geniuses. Secondly, there is an often unexamined assumption that the doing of mathematics is inherently beneficial to humanity, as a driver of 'progress' and so on. Such beliefs are commonly held by those with power, which helps to explain the unreasonable political effectiveness of what might be termed 'mathematical propaganda'. Belief in the intrinsic goodness of mathematics forms an integral part of the whole outlook of modernity.

'Images' are also social constructions for politicians, religious leaders, film stars, and others and there is a field of expertise in the art/science of fabricating such social constructions that can become more important than the 'real people' (whatever that may mean). Such activities have an obvious affinity with advertising material goods and the public relations industry systematised by Freud's nephew, Edward

Bernays, in the 1920s. Especially prominent in the advertising copy for mathematics are under-examined slogans such as 'mathematics for all' and 'mathematics helps you to think'. Algebra and calculus are products that have been sold hard, yet most people do not use school algebra to any significant degree, and calculus even less. And the assertion that they are essential to national achievement is undermined by looking at the works of civilisations predating their development as formal tools.

In Plato's parable of the cave, people look at shadows projected on the wall from an independently existing reality. In this book, we have joined in a general rejection of Platonism as a philosophy of mathematics. We suggest that the images perceived by people are human constructions, including those intentionally designed for ideological reasons.

And so to the core of our argument, which focuses on a cluster of images of mathematics and mathematics education, including images of mathematicians, mathematics learners and learning, mathematics teachers and teaching. These images influence the thinking and actions of mathematicians, scientists, philosophers of mathematics, mathematics learners and teachers, the general adult population, people with power to apply mathematics, and those with power to influence mathematics education.

It has often been commented that working mathematicians do not allow themselves to be distracted by philosophical considerations, even less by ethical and political issues (G. H. Hardy being an influential apologist for this position). If mathematicians were quarantined, their political and emotional detachment would not matter so much. Let us simply assert (the supporting evidence and arguments are scattered throughout the book) that when mathematicians put their thumbs on the scales of school mathematics, they can do a lot of harm (with honourable exceptions, of course, many figuring prominently in this book). The Bourbaki movement in mathematics, and its spillover into school mathematics (by no means entirely the fault of Bourbaki) represents the extreme case that may be characterised as confusing the foundations of mathematics (in the sense of the old philosophy of mathematics) with the foundations of mathematics education.

The institution of formal schooling is so familiar we forget how artificial it is as a cultural construction, removing children from their families during a large part of their development. For mathematics, a pervasive aspect of this artificiality is the chasm between what children experience in the mathematics classroom and what they experience in life. In school, too many people learn to fear and hate mathematics – to be more accurate, the interpretations and images of doing mathematics with which they are confronted and the demands placed upon them. Far too many individuals and groups of people, through classroom interactions and through testing, may have their self-images damaged as people who 'cannot do mathematics' and, by implication, as intellectually deficient.

In many ways, what could be termed natural rights of children, in particular sense-making and valorisation of the multiple forms of diversity, are violated. The example of Pythagoras (considered by a consensus of contemporary scholars to have been neither a mathematician nor a scientist) illustrates a failure to adhere to ethical standards of historical accuracy, insofar as that is possible. At the systemic level, work continues on a counternarrative to the Eurocentric myth of the origins and development of academic mathematics, that may be regarded as a manifestation of white intellectual supremacy.

What also deserves more penetrative research is the extent to which school mathematics, particularly in the early years, is foundational in forming people's worldviews – for example, that everything can be measured one-dimensionally and then ranked, that 'everything is linear', and that numbers as such have unimpeachable authority, no matter how flawed are the models that produce them. The fascination of mathematicians with the infinite may innoculate people against grasping the implications of living a finite life on a finite planet.

Whence do people get their images of mathematics? Mostly in school, but also out of school; mostly the latter tends to reflect and reinforce the former. People who failed to master abstract algebra, or fell at the early hurdle of fractions (when does anyone, really, *need* to compute 4/7 + 7/11??) are easily intimidated by what mathematics appears to them to be, yet have a feeling that is somehow of great significance and demands reverence. With immense political implications, a sense of the limitations of mathematical modelling is not generally nurtured in, or out of, schools, especially in relation to anything involving human complexity.

After school, a small minority pursue further studies within mathematics and are likely to be enculturated into the discipline. Many more use mathematics in their studies or work; in those cases they are likely to encounter and learn mathematics in context, often using specialised representations and formulations, rather than recollecting related, but decontextualised, elements of school mathematics. Increasingly, the mathematics needed for work is embedded in software.

The majority of people do not use significant amounts of formal mathematics, and the mathematics they use in 'everyday life' is learned in context. They remain open to the socio-cultural influences that shape images of mathematics in the media, in the very particular genre of books, plays, films about mathematicians, in the echo-chamber of asocial media (see the previous chapter in this volume), and so on. We suggest that most politicians, even those closely involved in educational policy and governance, are not much different from the general population and largely share their images of mathematics and mathematics education. They often have a minimal understanding of mathematics associated with a strong tendency to defer to mathematicians as authorities.

Accordingly, we envisage a programme of sustained research and analysis, building on the very considerable work already done.

The guiding framework for this effort would be that the framing of school mathematics shows a continuity from the images established in elementary school, developing progressively through later life into adulthood, and ultimately looping back into school mathematics. Contributing to the closing of this loop is the influence of those with political power, including mathematicians, who, to a considerable extent, shape school mathematics. In our opinion, mathematics education in schools will not fundamentally improve until this feedback loop is disrupted. One focus for the research program that we are advocating could be deciding which parts of the cycle are open to such disruption and how that might be achieved. We hope that more mathematicians will emerge from their ivory towers and recognise their consequential roles and ethical obligations in this project. In the same spirit, we welcome the philosophers of mathematics who have stopped endlessly mending their nets and actually put out to sea.

Meanwhile, we observe the manifestations of the vast chasm between the projected image of mathematics as the epitome of rationality and the collective irrationality of our species in failing to confront a confluence of existential crises.

All of these are human problems.

List of Figures

List of Tables

Index

About the Team

Alessandra Tosi was the managing editor for this book.

Adèle Kreager proof-read and indexed this manuscript.

Jeevanjot Kaur Nagpal designed the cover. The cover was produced in InDesign using the Fontin font.

Cameron Craig typeset the book in InDesign and produced the paperback and hardback editions. The main text font is Tex Gyre Pagella and the heading font is Californian FB.

Cameron also produced the PDF and HTML editions. The conversion was performed with open-source software and other tools freely available on our GitHub page at https://github.com/OpenBookPublishers.

Jeremy Bowman created the EPUB.

Raegan Allen was in charge of marketing.

This book was peer-reviewed by two anonymous referees. Experts in their field, these readers give their time freely to help ensure the academic rigour of our books. We are grateful for their generous and invaluable contributions.

This book need not end here...

Share

All our books — including the one you have just read — are free to access online so that students, researchers and members of the public who can't afford a printed edition will have access to the same ideas. This title will be accessed online by hundreds of readers each month across the globe: why not share the link so that someone you know is one of them?

This book and additional content is available at
https://doi.org/10.11647/OBP.0407

Donate

Open Book Publishers is an award-winning, scholar-led, not-for-profit press making knowledge freely available one book at a time. We don't charge authors to publish with us: instead, our work is supported by our library members and by donations from people who believe that research shouldn't be locked behind paywalls.

Join the effort to free knowledge by supporting us at
https://www.openbookpublishers.com/support-us

We invite you to connect with us on our socials!

BLUESKY
@openbookpublish
.bsky.social

MASTODON
@OpenBookPublish
@hcommons.social

LINKEDIN
open-book-publishers

Read more at the Open Book Publishers Blog

https://blogs.openbookpublishers.com

You may also be interested in:

Landscapes of Investigation

Contributions to Critical Mathematics Education

Miriam Godoy Penteado and Ole Skovsmose (Eds)

https://doi.org/10.11647/obp.0316

Advanced Problems in Mathematics

Preparing for University

Stephen Siklos

https://doi.org/10.11647/obp.0181

The Essence of Mathematics

Through Elementary Problemsk

Alexandre Borovik and Tony Gardiner

https://doi.org/10.11647/obp.0168

www.ingramcontent.com/pod-product-compliance
Ingram Content Group UK Ltd.
Pitfield, Milton Keynes, MK11 3LW, UK
UKHW022044041026
15591UKWH00013B/851

9 781805 11321